Multi Length-Scale Characterisation

Inorganic Materials Series

Series Editors:

Series Titles

Functional Oxides
Molecular Materials
Porous Materials
Low-Dimensional Solids
Energy Materials
Local Structural Characterisation
Multi Length-Scale Characterisation

Forthcoming Title

Structure from Diffraction Methods

Multi Length-Scale Characterisation

Edited by

Duncan W. Bruce
University of York, UK

Dermot O'Hare
University of Oxford, UK

Richard I. Walton
University of Warwick, UK

WILEY

This edition first published 2014

Registered office
John Wiley & Sons Ltd, The Atrium, Southern Gate, Chichester, West Sussex, PO19 8SQ, United Kingdom

For details of our global editorial offices, for customer services and for information about how to apply for permission to reuse the copyright material in this book please see our website at www.wiley.com.

Library of Congress Cataloging-in-Publication Data applied for.

A catalogue record for this book is available from the British Library.

ISBN: 9781119953197

Set in 10.5/13pt Sabon by Laserwords Private Limited, Chennai, India
Printed and bound in Malaysia by Vivar Printing Sdn Bhd

1 2014

Contents

Inorganic Materials Series Preface

Back in 1992, two of us (DWB and DO'H) edited the first edition of *Inorganic Materials* in response to the growing emphasis and interest in materials chemistry. The second edition, which contained updated chapters, appeared in 1996 and was reprinted in paperback. The aim had always been to provide the reader with chapters that while not necessarily comprehensive, nonetheless gave a first-rate and well-referenced introduction to the subject for the first-time reader. As such, the target audience was from first-year postgraduate student upwards. Authors were carefully selected who were experts in their field and actively researching their topic, so were able to provide an up-to-date review of key aspects of a particular subject, whilst providing some historical perspective. In these two editions, we believe our authors achieved this admirably.

In the intervening years, materials chemistry has grown hugely and now finds itself central to many of the major challenges that face global society. We felt, therefore, that there was a need for more extensive coverage of the area and so Richard Walton joined the team and, with Wiley, we set about a new and larger project. *The Inorganic Materials Series* is the result and our aim is to provide chapters with a similar pedagogical flavour but now with much wider subject coverage. As such, the work will be contained in several themed volumes. Many of the early volumes concentrate on materials derived from continuous inorganic solids, but later volumes will also emphasise molecular and soft matter systems as we aim for a much more comprehensive coverage of the area than was possible with *Inorganic Materials*.

We approached a completely new set of authors for the new project with the same philosophy in choosing actively researching experts, but also with the aim of providing an international perspective, so to reflect the diversity and interdisciplinarity of the now very broad area of inorganic materials chemistry. We are delighted with the calibre of authors who have agreed to write for us and we thank them all for their efforts and cooperation. We believe they have done a splendid job and that their work will make these volumes a valuable reference and teaching resource.

DWB, York
DO'H, Oxford
RIW, Warwick
October 2013

Preface

Inorganic materials show a diverse range of important properties that are desirable for many contemporary, real-world applications. Good examples include recyclable battery cathode materials for energy storage and transport, porous solids for capture and storage of gases, and molecular complexes used in electronic devices. Some of these families of materials, and many others, were reviewed in earlier volumes of the *Inorganic Materials Series*. When considering property-driven research in this large field, it is immediately apparent that methods for structural characterisation must be applied routinely in order to understand the functions of materials and thus optimise their behaviour for real applications. Thus 'structure–property relationships' are an important part of research in this area. To determine structure effectively, advances in methodology are important: the aim is often rapidly to examine increasingly complex materials in order to gain knowledge of structure over length scales ranging from local atomic order, through crystalline, long-range order to the meso- and macro-scopic.

No single technique can examine these levels of order simultaneously, and the chapters presented in this volume deal with recent advances in important methods that allow structural investigation of various classes of inorganic materials over a variety of length scales. This volume therefore examines some key methods required to characterise contemporary inorganic materials and to understand their use in practical applications. The five detailed chapters presented here cover Magnetic Properties, Thermal Methods, Atomic Force Microscopy, Gas Sorption and Dynamic Light Scattering.

The techniques discussed in this volume are diverse in nature, but in each case an excellent background and introduction is given, along with examples that illustrate the scope of the method, how it is applied and the information that it can be used to gain. Thus, three of the approaches (Magnetic Properties, Thermal Methods and Gas Sorption) deal with the bulk properties of materials, while AFM looks at structure on atomic length scales and Light Scattering deals with particle sizing. In particular, Light Scattering includes an introduction to, and discussion of, the new method of differential dynamic microscopy.

As ever, we approached a team of expert authors to contribute to this volume and we are delighted with the job they have done; we take this opportunity to thank them publicly. Taken together with *Local Structural Characterisation* and *Structure from Diffraction Methods*, the chapters in this volume describe a wide range of methods for the characterisation of inorganic materials, which we hope will find value as a reference source for those engaged in their study.

DWB, York
DO'H, Oxford
RIW, Warwick
October 2013

List of Contributors

Michael W. Anderson Centre for Nanoporous Materials, School of Chemistry, The University of Manchester, Manchester, UK

Pablo Cubillas Centre for Nanoporous Materials, School of Chemistry, The University of Manchester, Manchester, UK

Erika Eiser Cavendish Laboratory, University of Cambridge, Cambridge, CB3 0HE, UK

Michel B. Johnson Institute for Research in Materials, Dalhousie University, Halifax, NS, Canada

Philip L. Llewellyn Aix-Marseille University & CNRS, Laboratoire MADIREL, Marseille, France

Scott S. Turner Department of Chemistry, University of Surrey, Guildford, GU2 7XH, UK

Mary Anne White Institute for Research in Materials; Department of Chemistry; Department of Physics and Atmospheric Science, Dalhousie University, Halifax, NS, Canada

1

Measurement of Bulk Magnetic Properties

Scott S. Turner
Department of Chemistry, University of Surrey, Guildford, GU2 7XH, UK

1.1 INTRODUCTION

Measuring the response of a sample to an external magnetic field is a fundamentally important technique in inorganic materials chemistry, because of the close link between a material's magnetic properties and the number and orbital placement of its electrons. Chemistry can be considered the study of the manipulation of electrons, *e.g.* inducing electrons to move from one species to another in redox reactions or understanding how to manipulate electrons in order to hold atoms together by breaking and forming new bonds. From the point of view of coordination chemistry, great strides in understanding the links between magnetism and electron distribution advanced from the seminal work on the concepts of crystal field theory by Van Vleck and Bethe in the late 1920s.[1−3]

Crystal field theory, in which the ligand orbitals of a complex are modelled as negative point charges, leads to the splitting of valence orbitals at the metal, affecting how valence electrons fill the orbitals. This simple idea accounts successfully for many magnetic properties. The effect is a result of different magnitudes of electrostatic repulsion

Multi Length-Scale Characterisation, First Edition. Edited by Duncan W. Bruce, Dermot O'Hare and Richard I. Walton.

between electrons in the metal-based orbitals with specific directionality (and associated orbital momentum) and the distribution of the point charges.[4] However, this theory does not consider bonding and more sophisticated theories, such as ligand field theory[5] and molecular orbital theory, are often combined with magnetic measurements for this purpose. Magnetic data are simple to use to characterise samples as 'magnetic' or 'non-magnetic', to help improve materials that are technologically significant, such as those used as magnetic storage or as magnetic refrigerants (magnetocaloric devices).[6] However, a deeper analysis and understanding of magnetic data also sheds light on geometry, oxidation state, spin state and chemical bonding. Furthermore, the preparation and characterisation of new magnetic materials and the elucidation of structure–property relationships can be used to test predictions of new theoretical models. This is particularly true in the area of molecular or molecule-based magnets,[7,8] where the power of synthetic chemistry can be used to manipulate the nature and interaction of magnetic centres. For example, it is possible to prepare low-dimensional magnetic materials, which would otherwise be difficult or impossible to achieve using elemental materials. Magnetic data can be used in conjunction with other techniques that probe electron distribution, such as electronic spectroscopy and electron paramagnetic resonance (EPR) and used in conjunction with information about structure, such as that obtained from variable-temperature, single-crystal X-ray diffraction, magneto-structural correlations can be determined.

However, the theoretical study of magnetism is undoubtedly a difficult subject and often causes trouble among undergraduate students. The difficulty is often conceptual, although one issue is historical, since different authors use different unit systems and even different definitions of fundamental quantities. Another reason for difficulty is that in order to understand fully the origin of magnetic phenomena, an appreciation of quantum mechanics is required. It is fortunate that there are useful classical approaches that are more accessible, and as far as possible, such an approach will be taken here. A mathematical treatment of magnetism is unavoidable, and this chapter will present equations that explain the various physical interpretations. However, no detailed derivations will be given. For a much more comprehensive mathematical approach, the reader is referred to the excellent text on *Magnetism in Condensed Matter*.[9] There are many other first-rate texts that describe the general principles and theory of magnetism and a few are recommended to the reader.[10–13]

1.1.1 Purpose and Scope

The main purpose of this chapter is not to give an exhaustive overview of the theoretical underpinnings to magnetism; rather, it is to describe the physical techniques used by magnetochemists (and physicists) to determine the bulk magnetic properties of materials. Nevertheless, some background theory is necessary in order to provide clarity to subsequent sections and hopefully limit the need for the reader to refer to other texts. The discussion will explore a range of techniques, from the ubiquitous *Superconducting QUantum Interference Device* (SQUID) magnetometer, with which all magneto-experimentalists should be familiar, to more specialist ones such as the micro-SQUID. The development of the SQUID magnetometer was very important to the rapid development of the theory and to experimental work in magnetism, as it is relatively easy to use and provides accurate and very precise variable-temperature and magnetic-field measurements. Since the SQUID is the most common experimental technique, it will be described in the most detail and an example of raw data output and how the data are treated will be provided. Other techniques that can be used to provide magnetic information, such as nuclear magnetic resonance (NMR) spectroscopy, will be described briefly, although the focus will be on the specific experiments that give magnetic information and not on a detailed general description of the techniques themselves. The theoretical underpinning of these techniques can be found in other widely available – and more authoritative – sources. Finally, some complementary techniques will be described, such as electronic spectroscopy, where once again the focus will be on the information that can be extracted to complement magnetic data, rather than on the technique itself. The approach will be to describe very briefly the fundamentals of the method, describe how an experiment is conducted and, in some cases, show what data are obtained and how they can be used to inform the experimentalist about the magnetic properties. The text is intended to be accessible to a target audience of postgraduates and advanced undergraduates in the hope that it will encourage those who wish to know more about how magnetic properties are measured to enter this fascinating field.

1.1.2 The Origin of Magnetic Properties

1.1.2.1 The Classical View

The bulk magnetic behaviour of any material is caused by the presence of magnetic dipoles, which can be thought of as the magnetic equivalent of electric dipoles. Electric dipoles consist of two opposite charges separated by a finite distance, and magnetic dipoles likewise have two distinguishable ends (often called north and south poles) separated by a distance. However, the comparison is not perfect. For example, as we shall see, magnetic dipoles are associated intimately with electron angular momentum, whereas electric dipoles are not. The different categories of magnetism, such as diamagnetism, paramagnetism, ferromagnetism *etc*., as well as less well known types such as canted antiferromagnetism, emerge from the way in which a large ensemble of magnetic dipoles interact with one another and with their environment; specifically, from how they interact with an externally applied magnetic field. Magnetic dipoles, in the classical sense and on a macroscopic level, can be generated by a circuit or loop of moving charge, as shown in Figure 1.1. The resulting magnetic dipole is represented by a vector (just as with electric dipoles) orientated perpendicular to the current loop. The magnitude of the dipole, and therefore the length of the representing vector, is called the magnetic moment and given the symbol μ. The magnitude of μ is proportional to the area, A, prescribed by the loop and the charge passing a point in the loop *per* unit time – in other words, the current, I. Therefore, using SI units, we can define one unit of magnetic moment as that generated by a loop of area 1 m^2

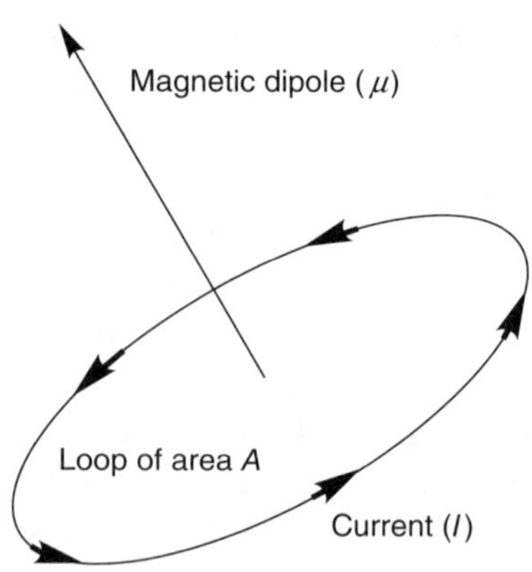

Figure 1.1 Generation of a magnetic moment from a current loop.

that carries 1 A of current, as given by Equation 1.1. Throughout this chapter, non-bold letters refer to the magnitude of a vector and bold letters to a vector quantity. The vector $\boldsymbol{A}$ has a magnitude, A, equal to the loop area and a direction parallel to the direction of the magnetic dipole. The current is a simple scalar quantity.

$$\mu = IA \tag{1.1}$$

Consider a magnetic dipole placed in an externally applied, homogeneous magnetic field with a defined direction. The current loop, and thus the magnetic dipole, will tend to orientate itself to become coincident with the direction of the external field, due to a propensity to minimise the energy of the magnetic moment. In a magnetic field, the dipole has energy, E, with a negative value equal to the dot or scalar product of the magnetic moment and external field, $\boldsymbol{H}$, vectors as given by Equation 1.2. According to Equation 1.2, the energy is minimised when $\boldsymbol{\mu}$ and $\boldsymbol{H}$ are coincident, for which α is zero and E becomes $-\mu H$.

$$E = -\boldsymbol{\mu} \cdot \boldsymbol{H} = -\mu H\cos\alpha \tag{1.2}$$

When α is not zero, the dipole rotates to minimise its energy. There must therefore be a force, $\boldsymbol{F}$, acting on the magnetic dipole. This has a magnitude equal to the cross-vector product of $\boldsymbol{\mu}$ and $\boldsymbol{B}$, as given by Equation 1.3 and shown in Figure 1.2.

$$F = \boldsymbol{\mu} \times \boldsymbol{H} = \mu H\sin\theta \tag{1.3}$$

A magnetic dipole that sits in a non-uniform (inhomogeneous) magnetic field is subject to a translational force. The energy is minimised

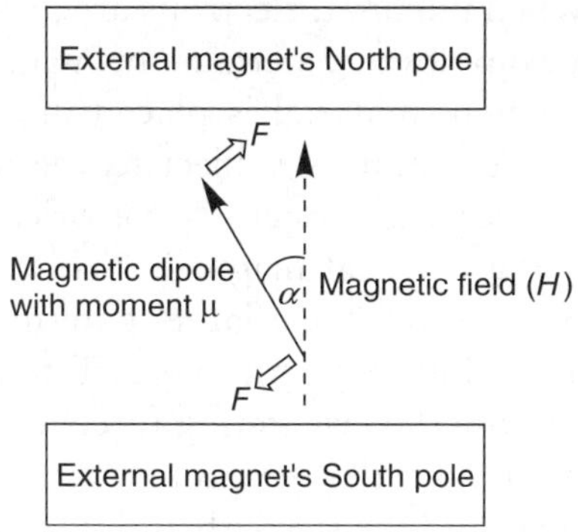

Figure 1.2 A magnetic moment, $\boldsymbol{\mu}$, in an externally applied magnetic field, $\boldsymbol{H}$. The energy of the magnetic moment is determined by its angle, α, to the magnetic field. The orientation force, F, acts on the magnetic moment to minimise its energy.

in Equation 1.3 if the dipole moves towards areas that contain larger magnitudes of $\boldsymbol{H}$, where the magnetic flux lines are most dense. When confined to a particular direction, this force is proportional to the magnetic field gradient, as described by Equation 1.4. In this equation, μ_z is the component of the magnetic dipole in the z direction and dH/dz is the field gradient. This effect is an important one, since the measurement of this translational force is the basis of some methods of determining the magnetic moment, as described in Section 1.2.

$$F_z = \mu_z \frac{dH}{dz} \tag{1.4}$$

1.1.2.2 The Quantum View

On a sub-atomic scale, the origin of magnetic moments can be visualised as arising from the movement of charge, although in this case it is associated with the motion of electrons in orbitals (which is linked to the spin and orbital momentum of the electrons). However, this way of thinking should only be used to help conceptualise the effect. Electron spin is commonly imagined as the electron particle spinning about an axis, rather like the Earth's rotation, but in reality it is actually a more nebulous quantity, with the momentum being generated internally.

Bulk materials evidently contain many atoms that have associated electrons, each having spin and orbital momentum. In cases where all electrons are paired, the sum of the momenta will cancel and no permanent magnetic moment is generated. However, external magnetic fields can still affect these materials (see the discussion on diamagnetism in Section 1.1.4.1). When unpaired electrons are available, the momenta are not cancelled and dipoles and permanent magnet moments will be present. When such a bulk material is placed in a magnetic field, the magnetic moment of each atom is subject to the orientating force, as described above. However, the propensity for orientation is moderated by the magnitude of the external magnetic field and the temperature. Clearly, according to Equation 1.3, a larger external magnetic field will result in a larger force. High temperatures will tend to randomise the moments, acting to oppose the orientating force. In addition, a material may have internal interactions that promote certain relative orientations of neighbouring dipoles, and the strength and type of these are relevant, as discussed below.

The enforced orientation of dipoles can be restated by saying that when a material is placed in a magnetic field, a magnetisation, M, is induced in the sample. The magnetisation is equivalent to the vector sum of the individual moments within a sample and therefore differs according to whether the sample is in or out of an external magnetic field. In order that the magnitudes of M can be compared, it is defined as the total magnetic moment *per* unit volume. Yet another way of stating this is that when any object containing electrons is placed in a magnetic field, the density of the magnetic lines of force (called the magnetic flux density or magnetic induction, B) within the object differs from the applied magnetic field, H, that surrounds the object. At this point it is necessary to warn the reader that the mathematical expression used for this idea is presented in several different ways in the research literature and textbooks; the issue is explained very nicely in an article by Quickenden and Marshall.[14] A significant problem is that it depends on which units are used (see Section 1.1.3 for more detail). The units used most commonly in science are the rationalised SI units (*Système Internationale d'unités*), but in magnetism the unrationalised cgs system is very widespread, which measures length in centimetres, mass in grams and time in seconds. The research literature is replete with data in cgs units and there is a firm reluctance in large portions of the scientific community to transfer to the use of SI – there are good reasons for this, but it does lead to confusion. Equation 1.5 is most often used for SI units, while Equation 1.6 is the corresponding equation for cgs units. In these equations, the external magnetic field, H, and the magnetisation, M, are treated as similar quantities with the same units. In other words, M is considered to be the 'additional magnetic field' induced internally that may support or oppose H, depending on whether it is negative or positive. One other confusion is that in cgs units, the permeability of a vacuum, μ_0, is dimensionless and equal to 1, so the reader will inevitably find equivalent equations that may or may not explicitly include this constant. In SI units μ_0, has a value of $4\pi \times 10^{-7}\ \mathrm{kg\,m\,s^{-2}\,A^{-2}}$.

$$\mathbf{B} = \mu_0(H + M) \tag{1.5}$$

$$\mathbf{B} = \mu_0(H + 4\pi M) \tag{1.6}$$

Another important quantity is the volume magnetic susceptibility, χ_v, which is a measure of how susceptible the dipoles are to reorientation as a function of the applied magnetic field, taken at some fixed temperature, *per* unit volume. A high susceptibility means that for a unit change in H, the magnetisation increases by a large amount. In other words, the dipoles are easy to orientate, requiring less force. A small value of χ_v means that the dipoles are more resistant to orientation, which leads to a small change in M *per* unit change in H. The value of χ_v can also be positive or negative, where an increase in H will increase or decrease M, respectively. This can be illustrated graphically by plotting the curve of M *versus* H, the gradient of which is the volume susceptibility according to Equation 1.7.

$$\chi_v = \frac{dM}{dH} \tag{1.7}$$

For most materials, under relatively small values of H, the plot of M *versus* H is linear, which means that χ_v is a constant, and therefore Equation 1.7 may be rewritten in the form of Equation 1.8. A magnetic investigation will routinely plot M *versus* H to determine the values of H for which linearity holds. Fields in this range can then be used to make variable-temperature measurements and so calculate magnetic susceptibility using Equation 1.8. Under these conditions, the magnetic susceptibility is independent of the magnitude of H.

$$\chi_v = \frac{M}{H} \tag{1.8}$$

1.1.3 The Units of Magnetism

As stated above, when beginning with Equations 1.5 and 1.6, the units in the magnetism literature are not consistent, and it can be argued that the situation is entirely confusing. It is clear from Equation 1.8 that the volume magnetic susceptibility, χ_v, is a dimensionless quantity, but in the cgs system it is not usually reported without units, being given as emu cm^{-3}, where 'emu', or electromagnetic units, is equivalent to cm^3. This clearly emphasises the relationship to volume. However, the quantity of interest to chemists is often magnetic susceptibility *per* mole, so the volume susceptibility is routinely converted as follows. The mass or gram susceptibility, χ_g, in units of $cm^3\,g^{-1}$ or emu g^{-1}, is obtained by dividing the volume susceptibility by the density in $g\,cm^{-3}$. This is then converted to molar susceptibility, χ_m, by multiplying by the molar mass in $g\,mol^{-1}$. As described in Section 1.3.1.3.2, the raw data output from a SQUID magnetometer (see Section 1.3.1) is calibrated to give mass magnetisation values rather than volume magnetisation. These data are converted to χ_m by multiplying by the sample's molecular weight and dividing by the sample mass and the applied external magnetic field. In the cgs system, the units of χ_m are reported as emu mol^{-1}, or often as the equivalent $cm^3\,mol^{-1}$. Clearly, in SI units χ_g is given in $m^3\,kg^{-1}$ and χ_m in $m^3\,mol^{-1}$. For more on this topic, see references [14, 15].

1.1.3.1 *Conversion of Units between SI and cgs Systems*

At first glance, the conversion between cgs and SI units seems straightforward: grams to kilograms, centimetres to metres. However, it is not quite that simple. As exemplified by Equation 1.6, the relationships may contain the constant π, since the permeability of free space is $4\pi \times 10^{-7}$ kg m s^{-2} A^{-2}. This reflects the fact that spherical symmetry is relevant in the origins of these equations ($4\pi r^2$ being the surface area of a sphere with radius r). However, while π is involved in the equation for B in cgs units, it may also appear in equations in SI units. In order to clarify the situation, Table 1.1 can be used to convert between SI and cgs units, so that literature values can be compared. For example, in order to convert a magnetic induction of 20 G to SI units, we multiply by 10^{-4} to give 2×10^{-3} T. Table 1.2 shows the values of some fundamental constants used in the text in both SI and cgs units. Note that in the cgs system, the Bohr magneton is 9.273×10^{-21} erg G^{-1} and in SI units

Table 1.1 Conversion between units of magnetism in the SI and cgs systems. The abbreviations for the units are: Oe, Oersted; A, Ampere; m, metre; G, Gauss; T, Tesla; Wb, Weber; emu, electromagnetic units. Note that H and M have the same units (within a unit system) and are treated as similar physical quantities (see text). The magnetic moment is often stated in units of the Bohr magneton, μ_B, by dividing throughout by that constant.

Quantity (symbol)	Units in cgs	Conversion factor	Units in SI
Magnetic field (H)	Oe	$10^3/4\pi$	A m^{-1} *or* J T^{-1} m^{-1}
Magnetic induction *or* magnetic flux density (B)	G	10^{-4}	Wb m^{-2} *or* T
Magnetisation *or* total moment per volume (M)	Oe	10^3	A m^{-1} *or* J T^{-1} m^{-1}
Magnetic moment (μ)	Bohr magnetons *or* erg G^{-1}	1 10^{-3}	Bohr magnetons *or* J T^{-1}
Volume susceptibility (χ_v)	dimensionless, but often written as emu cm^{-3}	4π	dimensionless
Mass *or* gram susceptibility (χ_g)	emu g^{-1} *or* cm^3 g^{-1}	$4\pi \times 10^{-3}$	m^3 kg^{-1}
Molar susceptibility (χ_m)	emu mol^{-1} *or* cm^3 mol^{-1}	$4\pi \times 10^{-6}$	m^3 mol^{-1}

Table 1.2 Fundamental constants used in the text, in SI and cgs units. Note that the erg, the unit of energy in the cgs system, is equivalent to $cm^2\,g\,s^{-2}$, and the Joule, J), the SI unit of energy, is equivalent to $m^2\,kg\,s^{-2}$.

Constant (symbol)	Value in cgs	Value in SI
Boltzmann constant (k_B)	1.3807×10^{-16} erg K^{-1}	1.3807×10^{-23} J K^{-1}
Bohr magneton (μ_B)	9.274×10^{-21} erg G^{-1}	9.274×10^{-24} A m^2 *or* J T^{-1}
Plank's constant (h)	6.626×10^{-27} erg s	6.626×10^{-34} J s

it is 9.273×10^{-24} A m^2. Using these values, the magnetic moment is $2.8279\sqrt{\chi_m T}$ Bohr magnetons in cgs units and $7.9774\sqrt{\chi_m T}$ Bohr magnetons in SI units, where T is temperature. However, it should be noted that these two expressions give the same numerical value for the magnetic moment, so the expressions found in Section 1.1.5.1, such as $\mu = g^2\sqrt{S(S+1)}$, are correct in either system of units.

1.1.4 Magnetism due to Paired Electrons

1.1.4.1 Diamagnetism

Diamagnetism arises from the interaction of an external magnetic field with paired electrons. This includes the paired electrons in core orbitals of materials with unpaired valence electrons. Therefore, all materials have some degree of diamagnetism, although in some cases the property is masked by another magnetic property (*vide infra*). When a purely diamagnetic material is placed in a magnetic field, a magnetisation is induced that opposes the external field. The magnetic susceptibility is therefore negative, although compared to other magnetic phenomena it is also relatively small. As shown in Figure 1.3, the susceptibility is independent of temperature.

In addition, in an inhomogeneous magnetic field, a purely diamagnetic material will be subject to a translational force, as given by Equation 1.3, which results in repulsion from the strongest parts of that field. Alternatively, we can state that a diamagnetic material expels the externally applied magnetic field. Since diamagnetism is present in all materials, when we wish to monitor another magnetic property, the underlying diamagnetic contribution must be subtracted from the total according to Equation 1.9. In this expression, χ_{total} is the total measured magnetic

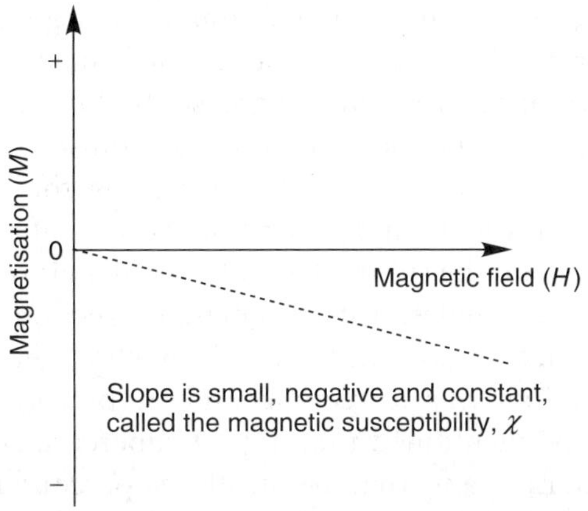

Figure 1.3 Magnetic susceptibility and magnetisation *versus* temperature for a diamagnetic material.

susceptibility, χ_{up} is that due to unpaired electrons and χ_p is that due to paired electrons. Diamagnetism is an atomic-scale consequence of Lenz's law, which states that an induced electromotive force (a voltage) always gives rise to a current whose associated magnetic field opposes that which induces it. This change is a requirement of the law of conservation of energy. Another way of stating this is that when the magnetic field interacts with paired electrons, the angular momentum of the electrons is altered in such a way as to create an opposing magnetic field.

$$\chi_{total} = \chi_{up} + \chi_p \tag{1.9}$$

After a sample is measured, the total magnetisation is obtained, from which the total susceptibility (usually *per* mole) can be calculated. In order to determine what part of the susceptibility is due to the unpaired electrons, we need some way of obtaining the value caused by the paired electrons. The best method is to measure a structurally equivalent sample that contains only paired electrons, although this can be impractical. A crude approximation can be made using the molecular weight of the sample, but the most common approach is to use Pascal's constants.[16,17] These are additive constants that give a good estimate of the diamagnetic contribution to the susceptibility arising from individual atoms, ions and some common ligands and other species. For many experiments, the diamagnetic contribution is a small correction to the total susceptibility and so the use of Pascal's constants is sufficiently accurate. It should also be noted that the sample is usually kept in a container during measurement, which might be a glass or quartz vial, a plastic tube or a gelatine capsule, and which may contain a fixative such as cotton wool or even glue or grease. It should be clear that the diamagnetic properties of the sample holder must also be taken into consideration. This is done simply by measuring the magnetisation of the blank holder at a few temperatures so as to confirm diamagnetism and obtain the value to be subtracted in later calculations. Finally, when measuring solution samples, a diamagnetic correction should be made for the solvent. Since the solvent volume is much larger than the volume of the magnetic solute, this correction can be substantial.

It should also be mentioned that Type I superconductors are perfect diamagnetic materials and that below the superconducting transition temperature, the induced magnetic field *exactly* opposes that of an external magnet. In other words, the external magnetic field is completely expelled from the interior of the superconductor. The classic experiment used to demonstrate this effect (the Meissner effect) shows the material hovering above a magnet. This is a spectacular manifestation of the repulsive force mentioned above. In fact, this is the same phenomenon seen in normal diamagnetic materials, except that the external magnetic field is not fully expelled.

When discussing some other types of magnetism in the following sections, it is assumed that the diamagnetism has already been taken into consideration. Therefore, the discussion relates only to the magnetism that originates from unpaired electrons.

1.1.5 Magnetism due to Unpaired Electrons

1.1.5.1 Curie Paramagnetism

The discussion will focus first on magnetic properties that result entirely from the spin angular momentum; Section 1.1.5.4 will then introduce deviations due to orbital momentum. Paramagnetic materials must have unpaired electrons; this is unavoidable in species carrying an odd number of electrons, *e.g.* radical species or Fe^{III} complexes, but can also arise in species with even numbers of valence electrons. It occurs when orbital filling according to the Aufbau principle, Pauli exclusion principle and in particular Hund's first rule, which maximises the total spin quantum number, leads to partially filled orbitals. The uncompensated angular momentum of the unpaired electrons produces magnetic dipoles and

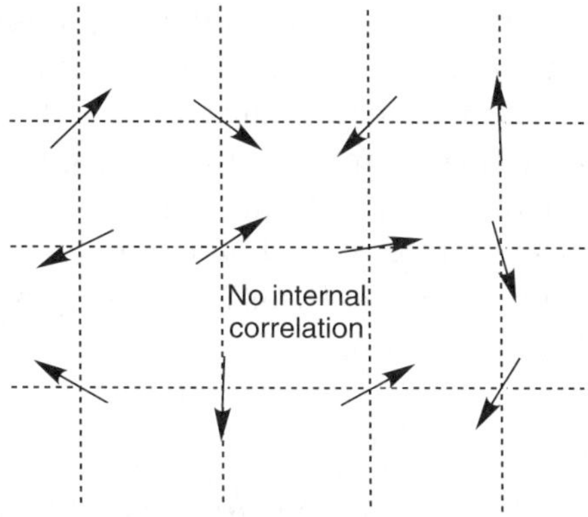

Figure 1.4 Random array of magnetic moments in a paramagnetic material. Each moment is assumed to have an identical magnitude reflected in the length of the representing vector, but this does not need to be true. Dotted lines represent the atomic lattice.

associated magnetic moments. For paramagnetic materials, there are no substantial internal interactions between the magnetic moments at each magnetic centre. In other words, the bulk material consists of an array of independent, uncorrelated magnetic moments in which the orientation of each dipole does not affect the orientation of neighbouring dipoles, as shown in Figure 1.4. We call this a dilute system.

When a dilute paramagnetic material is not in a magnetic field, the randomly orientated magnetic dipoles give a vector sum of the magnetic moments (the magnetisation) of zero. However, when it is placed in a magnetic field, as described in Section 1.1.2.1, there is a propensity for the magnetic moments to orientate parallel to the magnetic field. The magnetisation and magnetic susceptibility are positive, temperature dependent and larger in magnitude than they are in the case of diamagnetism. The dipoles are subject to thermal randomisation, so at higher temperatures, the magnetisation and susceptibility are smaller than they would be at lower temperatures, assuming the same applied magnetic field. Indeed, at sufficiently high temperatures, the randomised alignment of the dipoles can result in cancellation of any effect due to unpaired electrons and only the diamagnetic contribution remains.

The temperature dependency of the magnetic susceptibility for a dilute paramagnet can be described by the Curie law, which was proposed by Pierre Curie in 1895. This is given in Equation 1.10, where C is the Curie constant. The value of C depends on the total spin of the electrons and on a collection of fundamental constants given in

Equation 1.11. The derivation of this equation from first principles, *via* the Van Vleck equation, can be found in most texts on magnetism.[11] In Equation 1.11, N is Avagadro's constant, g is the electron g-factor and k_B is the Boltzmann constant. S is the total spin quantum number, which is the sum of the m_s quantum numbers – in other words, half the number of unpaired electrons. For example, for three unpaired electrons, $S = 3/2$, and for six electrons in a high-spin (see Section 1.1.7) octahedral metal complex, $S = 2$.

$$\chi_m = \frac{C}{T} \tag{1.10}$$

$$C = \chi_m T = \frac{Ng^2\mu_B^2}{3k_B} S(S+1) \tag{1.11}$$

Inputting the relevant constants in cgs units taken from Table 1.2 gives Equation 1.12, which is often how the equation is recalled. The electron g-factor is a dimensionless quantity, a simple ratio that relates the observed magnetic moment to the angular momentum quantum number and the fundamental atomic unit of magnetism. The expression for g is given in Equation 1.13, where h is Plank's constant and S is the spin angular momentum quantum number, as defined above.

$$C = \chi_m T = 0.12505 g^2 S(S+1) \tag{1.12}$$

$$g = \frac{h\mu_{\text{observed}}}{2\pi S \mu_B} \tag{1.13}$$

The Bohr magneton, μ_B, is a useful atomic unit of magnetic moment as it is the magnetic moment of a single 1s electron in hydrogen. This is given by Equation 1.14, where e is the electron charge and m_e the electron mass.

$$\mu_B = \frac{eh}{4\pi m_e} \tag{1.14}$$

For a free (not associated with an orbital) unpaired electron in which only the spin angular momentum contributes to the magnetic moment, the g-factor has a value of 2.0023. This is because the spin magnetic moment divided by the spin angular momentum is twice the orbital magnetic moment divided by the orbital angular momentum. The slight variation from exactly 2 is due to relativistic effects. Indeed, a g value of 2 is often used as a first guess when modelling magnetic data. However, the g-factor can take different values in real situations depending on the contributions from the spin and orbital momentum and is related to the spin–orbit coupling (see Section 1.1.5.4).

Adherence to the Curie law can be confirmed by plotting $1/\chi_m$ *versus* T, which, as shown in Equation 1.10, gives a straight line with a gradient of $1/C$. The extrapolated curve passes through the origin, as shown in Figure 1.5. So, for example, non-interacting (or 'dilute') Cr^{III} octahedral complexes have three unpaired electrons ($S = 3/2$) and, assuming $g = 2$, should give a straight line from which C or $\chi_m T$ equal to $1.8757\,\mathrm{emu\,K\,mol^{-1}}$ can be obtained. Another useful curve to plot is $\chi_m T$ *versus* T, which for a paramagnetic will be a straight horizontal line since C is constant. Any deviation from a straight line could indicate non-paramagnetic behaviour.

Older literature reports curves of magnetic moments, μ, against T rather than $\chi_m T$, and this form is still used by some non-specialists. The relationship between $\chi_m T$ and μ for $g = 2$ is given by Equation 1.15. The reason for the preference in reporting $\chi_m T$ is that the values are additive. For example, if a compound contains a mole of $S = 3/2$ centres in each mole of material that contributes $\chi_m T = 1.8757\,\mathrm{emu\,K\,mol^{-1}}$ (when $g = 2$), plus a mole of independent $S = 1/2$ centres with $\chi_m T = 0.375\,\mathrm{emu\,K\,mol^{-1}}$, then we will expect the total $\chi_m T$ to be the sum, *i.e.* $2.251\,\mathrm{emu\,K\,mol^{-1}}$. This develops from the fact that values of magnetic susceptibility are themselves additive. However, the values of magnetic moment are not additive, since they are related to the square root of the molar magnetic susceptibility, as shown in Equation 1.15.

$$\mu = \sqrt{8\chi_m T} \tag{1.15}$$

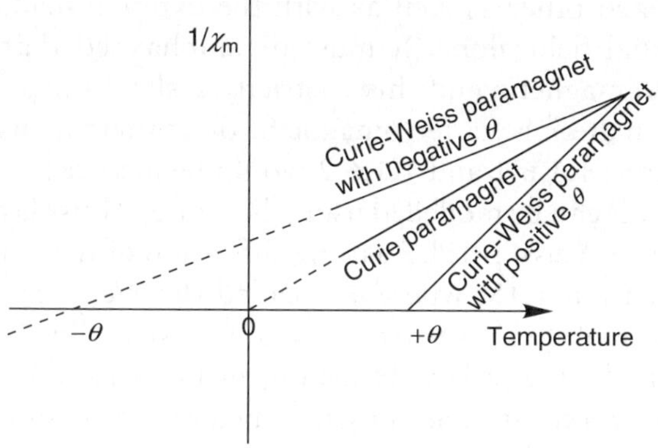

Figure 1.5 Plot of $1/\chi_m$ *versus* T for a Curie paramagnet and a Curie–Weiss paramagnet with a negative or positive Weiss constant, θ.

Note that Equation 1.15 has the magnetic moment in units of Bohr magneton. Even though it would be strictly correct to write μ/μ_B on the left side of Equation 1.15, it is normally written in the form given. Some useful expressions are also given in Equations 1.16 and 1.17. The former shows the relationship between the magnetic moment and S, noting that in this case μ_B has been written into the equation explicitly. As mentioned above, the value of μ is commonly reported in units of Bohr magneton, so this equation would then be correct without μ_B. Equation 1.17, in which n is the number of unpaired electrons ($S = n/2$), is a commonly used form of Equation 1.16, where μ is given in Bohr magnetons and g has been set to 2. Equations such as 1.16 and 1.17 are commonly known as the **spin-only formula**, since the magnetic moment is only a function of the electron spin momentum.

$$\mu = g\mu_B\sqrt{S(S+1)} \tag{1.16}$$

$$\mu = \sqrt{n(n+2)} \tag{1.17}$$

1.1.5.2 Curie–Weiss Paramagnetism

A paramagnetic material with no internal interactions between neighbouring magnetic moments is quite rare and real materials deviate from the Curie law. In these cases, the non-zero interactions can lead to neighbouring moments being more likely to align either parallel or antiparallel. In other words, there is a small coupling between neighbouring moments, which is a **ferromagnetic coupling** for parallel alignment or an **antiferromagnetic coupling** for antiparallel alignment. The magnetic dipoles interact with each other as well as with the external field, rather than with the external field alone. It must be emphasised that the materials are still paramagnetic and this is strictly a short-range coupling; to describe these materials as ferromagnetic or antiferromagnetic as discussed in Sections 1.1.6.1 and 1.1.6.2 would be incorrect.

These materials can be modelled using the Curie–Weiss law, which was detailed by Pierre Weiss in 1907 as a modification of the Curie law. This is given in Equation 1.18, where θ is called the Weiss or Curie–Weiss constant. A plot of $1/\chi_m$ *versus* T gives a straight line with gradient $1/C$, just as in the Curie law. However, in this case the extrapolated line crosses the x-axis at θ, as shown in Figure 1.5. It should be noted that the Curie–Weiss law only applies to temperatures that are much larger than the strength of any interaction. Deviations from the straight line commonly occur at low temperatures, where the magnetic dipole interactions are more significant. In addition, an occasional source of confusion is that the Weiss constant can be negative and yet has units of temperature (often Kelvin). However, it is the extension of the curve that is being considered, so there is no real experimental point with a negative value in Kelvin, despite expressions such as $\theta = -4\,\mathrm{K}$.

$$\chi_m = \frac{C}{(T - \theta)} \tag{1.18}$$

A common interpretation of θ is that a positive value indicates ferromagnetic coupling and a negative value antiferromagnetic coupling. While this interpretation is usually correct, there are other sources of deviation from the Curie law, such as when orbital angular momentum makes a significant contribution to the magnetic moment and so-called spin–orbit coupling occurs (detailed in Section 1.1.5.4). Therefore, caution should be used when associating the value of θ to specific interactions, unless it is certain that no other effects are present.

Finally, it should be emphasised that the Equations in Section 1.15 are strictly applicable to paramagnetic materials alone and as such are physically meaningless when any other type of magnetism is present. So relating $\chi_m T$ (or C) or the magnetic moment to the number of unpaired electrons can only be done when it is clear the sample is paramagnetic; this might be at sufficiently high temperatures, well above any critical temperature of moment ordering, for example (see Section 1.1.6).

1.1.5.3 Other Sources of Paramagnetism

1.1.5.3.1 Pauli Paramagnetism

Pauli paramagnetism can occur in materials that are metallic conductors of electricity, *i.e.* where the electrons are itinerant rather than localised. The magnetic susceptibility is small and positive – smaller than that of Curie or Curie–Weiss paramagnets. The susceptibility is also independent of temperature. In fact, this type of paramagnetism is of an order that competes with diamagnetism. For metals, the valence electrons are delocalised and are found in partially filled energy bands constructed from strong interatomic orbital overlap. These bands contain electrons that, without exposure to a magnetic field, are distributed equally with quantum numbers $m_s = \pm 1/2$, so that they are all paired. The highest occupied energy level is called the Fermi level, E_F. At this level the electrons sit just below a higher-energy, unoccupied level. Crucially, the unoccupied levels are very close in energy to E_F. Under some circumstances it could, therefore, be favourable energetically for some electrons to occupy the higher energy levels and thus increase the number of parallel spins according to Hund's rule, which maximises *S*. This is favourable by virtue of lowering electron–electron repulsion, but is also unfavourable because the electrons will occupy higher levels. The magnitude of the competing interactions determines how many electrons become unpaired. In fact, if there is a great propensity for this to occur, the result will be a ferromagnetic material, since there will be a strong energetic reason to have many unpaired electrons with parallel spins.

The probability of electron promotion is enhanced by having a large number of valence electrons. Electron promotion is also more likely if the electrons exist in narrow bands; this is the case with bands formed from 3*d* orbitals, since overlap between diffuse orbitals is poor. The two effects combine to give a high density of electrons at the Fermi level, so that more electrons are in a position to move to higher levels. When an external magnetic field is applied, the Zeeman splitting (see Section 1.5.1) leads to extra stability for electrons with spins parallel to the magnetic field. For many metals, this means ultimately that a modest percentage of electrons are promoted and the populations with $m_s = +1/2$ and $m_s = -1/2$ are no longer equal. The result is a small, permanent, magnetic moment and an extra contribution from this Pauli paramagnetism. However, for three *d*-block metals, Fe, Co and Ni, the energetics are such that a large number of electrons are promoted and their density is such that a strong ferromagnetic correlation exists between these electrons (see Section 1.1.6.1). These metals are thus the only common ferromagnetic elements in the Periodic Table.

There are also a number of examples where the nature of the electrons can be switched from being localised to being itinerant or where the density of the electrons is altered by some external perturbation. The effect is clearly observed in the change of magnetic properties and can be a method of monitoring metallic conductivity. Such examples include the conversion from Curie–Weiss to Pauli paramagnetism in $MNiO_3$ (M = lanthanide)[18] and, more dramatically, the switch between ferromagnetism and Pauli paramagnetism of a compound with the perovskite structure, $CaCu_3(Ru_{4-x}Mn_xO_{12})$, for a range of x.[19]

1.1.5.3.2 Temperature-Independent Paramagnetism

Temperature-independent paramagnetism (TIP) arises from the coupling of the ground state to high-energy excited states in the presence of a magnetic field.[11] As with Pauli paramagnetism, the magnetic susceptibility is most often similar in magnitude to that of diamagnetism, but with an opposite sign, although in some cases it can be significant. As its name states, this type of paramagnetism is invariant with changes in temperature. The effect is responsible for the small paramagnetism seen in species with a d^0 electron configuration, such as chromate(VI) or manganate(VII) ions. However, TIP can also be observed in compounds that do not have a diamagnetic ground state, and this should be taken into consideration as a correction in the same way as discussed for diamagnetism. For Curie and Curie–Weiss paramagnets, the effect is clearly seen in a deviation from linearity for the plot of $1/\chi_m$ *versus* T. Therefore, a Curie or Curie–Weiss equation plus an extra fitting parameter can be used when modelling paramagnetic data. This is shown in Equation 1.19 for a Curie–Weiss relationship. Clearly, as T increases, the Curie–Weiss term of Equation 1.19 becomes less significant and the TIP term more dominant. Therefore, the deviation from linearity is more obvious at higher temperatures.

$$\chi_m = \frac{C}{(T - \theta)} + TIP \tag{1.19}$$

1.1.5.4 The Orbital Contribution

The discussions in Sections 1.1.5.1–1.1.5.3 assume that the only significant contribution is from the spin of the electrons and ignore any contribution from the orbital momentum. Therefore, Equations 1.11 and 1.16 are written as being dependent only on S. In some cases, as will be shown, these are perfectly good relationships to use. However, an electron's orbital angular momentum may be significant, in which case equations that contain only the S quantum number will no longer be entirely valid. In such cases, the orbital and spin angular momentum may contribute to the magnetism independently, if they are weakly coupled, or else they may need to be treated differently due to spin–orbit coupling. In the former case, the total orbital angular momentum is given by the quantum number, L, which is the sum of the m_l values for individual electrons. Equation 1.11 is therefore incorrect and should be replaced by Equation 1.20, where the values of g are now distinguished and g_s is the g value for the spin contribution. However, note that g for the orbital contribution, g_L, is exactly 1, so this equation is often written without the inclusion of g_L. The corresponding equation for the magnetic moment is

Equation 1.21, with μ_B written explicitly.

$$\chi_m T = \frac{N\mu_B^2}{3k_B}\left(g_s^2 S(S+1) + g_L^2 L(L+1)\right) \tag{1.20}$$

$$\mu = \mu_B\sqrt{g_s^2 S(S+1) + L(L+1)} \tag{1.21}$$

Clearly if there is no orbital contribution and L is zero then Equation 1.20 reduces to Equation 1.11. For example, the orbital contribution from five unpaired electrons occupying separate d-orbitals will be zero, since the orbitals have orbital angular momentum quantified by $m_l = +2, -1, 0, -1, -2$. The sum of m_l values is, of course, zero and the d^5 species is expected to show a magnetic moment that is entirely a result of spin angular momentum. If two unpaired electrons are placed in degenerate d orbitals in a free ion, they will occupy the orbitals according to Hund's rules, which maximise S and L. Therefore, orbitals with $m_l = +2$ and $+1$ will be singly occupied and L will be non-zero, so an orbital contribution will be expected.

However, the orbital contribution may also be quenched when the orbitals are split in a crystal field. A simple, although not physically accurate, way to think about this is to consider that in order for an electron to contribute to the orbital angular momentum it must be able to orbit the atom's nucleus. In other words, an electron must be able to 'move' from its current orbital to an orbital that is identical to it in all respects except that it is at a different position about the atom. That is, the orbital in which the electron resides must be able to transform into an identical and degenerate orbital by a simple rotation about any axis. The other proviso is that the 'movement' of the electron does not violate the Pauli exclusion principle, so that each electron is still described by a different set of the four quantum numbers. For the d orbitals in a free ion, a rotation of 45° about the z-axis can transform the d_{xy} orbital into the $d_{x^2-y^2}$ orbital, so an orbital contribution from an electron in one of these orbitals is expected. Indeed, the m_l values for these orbitals are ± 2. However, if both the d_{xy} and $d_{x^2-y^2}$ orbital are singly occupied with electrons of the same spin, the electrons can contribute no orbital momentum, as electron movement would violate the Pauli exclusion principle. In the free ion, the d_{z^2} orbital is unable to transform into any of the other orbitals, since it has a different shape, so d_{z^2} electrons do not contribute to the orbital angular momentum. It is not a coincidence that m_l is zero for the d_{z^2} orbital. The d_{xz} and d_{yz} orbitals have m_l values of + and -1, and the same arguments apply.

The situation is not the same when a crystal field is present. For example, in an octahedral crystal field, the *d* orbitals are split into a lower set consisting of the d_{xy}, d_{xz} and d_{yz} orbitals, with t_{2g} symmetry, and a higher-energy e_g set consisting of the d_{z^2} and $d_{x^2-y^2}$ orbitals. Since there is a stipulation that the transforming orbitals must have the same energy, the splitting affects which orbitals can contribute to the orbital momentum. The members of the degenerate t_{2g} set of orbitals can be converted into one another by rotation about an appropriate axis. For example, a d_{xy} orbital can be converted to a d_{yz} orbital by a 90° rotation about the common *y*-axis. However, since the e_g orbitals have different shapes, they cannot be converted into one another. Therefore, an electron in a d_{z^2} or $d_{x^2-y^2}$ orbital does not contribute directly to the orbital angular momentum. In addition, unlike in the free ion, the d_{xy} orbital can no longer transform into the $d_{x^2-y^2}$ orbital, so this source of orbital contribution is removed.

The above argument can be restated using term symbols, which reflect the orbital occupancy of an electron. It is not the objective of this chapter to discuss the detail of what term symbols are and how they are determined, and several standard texts are available that have clear descriptions.[20] A contribution from orbital angular momentum is expected when the ground term is a *T* state, *e.g.* a d^1 species. If the ground state has symbol A_1 (orbital singlet from an *S* free-ion term), A_2 (orbital singlet from an *F* free-ion term) or *E* (doublet from a *D* free-ion term), there is no *direct* contribution to the orbital momentum. However, the situation is not as clear-cut as this – hence the use of the word 'direct'. If the ground state is an A_2 or *E* term, there is always an excited state *T* term of the same spin multiplicity. According to the selection rules of electronic spectroscopy (see Section 1.5.2), electron transfer may occur between states with the same spin multiplicity. An alternative way to put this is to say that the A_2 and *E* ground states can mix with the higher *T* states, which will introduce an amount of deviation from spin-only magnetic moments by introducing orbital momentum from the excited states. It is possible to quantify this using Equation 1.22 for the deviation of the *g*-factor from its free-electron value. In Equation 1.22, Δ is the crystal field splitting parameter found from electronic spectra (see Section 1.5.2) and the constant α is 2 for the *E* ground state and 4 for the A_2 ground state. Clearly, a larger crystal field splitting will reduce the orbital contribution, since excited states will be higher in energy and more inaccessible. λ is related to the spin–orbit coupling constant, ζ, by Equation 1.23, where ζ is always positive for shells less than half full (d^1–d^4) and negative for shells more

Table 1.3 A comparison of spin-only and experimental values of $\chi_m T$ for octahedral high-spin complexes with electron configurations from d^1–d^9.[21]

Number of *d* electrons	Ground term in an octahedral field (first excited term)	Spin-only value of $\chi_m T$ ($cm^3\,K\,mol^{-1}$) from Equation 1.12	Expected orbital contribution	Experimental range of $\chi_m T$ ($cm^3\,K\,mol^{-1}$) at 300 K
1	T	0.375	Yes	0.36–0.41
2	T	1.000	Yes	0.98–1.20
3	$A_2(T_2)$	1.875	Yes – *via* mixing to give lower $\chi_m T$	1.71–1.90
4	$E\,(T_2)$	3.001	Yes – *via* mixing to give lower $\chi_m T$	2.88–3.00
5	A_1	4.377	No	4.06–4.50
6	T	3.001	Yes	3.12–3.92
7	T	1.875	Yes	2.31–3.38
8	$A_2(T_2)$	1.000	Yes – *via* mixing to give higher $\chi_m T$	1.05–1.90
9	$E\,(T_2)$	0.375	Yes – *via* mixing to give higher $\chi_m T$	0.45–0.55

than half full (d^6–d^9). This means that values of g can be less than or greater than 2 and commonly range from about 1.7 to 2.2. Table 1.3 shows a comparison between the values of $\chi_m T$ for the spin-only case and typical experimental ranges of high-spin complexes (with relatively small crystal field splitting energies). For reference, the ground-state term symbol and, if relevant, the excited-state term symbol are shown.

$$g = 2.0023\left(1 - \frac{\alpha\lambda}{\Delta}\right) \tag{1.22}$$

$$\lambda = \pm\zeta/2S \tag{1.23}$$

A feature of materials with significant spin–orbit coupling is a strong temperature dependence of the magnetic properties. This occurs because T states are split in an external magnetic field and, as the temperature is lowered, some levels are depopulated. Care must be taken, therefore, when attributing a variable temperature change in $\chi_m T$. Spin–orbit coupling also increases with increasing atomic weight of the metal centre. For example, f-block orbitals have a large contribution from spin–orbit coupling and small crystal-field effects. Therefore, it is generally true that f-block magnetic materials do not obey the Curie or Curie–Weiss laws. In these cases, the S and L quantum numbers may not be appropriate and the total orbital quantum number, J, must be used, as in Equation 1.24.

The appropriate value of g is found using Equation 1.25. For less than half-filled subshells, the ground state is given by $J = L - S$, and for more than half-filled subshells by $J = L + S$.

$$\chi_m T = \frac{N\beta^2}{3k_B} g_J^2 \left(J(J+1)\right) \tag{1.24}$$

$$g_J = \frac{3J(J+1) + S(S+1) - L(L+1)}{2J(J+1)} \tag{1.25}$$

1.1.6 Magnetic Materials with Long-Range Order

1.1.6.1 Ferromagnetism

Like paramagnets, ferromagnetic materials must have unpaired electrons, but unlike with paramagnets, the neighbouring magnetic moments interact strongly and in a long-range manner. The interaction is such that the magnetic moments tend to align parallel with each other, as shown in Figure 1.6. As a consequence, when a ferromagnet is placed in a magnetic field, the magnetic moments orientate to align with the field, and this alignment is further promoted by the internal interactions.

Therefore, at a specific temperature and external field, the magnetisation and magnetic susceptibility of a ferromagnet are always much larger than those of a paramagnet. However, temperature still has a randomising effect and there is competition between this and the magnitude

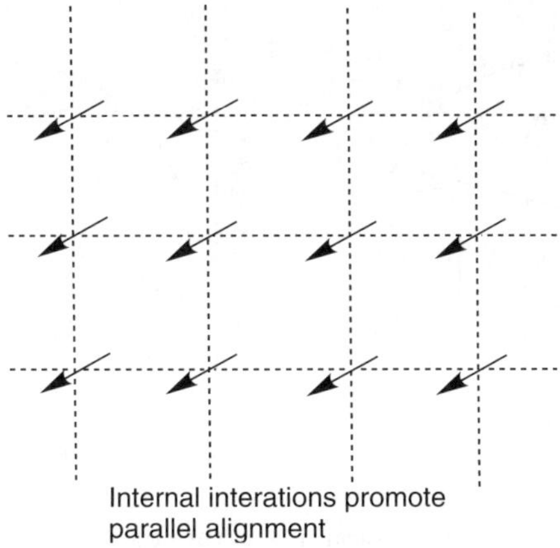

Figure 1.6 Ordered array of magnetic moments in a ferromagnetic material.

of the interaction between the magnetic moments. Above a sufficiently high temperature, called the Curie temperature, T_C, the thermal effects become a dominant, randomising factor that influences the moment's orientation. Therefore, above T_C the material acts like a paramagnet (although short-range, ferromagnetic coupling may still be evident) and will obey the Curie–Weiss law, as given in Equation 1.18, with a positive Weiss constant.

Below T_C, the behaviour is dictated by the internal interactions that tend to align the magnetic moments. However, below T_C and in the absence of an external magnetic field, the material still may not exhibit long-range magnetic order, due to the presence of a randomly orientated domain structure. In this case, the structure consists of areas (or domains) of equivalent moment alignment, but neighbouring domains are aligned randomly with respect to each other, as in Figure 1.7a. A small external field is required in order to orientate the domains parallel to each other and thus achieve long-range ferromagnetic order. This is equivalent to stating that the domains that are aligned with the magnetic field grow at the expense of non-aligned domains, as seen in Figure 1.7b. Perfect ordering can only occur at zero Kelvin, however, since the randomising thermal effects are diminished only below T_C, rather than eradicated.

In addition to a very large and positive magnetisation, ferromagnets have two other defining characteristics that are important for their technical applications. First, plotting a curve of magnetisation *versus* external magnetic field – recall that this is the curve with magnetic susceptibility as the gradient – shows an initial large rise in magnetisation at small or moderate values of the magnetic field. In other words, the external

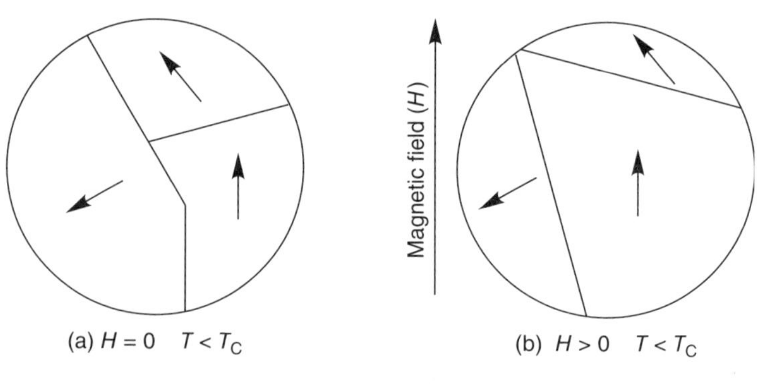

Figure 1.7 (a) Schematic of the domain structure of a ferromagnetic material with no externally applied field, *H*. (b) As *H* is applied, the domains that align parallel with the field grow larger.

field easily induces a large magnetisation. As stated previously, the magnetisation is the vector sum *per* unit volume of the individual magnetic moments and, since there is only a finite number of electrons *per* sample volume, the magnetisation has a saturation value, M_{sat}. With ferromagnets, this can be reached under relatively small values of the external field. In fact, since paramagnets also have a finite number of electrons, they too have a saturation magnetisation. However, paramagnets have small interactions between the moments and are more prone to thermal effects, so in practice very large external fields are needed to achieve saturation. The value of the saturation magnetisation is related to the number of unpaired electrons by Equation 1.26.

$$M_{sat} = Ng\mu_B S \tag{1.26}$$

For one electron *per* mole ($S = 1/2$), the contribution from the electron spin to M_{sat} with $g = 2$ is $5585\,\mathrm{cm^3\,G\,mol^{-1}}$. Therefore, it is common practice to plot magnetisation against field and present the magnetisation values divided by 5585, thus effectively plotting 'number of electrons' *versus* field.

The second, extremely important, property of ferromagnets is that they display magnetic hysteresis, which is a consequence of their domain structure. This is also revealed by plotting magnetisation *versus* external field. In this case, the magnetic field is cycled from zero up to a point that ensures magnetic saturation, then reduced back to zero and reversed (the poles are switched) until saturation is again achieved in the opposite orientation. Finally, the magnetisation loop is closed by increasing the field through zero and back to the maximum positive value. An example of a hysteresis loop is shown in Figure 1.8.

The form of the loop is directly related to the domain structure, as the external magnetic field causes the domains to align with each other, or in other words to grow in size *via* the movement of the domain boundary. But there is a resistance to movement of the boundary, which appears as irreversibility. The presence of hysteresis is relevant to the use of ferromagnets in data storage applications, since at a specific value of the external magnetic field the sample can exist in one of two magnetised states. As shown in Figure 1.8, at zero external field the sample can have magnetisation M_r or $-M_r$, depending on its magnetic field history. In other words, it is possible to store binary information in the form of a positive or negative magnetisation, with the values interpreted as 0 and 1 bits, respectively. Two important parameters are obtained from the hysteresis loops: a remnant magnetisation, M_r, and a coercive field, H_c.

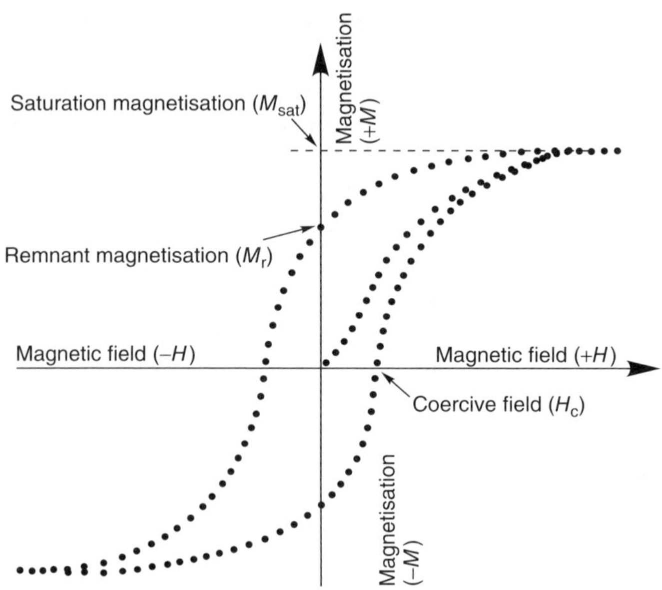

Figure 1.8 Form of a hysteresis loop displayed by a ferromagnetic material. See text for details.

Greater resistance to the relative movements of domains will give higher values for both of these parameters. A large value of M_r is often desired, since this means that there is a large difference between the two magnetisation values used to store the data, making them easier to read. A high H_c is also often required, because this is the value of the magnetic field that will remove any remnant magnetisation. In other words, once the data are stored, they can be deleted if the storage medium is subjected to a field at or above H_c. This can be verified experimentally by ignoring the warnings not to carry magnetically stored data (*e.g.* credit cards) too close to an NMR machine. Even with the efficient magnetic shielding in modern NMR machines, they are usually surrounded by a marked-off area displaying warnings of the effects of their magnetic field.

1.1.6.2 Antiferromagnetism

Antiferromagnetic materials have unpaired electrons and a strong interaction between the resultant magnetic moments. This interaction leads to antiparallel alignment, as shown in Figure 1.9. Thermal randomisation of the moments means that above a specific temperature, called the

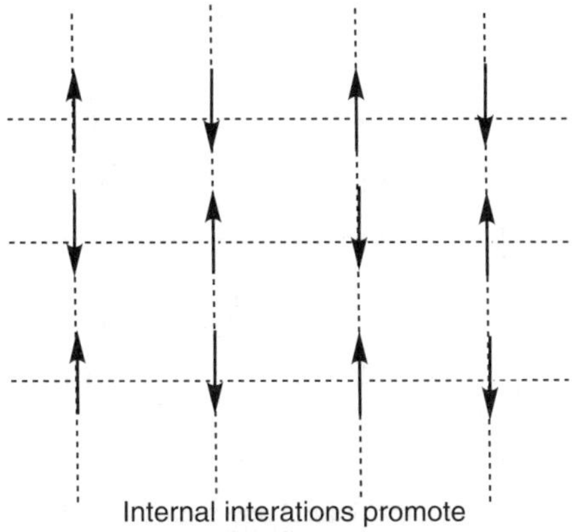

Figure 1.9 Fully ordered array of magnetic moments in an antiferromagnetic material.

Néel temperature, T_N,[22] the material behaves like a paramagnet and the behaviour is modelled by the Curie–Weiss law (Equation 1.18) with a negative value of the Weiss constant. Below T_N, the antiparallel alignment of the magnetic moments will result in a magnetisation of zero if realised fully. However, in practice, finite values of magnetisation are always seen, which are lower than the values predicted for paramagnets. This is because the randomisation afforded by any temperature above zero Kelvin results in the incomplete antiparallel alignment of the magnetic moments.

1.1.6.3 Ferrimagnetism

Ferrimagnetic materials display aspects of both antiferromagnetism and ferromagnetism. As with antiferromagnets, the neighbouring magnetic moments are coupled in an antiparallel alignment. However, the neighbouring moments have different magnitudes, such that their vector summation gives a non-zero magnetisation. The arrangement of magnetic moments is shown in Figure 1.10. This situation occurs in materials that have neighbouring atomic centres with different numbers of unpaired electrons.

For example, a series of molecular magnets that have chains of alternating Mn^{II} (five unpaired electrons, $S = 5/2$) and Cu^{II} (one unpaired electron, $S = 1/2$),[23–25] leads to the antiparallel alignment of a moment one unit in length with a moment five units in length. An

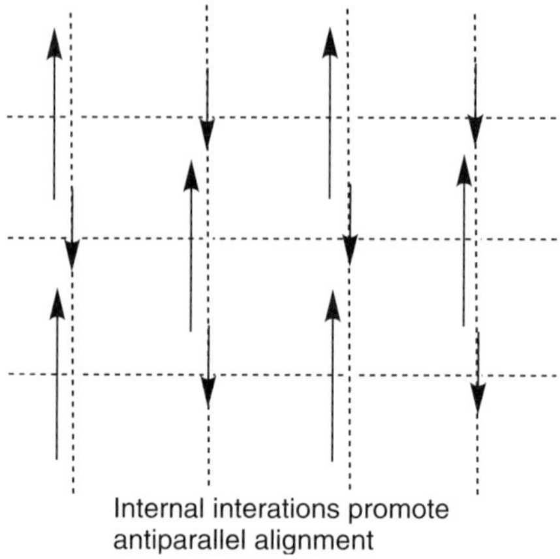

Figure 1.10 Fully ordered array of magnetic moments in a ferrimagnetic material.

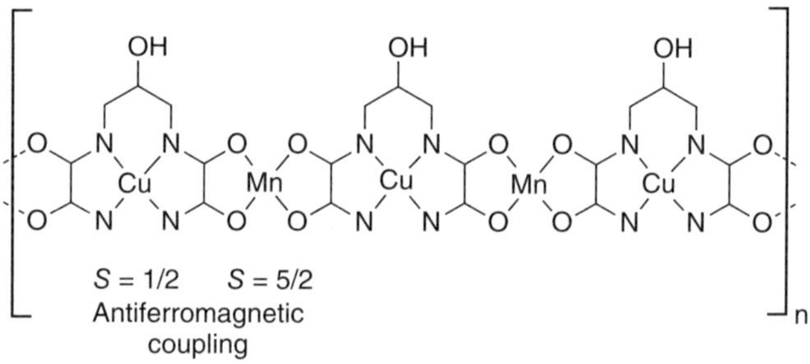

Figure 1.11 Atom connectivity in a Mn^{II}/Cu^{II} ferrimagnet, synthesised by Kahn and co-workers.[23,25] The octahedral coordination about the Mn site is completed with water molecules.

example of a Cu/Mn chain is shown in Figure 1.11, where interchain interactions are similarly antiferromagnetic between different metals. The result is an array of moments that are effectively four units in length, orientated in the same direction as the moments associated with the $S = 5/2$ centre. This is equivalent to the ferromagnetic spin structure shown in Figure 1.7.

Therefore, the macroscopic behaviour of ferrimagnets is similar to that of ferromagnets, including the presence of a domain structure and hysteresis. In fact, synthesising ferrimagnets is often a good tactic for making materials with ferromagnetic-like properties, as it is easier to engineer antiferromagnetic coupling than it is to arrange ferromagnetic coupling. As with other ordered states, ferrimagnetic materials act like paramagnets above a specific temperature. Above the critical temperature, the materials can be modelled with the Curie–Weiss law, with a negative Weiss constant that reflects the antiferromagnetic nature of the short-range coupling. Many ferrites – oxides containing Fe^{II} – show ferrimagnetism, such as magnetite (Fe_3O_4) and maghaemite (γ-Fe_2O_3). Magnetite has an inverse spinel structure and is ferrimagnetic by virtue of having Fe^{II} at two different sites – with octahedral and tetrahedral geometry – together with Fe^{III} at octahedral sites (Figure 1.12). It is the combination of the antiferromagnetic coupling between the different Fe^{II} sites and the ferromagnetic coupling between the Fe^{II} and Fe^{III} octahedral sites that leads to ferrimagnetism and to application in magnetic storage devices.

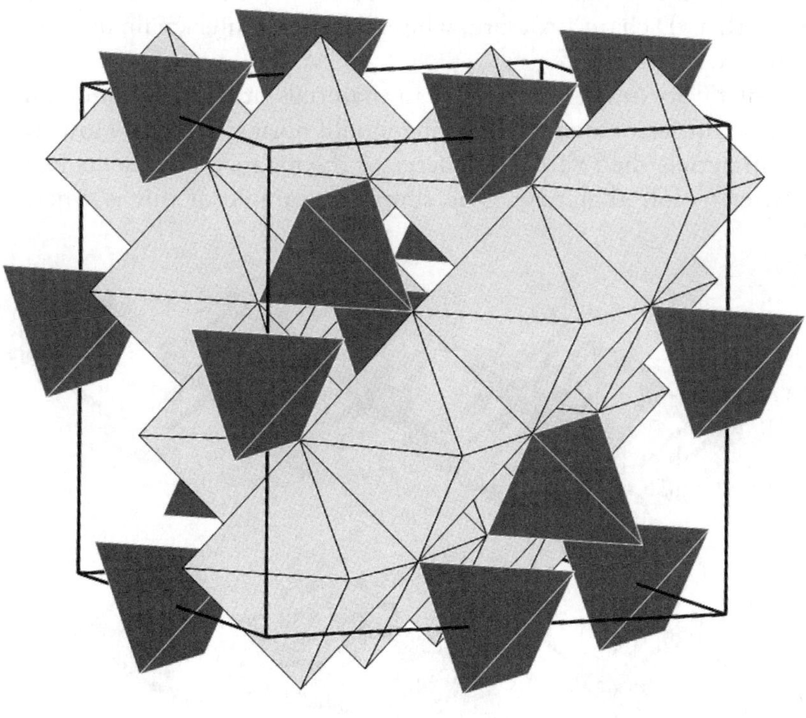

Figure 1.12 Structure of magnetite, Fe_3O_4, showing the octahedral (light grey) and tetrahedral (dark grey) Fe sites. The oxygen atoms are at the corners of each polyhedron.

1.1.7 Brief Notes on Other Types of Magnetism

Many other categories of magnetism are known. These are again defined by their behaviour when in a magnetic field. A brief overview of some of them is given here, although the list is not intended to be exhaustive. Those mentioned reflect the author's own bias and relate to some specific points that will be made later in the chapter. For example, metamagnetic materials[26] have a phase change induced by a magnetic field. This can occur in materials that have competing ferromagnetic and antiferromagnetic interactions. Antiferromagnetic interactions often dominate until a sufficient magnetic field is applied, and then bulk ferromagnetism is observed. For example, $FeCl_2$ adopts a layered structure (Figure 1.13) in which the intra-layer coupling is ferromagnetic and the inter-layer coupling is antiferromagnetic.[27] Below a T_N of 24 K, the material acts like a bulk antiferromagnet, but it becomes ferromagnetic above an external field of 1 T. The effect has also been observed in Fe^{II} coordination polymers with a 1D chain structure, where intra- and inter-chain interactions compete with each other.[28]

Magnetic frustration[29] occurs in materials at the low temperatures at which antiferromagnetic ordering would normally occur. However, in these materials the relative geometry of the magnetic moments leads to a barrier to full alignment. The simplest example of this is where the

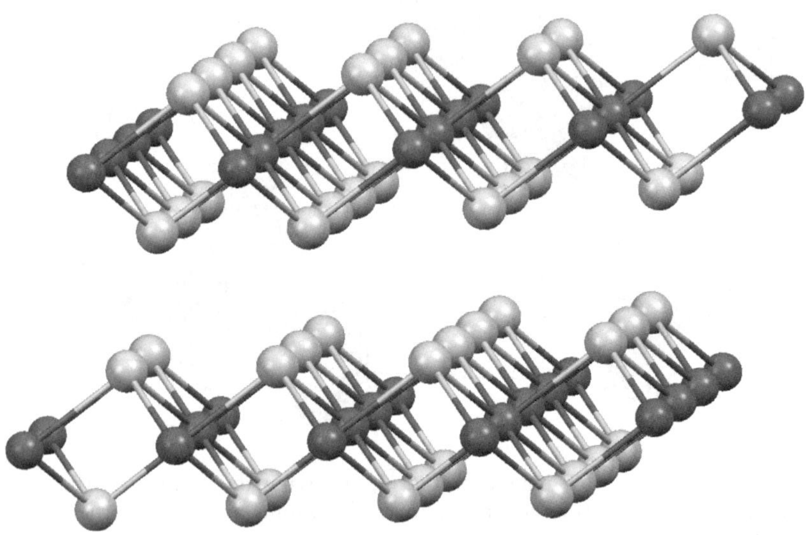

Figure 1.13 Layered structure of $FeCl_2$. Dark grey atoms are Fe, lighter grey are Cl.

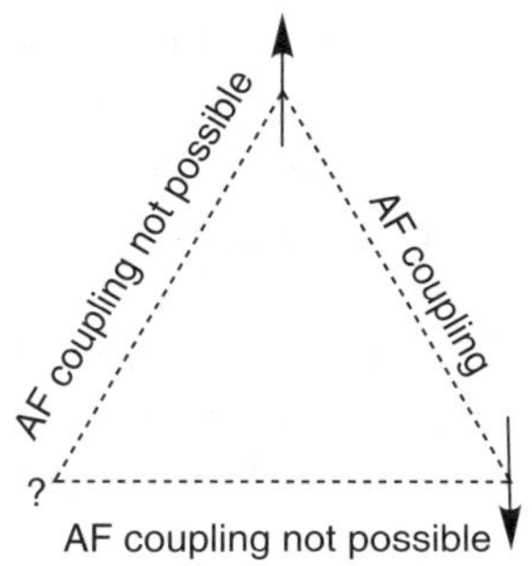

Figure 1.14 Triangular arrangement of magnetic moments with strong antiferromagnetic (AF) coupling. Full compliance of the coupling is not possible.

magnetic moments occur at the corners of an equilateral triangle, as shown in Figure 1.14. Within this geometry, only two sites can attain antiferromagnetic alignment and the third site is unable to couple simultaneously with these.

Extended lattices based on geometric frustration can be made, such as the well-known Kagome lattices[30,31] and pyrochlore structures.[32] The result of the extended competing interactions is that unusual magnetic ground states are found. An example of such a state is the so-called spin-glass state,[33] in which the magnetic moments exist in a random pattern. Unlike paramagnets, which also have random orientation of the moments, a spin glass is a fixed, ordered state. As with other ordered states, there is a temperature below which a phase transition to a spin glass occurs, the freezing temperature, T_f, and so it is very much different from the random dynamic pattern found in paramagnets. Another, somewhat exotic state is the spin liquid, in which the magnetic moments interact strongly but are still able to fluctuate at very low temperatures.

Superparamagnetism[34–36] is a type of magnetism seen in ferromagnetic and ferrimagnetic nanoparticles, when their size is of the order of 1–10 nm. This is particularly important since most methods of magnetic storage use particulate magnets in their construction. The behaviour is also seen in high-spin-state molecules, such as those found in the area of single-molecule magnets[36,37] (also called molecular nanomagnets); it results from the bulk materials being inhomogeneous. These ferro- or ferri-magnetic particles or molecules can themselves order at a certain temperature, but the interactions between these clusters are paramagnetic. Therefore, even below the particle's ordering temperature, there is sufficient thermal energy to change the overall magnetisation of the bulk material. The resulting fluctuation leads to zero overall magnetisation, just as in the domain picture of a ferromagnet, but in this case the paramagnetism results from a random ordering of ordered species, rather than from the atomic moments in 'normal' paramagnetism. Therefore, the effect is dubbed 'superparamagnetism'. At lower temperatures, below a so-called blocking temperature, T_B, the particles exhibit single-domain ferromagnetic behaviour. In the area of molecular nanomagnets, the latest research aims to maximise the number of coupled spins in the molecules and to introduce anisotropy in them so that there is a preferred direction of magnetisation, thus promoting irreversibility. A prospective use of these materials is in magnetic cooling.[38]

Spin-transition (also called spin-crossover) compounds have labile electronic states, switchable between high-spin (HS) and low-spin (LS), as driven by an external stimulus such as temperature, pressure or electromagnetic irradiation.[39] The most studied compounds are neutral and cationic Fe^{II} or Fe^{III} complexes with octahedral coordination, where all or some of the complex ligands are nitrogen-donor heterocycles. For example, octahedral Fe^{II} complexes can be induced to change reversibly, sometimes with hysteresis, between diamagnetic LS ($S = 0$, $^1A_{1g}$) and paramagnetic HS ($S = 4/2$, $^5T_{2g}$) states. The major interest is in using this bistability as the active component in molecular-based switches, data-storage applications and displays.[40] The transition can be monitored by following variable temperature electronic spectra, since the colour and thus the absorption characteristics of the LS and HS complexes are different. The HS to LS conversion also leads to a decrease in metal–ligand bond lengths of up to 10% and to changes in bond angles, since the two states have different numbers of electrons in the antibonding e_g^* orbitals. Of more relevance to this chapter, the transition can be monitored by following the magnetisation *versus* temperature. As the temperature is lowered, the LS state becomes populated in preference to the HS state, and as a consequence the number of unpaired electrons decreases, as reflected in a decrease in $\chi_m T$. A crucial parameter in designing spin-transition materials is the energy of the *d*-orbital splitting, which is 'lost' when in the HS state. This must be of a similar magnitude to the spin-pairing energy 'lost' when in the LS state – the most stable state wins out. For this reason, researchers tend to focus on a small range of metal ions and types of ligand, since these factors impact on the *d*-orbital splitting. For example, the majority of compounds contain Fe as the metal ion with heterocyclic *N*-donor ligands or a mixture of *N*- and *O*-donor ligands. A typical example is the cationic Fe^{II} complex of 2,6-di(pyrazol-1-yl)pyridine, the $[BF_4]^-$ salt of which has a sharp

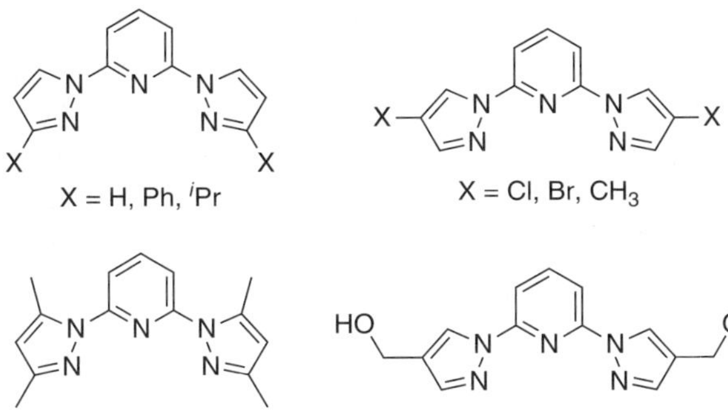

Figure 1.15 The ligand 2,6-di(pyrazol-1-yl)pyridine (top left with X = H) and some of its derivatives that have been investigated in attempts to promote spin-transition behaviour.

spin transition at 261 K.[41] As a result of this discovery by Halcrow and co-workers, a large number of substituted derivatives of the ligand have been investigated, a subset of which is shown in Figure 1.15.[42,43]

1.1.8 General Considerations for the Measurement of Magnetic Properties

Bulk magnetic moments are measured *via* two different effects. It is possible to monitor the force or torque that a moment experiences in an inhomogeneous magnetic field (see Section 1.1.2.1). Alternatively, the current or voltage induced in a simple conducting wire, called a pickup coil, can be monitored by a nearby changing magnetic moment. The force and torque methods were the first to be exploited, and are still used for teaching or where excellent accuracy is not required. The principle is to monitor the force in terms of a sample movement with respect to a fixed magnet (see the Gouy method in Section 1.2.1 and the Faraday method in Section 1.2.3) or to fix the sample and monitor the movement of the magnet (see the Evans method in Section 1.2.2).

Most modern methods use variations on induction. Induction is achieved either by steadily moving a sample through a set of pick-up coils, as in the SQUID magnetometer (see Section 1.3.1), or else by vibrating the sample, as in the vibrating sample magnetometer (VSM) (Section 1.3.4). The pick-up coils are often made of copper wire, but the SQUID uses a much more sensitive set of superconducting pick-up coils, which are in turn coupled to a SQUID detector. In addition to bulk magnetic measurements, there are experiments of particular use for the measurement of the magnetic effects *via* resonance spectroscopy. These are made possible by the fact that a magnetic field is able to change electronic energy levels; the techniques measure the local effects of magnetic fields rather than the bulk magnetisation. Such techniques include the Evans NMR method (Section 1.4), EPR (Section 1.5.1) and Mössbauer spectroscopies (Section 1.5.4).

1.2 MAGNETIC MEASUREMENT BASED ON MEASURING A FORCE OR TORQUE

1.2.1 The Gouy Balance

Many of the early methods of measuring magnetisation depended on measuring the translational force resulting from the interaction between an inhomogeneous field and the magnetic moment within the sample, as given by Equation 1.4. One such method used the Gouy apparatus,[44,45] which is probably still familiar to many students, since it is often used in undergraduate teaching laboratories. The method was invented by a French scientist, Gouy (1854–1926). It is most suitable for finely powdered samples but can also be used with liquids. A sample is placed into a narrow, cylindrical glass or quartz tube; if a powder is used, it must be compacted carefully. As will be seen below, the amount of sample in the magnetic field is taken into consideration, so it is important that there are no voids from inefficient packing of the sample. The sample tube is attached to a sensitive balance and suspended between the north and south poles of an electromagnet or a permanent magnet with a typical maximum field of a few tenths of a Tesla. Importantly, the sample tube is positioned so that its base sits between the poles of the magnet, where the magnetic field is uniform, but the uppermost part of the sample sits in a region of space where the magnetic field is zero. A simplified schematic of the apparatus is show in Figure 1.16. In this way, the whole sample sits in a field gradient of an inhomogeneous field. As noted in Section 1.1.5, paramagnetic samples are attracted to areas of high field, and therefore there will be a force on the sample that can be detected as an increase in sample weight. By contrast, diamagnetic samples experience a weight decrease, with a much smaller magnitude.

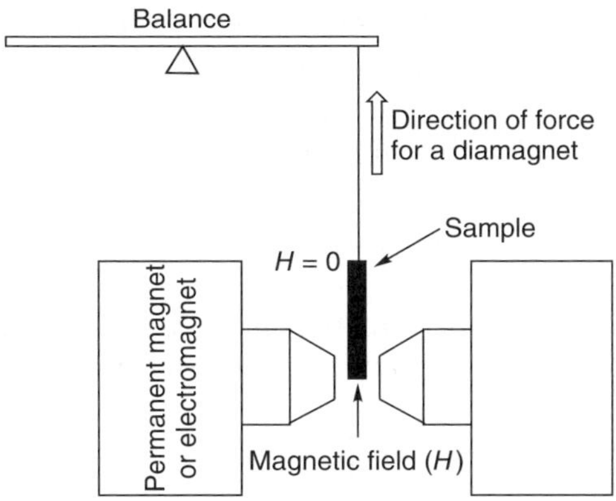

Figure 1.16 Schematic of the apparatus used in the Gouy method.

Practically, the magnet, sample and balance are enclosed in a box with a door for access. The function of the box is to exclude air currents that might affect the balance readings. For the same reason, the apparatus should be positioned so as to avoid any vibration, *e.g.* on a sturdy bench. The balance reading is noted before, w_0, and after, w_H, the magnetic field is applied, so that the force exerted is proportional to $w_H - w_0$. For a paramagnetic sample, this difference will be positive ($w_0 < w_H$), while for a diamagnetic sample it will be negative ($w_0 > w_H$). Assuming that the magnetic field, H, varies as a function of length, z (parallel to the sample tube), the z component of the force, f_z, *per* unit volume of the sample is given by Equation 1.27, where M is the magnetisation *per* unit volume and μ_o is the vacuum permeability in the required units. The right side of Equation 1.27 is used to convert M into terms involving susceptibility.

$$f_z = \mu_0 M \frac{dH}{dz} = \chi_v \mu_0 H \frac{dH}{dz} \tag{1.27}$$

We can also calculate the work done, dw, by this force when a small volume of the sample dV moves out of a position with field strength H_1 to one with H_2; this is given by Equation 1.28.

$$dw = dV \int_{H_1}^{H_2} f_z dz = dV \chi_v \mu_0 \int_{H_1}^{H_2} H \frac{dH}{dz} dz = \frac{1}{2} dV \chi_v \mu_0 (H_2^2 - H_1^2) \tag{1.28}$$

When a sample with cross-sectional area, A, moves by an amount $|\Delta z|$, this corresponds to the movement of a sample volume, $A|\Delta z|$. Assuming that this volume moves from a region of zero field strength (so H_1 can be set to zero) to a region of a non-zero field strength ($H_2 = H$), the work done is given by Equation 1.29. This is equated to the force times the distance moved, also shown in Equation 1.29.

$$w = \frac{1}{2}A|\Delta z|\chi_v\mu_0 H^2 = f|\Delta z| \tag{1.29}$$

A simple rearrangement of Equation 1.29 gives an expression for the force on the tube resulting from its interaction with the magnetic field, shown in Equation 1.30.

$$f = \frac{1}{2}A\chi_v\mu_0 H^2 \tag{1.30}$$

In making the measurements, the weights are determined in the absence and presence of the magnetic field. The weights are converted to forces (in Newtons if using SI units) by multiplying them (in kg if using SI units) by the acceleration due to gravity in ($9.81\,\mathrm{m\,s^{-2}}$), G in Equation 1.31. A capital letter G is used here to avoid confusion with the symbol for the g-factor above.

$$G(w_H - w_0) = f = \frac{1}{2}A\chi_v\mu_0 H^2 \tag{1.31}$$

Simple rearrangement gives the final equation for the volume magnetic susceptibility, Equation 1.32, which can then be converted to molar susceptibility as outlined in Section 1.1.3.

$$\chi_v = \frac{2G(w_H - w_0)}{A\mu_0 H^2} \tag{1.32}$$

In fact, the normal practice is to determine a constant of the apparatus for the value of $2G/A\mu_0 H^2$ by measuring a material with a known susceptibility. Common standards for paramagnetic samples are $Hg[Co(NCS)_4]$ and Tutton's salt, $(NH_4)_2[Mn(SO_4)_2] \cdot 6H_2O$; water is often used as calibrant for diamagnetic materials.

1.2.2 The Evans Balance (or Inverse Gouy Method)

The Evans balance[46] is a more modern method than the Gouy balance and can also be found in undergraduate laboratories. It relies on an inversion of the Gouy method, whereby a fixed sample is used and a change

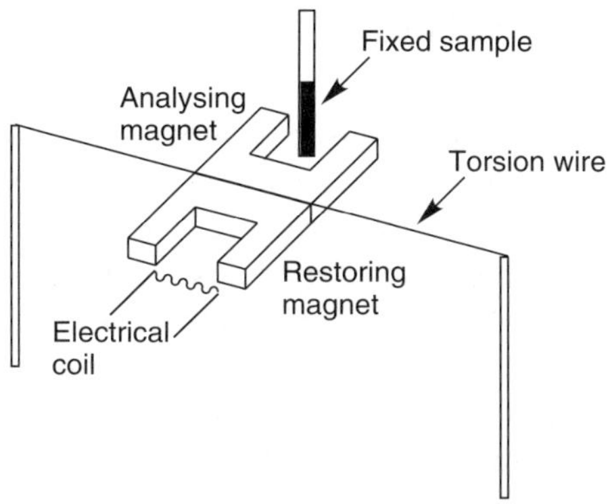

Figure 1.17 Schematic of the Evans balance.

in the weight of the magnet is effectively measured. An advantage of this method is that the equipment is simpler and smaller, since there is no need for an accurate balance. As shown in Figure 1.17, the apparatus can consist of two magnets joined back to back on a torsion wire fixed across two points. Sometimes the configuration has the two magnets at each end of the wire, but either method requires that, in the absence of the sample, the system is kept in balance and any twist or torsion in the wire can be measured, which is normally done using an optical sensor.

Between the poles of one magnet (the restoring magnet) sits an electrical coil. Initially the space enclosed by the other magnet (the analysing magnet) is empty. When a magnetisable sample is placed in a fixed position between the poles of the analysing magnet, the resultant attraction or repulsion is translated into a twist of the wire. The current required in the electrical coil to restore the assembly to its original balanced position is proportional to the force between the sample and the analysing magnet.

As with the Gouy method, the Evans balance must be used in conjunction with a calibration standard of known susceptibility. For one experiment, three readings are needed: one of the empty diamagnetic tube, R_0), one of the same tube filled with the chosen calibrant and one of the same tube filled with the sample, R_s). A machine calibration constant is determined from R_0 and the calibrant reading to give the correct susceptibility. The sample's mass susceptibility is then given by Equation 1.33, where L is the length of the sample, m is its mass and C is the calibration constant.

$$\chi_g = \frac{CL(R_s - R_0)}{m} \tag{1.33}$$

1.2.3 The Faraday Balance

The Faraday magnetometer[46] uses a similar approach to the Gouy balance, in that the apparatus measures the force experienced by the sample in a magnetic field. However, there is a difference in the origin of the magnetic field in which the sample sits. As shown in Section 1.2.1, the sample in a Gouy balance rests in an inhomogeneous field, since one end is placed between the poles of a homogeneous magnetic field and the other is kept away from the magnet, in a region of zero field. In the Faraday balance, the inhomogeneous field is created by the appropriate shaping of the pole pieces of the magnet. From Equation 1.27 on the force exerted on a moment in an inhomogeneous field, the pole pieces are shaped such that there is a region where H multiplied by $\mathrm{d}H/\mathrm{d}z$ is a constant. The sample is placed in this region of the field. Therefore, the force is independent of the volume (density) of the sample and rigorous packing of the sample – crucial in the Gouy method – is not required. The effect of measuring χ_g directly is that reproducibility is also improved. The force, again measured by a weight change, is measured by a torsion balance that is very sensitive to small changes in weight. A third advantage the Faraday method has over the Gouy balance is that much smaller samples can be measured: a few milligrams as opposed to at least a hundred times that.

1.3 MAGNETIC MEASUREMENT BASED ON INDUCTION

1.3.1 The DC SQUID Magnetometer

Currently, the SQUID is the most sensitive means of measuring magnetic moments. Therefore, a considerable amount of this chapter will focus on the technique. Essentially, a SQUID acts as an amplifier to convert small magnetic fields into a relatively large electrical response. Often the magnetometer as a whole, including the external magnet, electronics, cooling system and even the controlling computer, is referred to as the 'SQUID', but this is actually the name of the detector that gives the magnetometer its sensitivity; this is rather akin to referring to the clock tower of the British Houses of Parliament as 'Big Ben' when in fact that is the name of the bell inside the tower. Typical modern SQUIDs, coupled to sensitive electronics, are capable of measuring magnetic fields down to about 10^{-18} T. To put this into perspective, it is many orders of magnitude lower than the Earth's magnetic field, which at its minimum is approximately 3×10^{-5} T. Incredibly, SQUIDs are also sensitive enough to detect magnetic fields that are created during biological processes, such as those created by charge flowing across synapses in the human brain. This fact has worked its way into popular media as the premise for a device used to record dreams and memories in the (fictional) 1995 film *Strange Days*.[47] (Un)fortunately we cannot do this quite yet.

The magnetometer has several components, one of which is a cryostat that controls the flow of liquid helium in order to vary the temperature of the sample, which, under reduced pressure, can typically reach a base of 1.8 K. A typical measurement covers the range 300 to 2 K, but some machines have heaters that allow temperatures up to 800 K to be reached.[48] The use of liquid helium means that the magnetometers are expensive to run, although modern machines mitigate against this by including integrated helium recycling units. The liquid helium is also used to cool an external magnet (external to the sample), which operates at low temperatures because it is made using loops of superconducting wire. An advantage of using a superconducting electromagnet is that large fields, typically up to 7 T, can be generated without resistive heating. The only limit to the size of the field is the critical current of the superconductor, above which superconductivity is lost. Once a current has been induced to flow in a superconducting material, it does not degrade, since there is no resistance to its flow. This current persistence means that once the magnetic field reaches the appropriate value, the driving voltage can be switched off, without loss of the field.

The next part of a SQUID is a detection coil – a closed loop of superconducting wire – again cooled below its operating temperature by liquid helium. One end of the wire is coiled about the sample, and it is this that detects the sample's magnetic moment. During an experiment, the sample is moved through the wire's loops and experiences a changing magnetic moment. This induces a varying current in the detection coil. The loop is wound around the sample space in a configuration called a second-order gradiometer (literally a detector of field gradients), shown about the sample in Figure 1.18. Due to the counterwound loops, this configuration means that a current is not induced in the coils by a constant magnetic field (*i.e.* that which magnetises the sample in the first place). Therefore, the detector coil is only sensitive to the moving sample and, potentially, the sample holder. However, the sample holder is designed to be a long, homogeneous piece of material, so that its moment does not vary with its movement and it produces no signal, providing that the sample extends well beyond the ends of the coil during measurement. This set-up has consequences for how samples are mounted onto their sample holders, as described in Section 1.3.1.3.

The other end of the pick-up coil is close to the SQUID detector itself, which detects the changing magnetic field generated by the current in the coil. Figure 1.18 shows a schematic of this arrangement. The reason that the sample is not placed directly next to the SQUID detector (although refer to Section 1.3.3 to see when it can be) is that the SQUID detector is sensitive to the constant external magnetic field. Indeed, SQUIDs are also sensitive to stray magnetic fields, *e.g.* from nearby electronic equipment. Therefore, the SQUID detector is isolated magnetically from these other magnetic fields using a superconducting shield. This shielding relies on the perfect diamagnetism of a superconductor, which can completely expel magnetic fields below a certain magnitude, as outlined briefly in Section 1.1.4.1. The configuration thus allows the SQUID detector to measure the magnetic field only from the pick-up coils; the field is itself generated only by the sample.

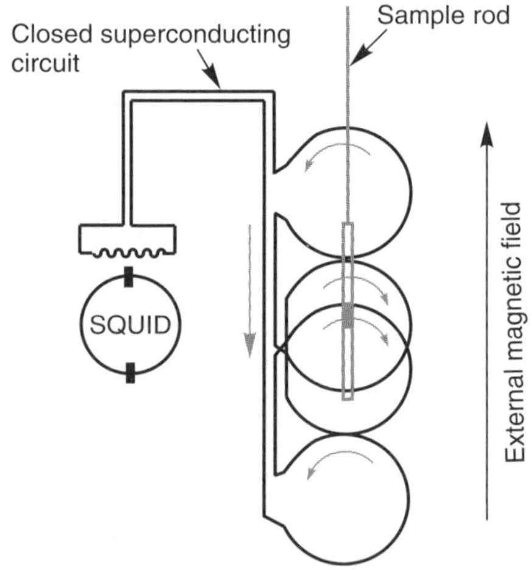

Figure 1.18 Schematic of a second-order gradiometer pick-up coil coupled to a SQUID detector. Grey arrows show the direction of the pick-up coil's turning.

1.3.1.1 The Josephson Effect

SQUIDs usually consist of a ring of superconducting material into which one or two poorly conducting thin sections have been inserted, with widths of the order of a few tens of Ångströms. Generally, the sections can be either insulating or metallic, or simply narrower sections of the superconducting material, so long as they restrict current flow; they need only to act as a barrier to conductivity. A radio-frequency (RF) SQUID contains one barrier and is therefore cheap to produce, but it is less sensitive than a DC SQUID with two barriers. Section 1.3.1.2 contains a schematic of a single DC SQUID detector (Figure 1.20). The barriers, or Josephson junctions, lead to the occurrence of the Josephson effect at the superconductor–barrier–superconductor interface. Despite the insertion of a barrier into the superconducting loop, a supercurrent still flows from one side to the other. Supercurrents are transported by a loose pair of associated electrons called a Cooper pair, according to the famous BCS theory.[49] The Cooper pair can pass through the barrier if it is thin enough. This is a quantum effect, called tunnelling, which was predicted to occur in superconductors by Josephson in 1962;[50] Josephson was awarded the 1973 Nobel Prize in Physics for this prediction. The first experimental observation of the Josephson effect was probably in 1963 or 1964,[51,52] although it had been observed previously but was considered to result from imperfections in the barrier construction.

Cooper pairs can also be described in terms of a quantum collective wave or condensate, in which the electron pairs adopt the same phase and energy. The current is able to flow across the Josephson junction because the wave's tail effectively extends across the barrier to the other side. In this way, the wave on one side of the barrier couples to the wave on the other side and they interfere. Furthermore, as the two superconducting regions are distinct, the phases of the Cooper pairs in the two regions are not the same ($\phi_1 \neq \phi_2$). Current that passes across the barrier is thus dependent on a phase difference between the Cooper pairs on each side of the barrier. The super-current, I, is shown in Figure 1.19 and given by Equation 1.34. I_0 is a critical current, a defining characteristic of the barrier, and δ is the phase difference across the junction. If I_0 is less than the critical current of the main superconducting ring then a maximum current flows across the gap, equal to I_0, when the phase difference is $\pi/2$.

$$I = I_{\mathrm{o}} \sin \delta \tag{1.34}$$

$$\delta = \phi_1 - \phi_2 \tag{1.35}$$

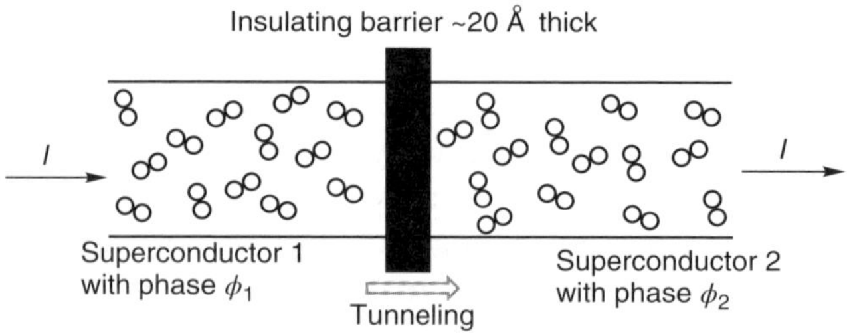

Figure 1.19 Schematic diagram of a single Josephson junction. Cooper pairs of electrons are represented as paired circles.

1.3.1.2 Mechanism of Detection by a SQUID

SQUIDs are commonly made out of niobium and niobium-based alloys, although high-T_C SQUIDs using derivatives of copper oxides are becoming commercially relevant. In the DC SQUID detector, shown in Figure 1.20, a bias current, I_B, passes through from A to D. Since there is no resistance difference in path ABD *versus* ACD, this current splits into the two branches equally, with a current of $I_B/2$ travelling though each path. When the magnetic field, H, from the pick-up coil is applied

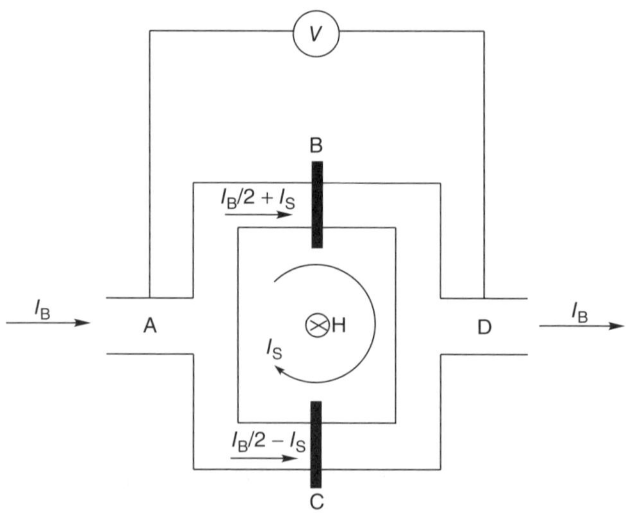

Figure 1.20 Schematic diagram of a DC SQUID detector. See text for an explanation of the mechanism of operation.

perpendicular to the ring, a small shielding current, I_S, is induced to flow around the ring. The current is such that it counters the magnetic field. As shown in Section 1.1.4.1, this is a feature of the Meissner effect, the diamagnetic response of superconductors. In Figure 1.20, it can be seen that the direction of the flow of the current, I_S, is the same on the B side of the SQUID but is opposite on the C side. Thus the current is different in the different branches of the SQUID, either $I_B/2 + I_S$ or $I_B/2 - I_S$. However, these currents ultimately combine so that the output current is invariant, at I_B.

This is equivalent to stating that the magnetic flux from the pick-up coil changes the phasing in the ABC part of the SQUID compared to the DBC part and that this is what moderates the current through each barrier. Normally the magnitude of I_S, induced in a superconductor is sufficient completely to cancel the magnetic flux so that it becomes zero within the ring of the SQUID loop. However, the critical currents of the weak links prevent this from happening and I_B is such that the critical current is always reached and the SQUID is said to operate in a resistive mode, in order to limit the magnitude of I_S. When the critical current is reached at either Josephson junction, a stable voltage, V), can be measured across the junction.

The final bit of information needed to follow the operation of a SQUID is that the magnetic flux in the hole of a loop of superconducting material is quantised, *i.e.* it can exist only in quanta of $nh/2e$, where h is the Plank constant, e is the fundamental electronic charge and n can take values of 0, ±1, ±2, ±3 *etc.* The value of $h/2e$ is 2.07×10^{-15} Wb (Weber), as compared to magnetic flux density, B, which is measured in $\mathrm{Wb\,m^{-2}}$ (*i.e.* Tesla; also see Table 1.1). Clearly, if the field is fully expelled then $n = 0$, but as described above this case does not occur and n must take one of the other values.

In a situation where the magnetic flux exceeds half the magnetic flux quantum but is less than a full quantum, it is favourable energetically to force n to be +1. In other words, the system must act in such a way as to increase, rather than decrease, the magnetic flux within the loop. This is done by reversing the screening current so that it enhances the magnetic flux and quantisation is satisfied. In other words, I_S reverses each time the flux increases by half-integer multiples of the flux quantum. The result is that the output voltage across the SQUID oscillates, while the value of current passing though the SQUID is constant. As such, the voltage is a periodic function of the flux in the SQUID. The fact that this measures the flux in half-quantum units gives the SQUID its sensitivity. Therefore, the SQUID acronym refers to the fact that Superconducting materials are required, a QUantum phenomenon (tunnelling) and the Interference of Cooper-pair waves are all necessary for the Device's operation.

1.3.1.3 Notes on Practical SQUID Operation

The following information is given to provide some very general guidance on using a SQUID magnetometer. The machines are operated with computers that can be programmed with a set of simple instructions, similar to simple BASIC programming. A list of sequential steps is written to instruct the machine when and how to change the external magnetic field and/or the temperature, then how long to wait for these values to settle, when to measure the sample's moment and finally which data to store for later analysis. Each measurement can also be repeated, by using program loops, to improve signal-to-noise ratios (SNRs). In this way, programs can be run to make measurements without the need for constant attention by the operator, even for several days.

The sensitivity of the SQUID is not only useful for measuring small magnetic moments but also means that in practice it is possible to use as little as 1 or 2 mg of sample for a clear noise-free signal, although 10 mg or above is used commonly. The sample itself should not exceed a length of about 10 mm, so that it sits wholly within the pick-up coils. As with all scientific work, it is also important for the sample preparation area to be clean, since a typical laboratory contains many sources of magnetic contamination. An often overlooked issue, even before the magnetic properties of a sample are measured, is the use of metallic spatulas. Scraping a sample from the base of a round-bottomed flask or a sintered glass filter can introduce contaminants from small metallic shavings. It is thus advised that experimentalists use plastic spatulas.

Powdered samples can be mounted in a diamagnetic sample holder, which often consists of a gelatine capsule fixed inside a plastic straw. Fixing the sample can be done by using a gelatine capsule that fits very tightly inside the straw. Alternatively, cotton wool, empty gelatine capsules above and below the sample or a smaller collapsed straw wedged in above and below the sample can be used. The base of the straw is then capped with non-magnetic tape, which mitigates against the sample moving in the straw during measurement, or even falling into the sample space. Clearly, the sample space must be kept free of contamination from foreign materials, with powdered samples – probably the most commonly measured – posing a particular threat. The unclosed end of the straw is then attached to a sample rod consisting of a metal

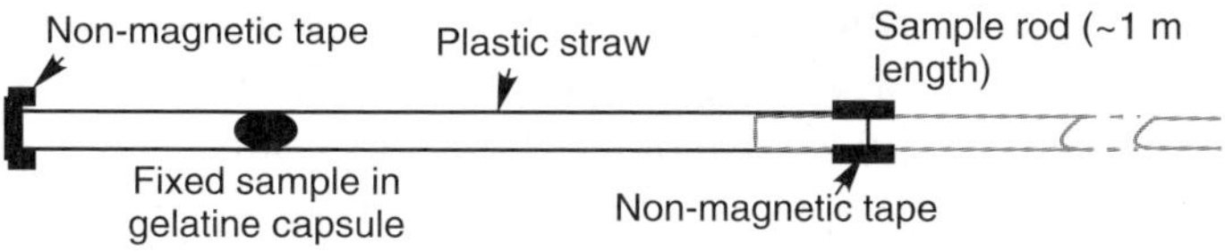

Figure 1.21 Schematic of the mounting of a sample to the sample rod in a SQUID.

rod about 1 m in length with a plastic lock nut at the top, assembled as shown in Figure 1.21. Some more sophisticated sample rods even contain built-in fibre-optic cables so that samples can be illuminated with an external light source during magnetic measurements, *e.g.* to monitor light-activated spin transitions.[53]

The sample holder must perform well across the experimental temperature range, which can be problematic when using plastic mounts at elevated temperatures. Also, as explained in Section 1.1.4.1, the sample holder may contribute a background signal, which must be measured and accounted for. However, as noted in Section 1.3.1, a long sample holder, such as a plastic straw, should not produce a signal, assuming that it is homogenous along its length. Small signals are seen from changes in such parameters as thickness, mass and density along the length of the straw. It is also crucial that the sample be held rigidly, in order to prevent any unwanted sample motion. This point is even more crucial when making measurements on single crystals or arrays of crystals, where magnetic susceptibilities might be anisotropic. In other words, the results may depend on the orientation of the crystal with respect to the external magnetic field. If the sample is air-sensitive or is liquid, it should first be sealed into a tube, such as a quartz NMR or smaller EPR tube. This can then be fixed inside the straw. A very brief discussion of EPR is given in Section 1.5.1 and a more detailed one can be found in Chapter 4 of *Local Structural Characterisation*.[54]

At this stage, the sample position relative to the sample rod's locking nut can be adjusted so that the sample is placed at roughly the centre line of the external magnetic field. This can be done outside the magnetometer by comparison with samples run previously. This process can be quite important, especially for very weakly magnetic materials or very small samples that might be difficult to locate once in the body of the magnetometer. Finally, it is common to make sure there are a few small holes in the side of the straw, which aids in removal of oxygen when introducing the sample. The sample is then moved into the body of the apparatus *via* an air lock system. The sample space of a SQUID magnetometer has a low-pressure helium atmosphere and contamination with oxygen (since it is paramagnetic) or water vapour (since it will freeze at the operating temperature) from air must be avoided. For this reason, the sample space is isolated from the atmosphere with a valve, above which sits a chamber through which air can be removed and replaced with helium. The sample is placed in this chamber, with the sample rod sticking out above the machine, and the sample space is purged – typically three times, through a cycle that evacuates the space and replaces the air with helium at the sample space pressure. The valve is opened and the sample rod is pushed down so that the sample moves into the body of the magnetometer. The locking nut at the top of the sample rod is then connected and locked to a step motor used to move the sample through the pick-up coils (see Section 1.3.1). If oxygen should leak into the sample chamber, this will be seen in the data as an antiferromagnetic transition for molecular oxygen at 43 K and an extra paramagnetic signal above this temperature.

A note of caution: purging the vacuum space may cause the sample holder to explode at this stage and contaminate the air-lock, but this can be observed through a window in the side of the chamber. Although this is a very rare occurrence, it is more convenient to have to clean out the air lock than the sample chamber. But, just in case, the sample chamber may be cleaned by first increasing the temperature to 300 K and allowing it to equilibrate.

Now that the sample sits in the body of the magnetometer, roughly at the centre line of the magnet, an accurate centring procedure can be carried out. The step motor moves the sample through the pick-up coils for a single measurement over the whole range of the sample motion. The resulting signal, an example of which is given in Figure 1.22, shows where the centre of the sample is with respect to the magnet. In Figure 1.22, the sample sits where the maximum signal occurs, at 3 cm. A centring scan is commonly carried out using a small external field (*e.g.* 0.01 T) and at a temperature that provides a decent sample signal. Any required positional displacement is done *via* the stepping motor and the centring scan is redone until the sample is in the correct position.

The next step is to write the sequence of instructions for the magnetometer to follow. A common experiment is to measure samples in very low magnetic fields, *e.g.* when trying to assess the type of a superconducting sample, and it is, therefore, important to realise that superconducting magnets can trap remnant magnetic fields. This can happen if a large magnetic field has been used in previous measurements and means that any applied magnetic field may not be what is expected. A magnetometer can have different ways of applying magnetic fields. One method is the

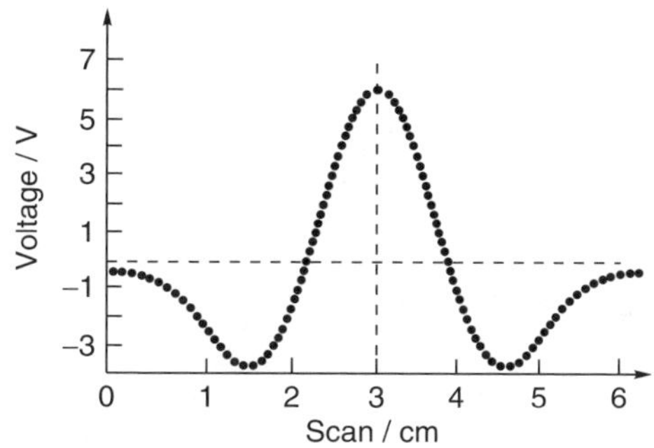

Figure 1.22 Example of a centring scan for a SQUID magnetometer.

'oscillate mode', in which the field oscillates about the required field at decreasing amplitudes – this minimises remnant fields. A 'no overshoot mode', as the name suggests, simply changes the field to the desired value without overshooting it. This is clearly important if the sample is sensitive to the history of its exposure to magnetic fields, as are ferro- and ferrimagnets. It has been noted that the magnetic field will relax (decay) to its equilibrium value after each field change, so it is recommended that a wait time of about 10 minutes after each change be included in the program. However, in the third mode, called 'hysteresis mode', the current through the external magnet's coils is left on so that rapid field changes can be made and hysteresis loops can be quickly measured. Another important consideration is how the temperature is set. A quick set of measurements can be made by sweeping the temperature up from its base value (commonly 2 K) at a particular scan rate. The downside to this method is that the temperature may change during a measurement. Therefore, a more measured stepping sequence, including a settling time, can be used for more accurate measurements.

1.3.1.3.1 Example of a Measurement Sequence for a ZFC/FC Measurement

A common experiment that is used to monitor irreversibility in ferro- and ferrimagnets (see Sections 1.1.6.1 and 1.1.6.3) and to determine the T_C is to collect a zero-field-cooled/field-cooled (ZFC/FC) data set. The temperature is first set well above the T_C of the sample, so that it is in the paramagnetic regime, and the magnetic field is set to zero, taking any remnant field into consideration. The requirement that $H = 0$ can be checked, once the sample has been centred, by measuring the sample and observing a low output from the SQUID. The sample is cooled through

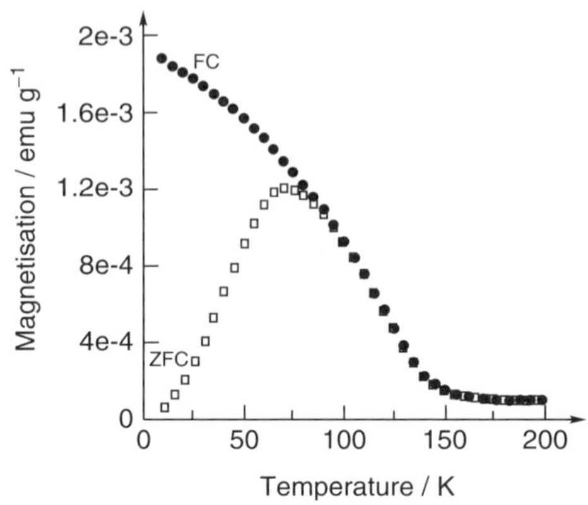

Figure 1.23 Typical curve of magnetisation *versus* temperature for zero-field-cooled (ZFC) and field-cooled (FC) measurements of a ferromagnet. T_C is approximately 70 K.

the critical temperature to the lowest available temperature in the zero magnetic field. After a stabilisation time, a small magnetic field is applied (*ca* 0.01 T) and the sample is measured as a function of temperature as it warms to the highest measurement temperature, giving the ZFC data set. The sample is then cooled in the same applied field to the base temperature and the FC data set is measured while it is warmed again. A typical plot of both the FC and the ZFC data for a ferromagnetic material is shown in Figure 1.23. The T_C is taken at the point where the two measured curves separate, where irreversibility occurs.

1.3.1.3.2 Example Data Set Taken from a SQUID

Table 1.4 shows a few lines of a data set taken from a SQUID magnetometer. The column headed 'SQ' indicates that longitudinal pick-up coils were used in the measurement. Some systems are also able to use transverse coils, the use of which will not be expanded upon in this chapter. The next two columns contain the temperature and the error in temperature in Kelvin (this is clearly a variable-temperature rather than variable-field experiment). The external field, given in Gauss, is next; for a variable-temperature experiment, this column is constant. The 'EMU' column is the measured magnetisation and the STD column contains the standard deviation of the emu. As mentioned above, more than one measurement is usually taken at each fixed temperature or field, and the standard deviation is calculated from the combination of these measurements. The final two columns are other indications of the error in the

Table 1.4 Sample data set from a SQUID magnetometer. The date and time columns, which are usually part of the output, have been removed.

SQ	TEMP	dTMP	FIELD	EMU	STD	REG	%ERR
L	2.00	0.00	1000.00	5.4263e-3	5.43e-7	0.962	1.0e-2
L	2.25	0.00	1000.00	4.9438e-3	3.17e-6	0.963	6.4e-2
L	2.50	0.00	1000.00	4.5452e-3	1.83e-6	0.964	4.0e-2
L	2.75	0.00	1000.00	4.1973e-3	1.76e-6	0.964	4.2e-2

data. The REG column is a goodness-of-fit parameter for the fitting of a theoretical curve to the experimental signal shown in Figure 1.22, a perfect fit having REG = 1. It is this curve fitting that yields the emu value from the maximum of the curve. Finally, %ERR is a percentage error calculated using the emu and STD columns.

The emu values are easily converted to molar magnetic susceptibility by using Equation 1.36, where RMM is the molecular mass of the sample in g mol^{-1}. From Equation 1.36, the diamagnetic corrections can be carried out and $\chi_m T$ and $1/\chi_m$ calculated and plotted, as described in Section 1.1.5.1.

$$\chi_m = \frac{\text{emu}}{\text{FIELD in Gauss}} \times \frac{\text{RMM in g mol}^{-1}}{\text{Sample mass in g}} \tag{1.36}$$

1.3.2 AC Magnetometry

In AC magnetic measurements, a small sinusoidal alternating magnetic field, where the north and south poles swap over many times *per* second, is superimposed onto a DC field and applied to a sample. This is achieved using an alternating current through a coil that sits around the sample. The AC field leads to an alternating induced magnetisation in the sample, as the magnetic moments are subject to an alternating alignment force, given in Equation 1.3. This magnetisation is in turn detected by either a normal induction coil or a coil coupled to a SQUID detector, as outlined in Section 1.3.1.2. As the DC field is changed, different parts of the M *versus* H curve can be probed, which ultimately yields different magnetic susceptibilities. A distinct difference between this method and the DC (or fixed external field) methods is that the magnetisation is measured as a function of a changing field. In other words, the derivative dM/dH is measured directly. This is, of course, the magnetic susceptibility from Equation 1.7, which is measured without the need for extra experiments to determine a field range with a linear response in M. This also means that, unlike in the DC induction methods, the sample does not need to move in order to be measured by the pick-up coils. Instead of moving the sample, this method uses a changing magnetic field to alter the sample's moment. The utility of this technique also relies on the fact that the frequency of the alternating field can be adjusted easily. Since the measurements are time dependent – a function of the AC frequency – this also provides information about the dynamics of the magnetisation process; information that cannot be obtained from DC measurements. Simply put, it is possible to determine whether the change in magnetisation is able to follow the changing applied field as the rate of change is altered.

Since the applied magnetic field follows a sinusoidal curve, the applied AC field at time t is given by H_{AC} in Equation 1.37, where H is the applied field and ω is the frequency of the oscillation. Figure 1.24 illustrates this further.

$$H_{AC} = H \sin(\omega t) \tag{1.37}$$

At very low frequencies, the measurement is similar (as near as possible) to a DC measurement and the magnetisation, M_{AC}, is given simply by Equation 1.38. However, at higher frequencies the induced magnetisation may or may not follow H_{AC} exactly and the lag is described by

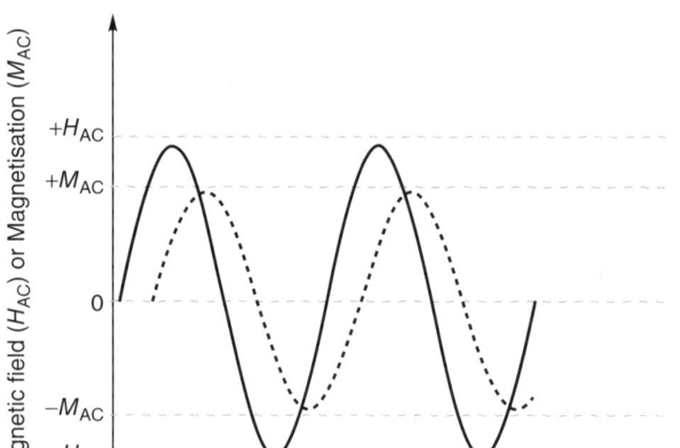

Figure 1.24 Oscillating AC field (solid line) and corresponding magnetisation (dotted line), showing the effect of a phase shift or 'lag'.

Equation 1.39 through a lag angle or phase shift, ϕ.

$$M_{AC} = \frac{dM}{dH} H \sin(\omega t) = \chi H_{AC} \tag{1.38}$$

$$M_{AC} = M \sin(\omega t - \phi) \tag{1.39}$$

Equation 1.39 may also be rewritten as Equation 1.40 in terms of magnetic susceptibility by using the relationships in Equations 1.41 and 1.42. As can be seen from Equation 1.42, the magnetic susceptibility is split into a real, χ', and an imaginary, χ'', component. In the limit of low frequencies, the real component is equivalent to the slope of M *versus* H in a DC field. A non-zero, imaginary component indicates that a lag is experienced by the sample. A similar set of relationships is applicable to other AC techniques, such as AC conductivity measurements, where the real and imaginary components of resistivity (or conductivity) can be examined. A typical experiment uses a small driving field of a few Gauss and measurements can be taken of magnetisation *versus* temperature, *versus* driving frequency and *versus* the DC and AC fields.

$$M_{AC} = \chi' \cos(\omega t) + \chi'' \sin(\omega t) \tag{1.40}$$

where

$$\chi' = \frac{M}{H} \cos(\phi) \quad \text{and} \quad \chi'' = \frac{M}{H} \sin(\phi) \tag{1.41}$$

$$\chi = \chi' + i\chi'' \tag{1.42}$$

The AC technique is particularly useful for determining the T_C of ferro- and ferri-magnetic materials accurately. As this type of material is cooled through T_C, the internal interactions between magnetic moments become significant and there is a resistance to the motion of the domain walls and an irreversibility of movement, as described in Section 1.1.6.3. As a consequence of this irreversibility, the sample magnetisation is unable to exactly follow the drive field as it oscillates. The T_C is taken at the temperature where the imaginary component of the susceptibility becomes non-zero.

Materials with spin-glass behaviour, described in Section 1.1.7, are normally characterised using AC techniques, since they can exist in irreversible magnetic states or long-lived states that are less stable than the most stable state (*i.e.* metastable). As with other ordered systems, the spin-glass state is only present below a certain temperature, above which it would be paramagnetic. This temperature can be determined by measuring χ' as a function of temperature, with a maximum being found at the critical temperature. Interestingly, the critical temperature is dependent on the frequency of the drive field, and this feature is distinctly associated with spin glasses. In addition, below the critical temperature the imaginary component of the magnetic susceptibility is non-zero, due to the irreversibility (resistance to motion) of the magnetic moments.

The AC technique is also crucial in investigating the phenomenon of superparamagnetism, as described in Section 1.1.7. Above the blocking temperature, the imaginary component of the susceptibility is small and the real component follows the Curie law as predicted for a paramagnet. The imaginary component then peaks at the blocking temperature, T_B. Since T_B depends on the measurement frequency, the peak of the imaginary component occurs at different temperatures for different frequencies.[36]

1.3.3 The Micro- (and Nano-) SQUID

For samples that require a very large sensitivity, even above that of a 'normal' SQUID, micro- or nano-SQUIDs can be used.[55] These methods utilise the full sensitivity of the SQUID instead of coupling the sample *via* a gradiometer coil, as discussed in Section 1.3.1. Here the sample is positioned directly on a SQUID detector or an array of detectors with micro- or nanometre dimensions; such SQUIDs are prepared employing the lithographic techniques used in the semiconductor industry. The usefulness of these techniques lies in their ability to study small single crystals or even single molecules. The main issue, as mentioned in Section 1.3.1.2, is that the magnetising field not only influences the sample but is also detectable by the SQUID. In order to solve this issue, the external magnetic field is strictly applied parallel to the SQUID plane rather than perpendicular to it (or at some other angle). Furthermore, the SQUID is manufactured to be extremely thin, which prevents flux trapping in the superconducting material.

1.3.4 The Vibrating Sample Magnetometer (VSM)

As with other methods, in the vibrating sample magnetometer (VSM)[56] the sample is magnetised using a homogenous external magnetic field, but the defining characteristic of this method is that the sample is oscillated in a sinusoidal manner through the induction pick-up coils. The vibration is typically induced by attaching the sample to a piezoelectric material that vibrates when a current is passed through it. In early designs, the sampled movement was generated using a loudspeaker set to vibrate at a defined frequency and the magnetic flux seen by the pickup coils changed in a sinusoidal manner with time. The resultant alternating voltage is processed by a lock-in amplifier, with the piezoelectric vibration used as a reference, and the result is a good signal-to-noise ratio. Unlike the Gouy and Faraday methods, the VSM can measure single crystals easily, since the apparatus is not designed to respond to a force – the sample is in a homogeneous field, so the crystallographic axes can be kept stationary with respect to the magnetic field.

1.4 THE EVANS NMR METHOD

The molar magnetic susceptibility of soluble paramagnetic materials can be measured by means of NMR spectroscopy.[57,58] This is quite convenient as NMR instruments are much more common than SQUIDs or even Gouy balances. In this method, the compound is dissolved in a suitable solvent that gives a sharp singlet in its hydrogen NMR spectrum. The solution is placed into an NMR tube along with a co-axially mounted sealed capillary containing the pure solvent. A paramagnetic compound causes the solvent resonance peaks to shift from the normal position. As a result, the NMR spectrum will contain two resonance peaks, the first due to the solution and the second due to the pure solvent. The frequency difference between the two peaks is then related to the magnetic susceptibility of the solute. The mass susceptibility of the sample is given by Equation 1.43, where Δf is the frequency difference of the two lines in Hertz, f is the frequency of the radio waves at which the spectrum is recorded, m is the mass of the sample *per* cm^3 of solution, χ_0 is the mass susceptibility of the solvent and d_0 are d_s are the densities of the solvent and solution, respectively.

$$\chi_g = \frac{3\Delta f}{2\pi f \mathrm{m}} + \chi_0 + \frac{\chi_0(d_0 - d_s)}{m} \tag{1.43}$$

1.5 BRIEF NOTES ON COMPLEMENTARY TECHNIQUES

1.5.1 Electron Paramagnetic Resonance (EPR)

EPR[59] is often referred to in older literature as 'electron spin resonance' (ESR); they are identical techniques. As described throughout Section 1.1.5, the *g*-factor is often estimated when modelling magnetic data or correlating the number of unpaired electrons with magnetic information. This can also be estimated if the orbital contribution is known using Equation 1.22, or *g* can be set as a fitting parameter if there is an available theoretical model. However, EPR spectroscopy is useful as it gives a direct measure of *g*-values. EPR spectroscopy has been described as being similar to NMR spectroscopy except that it looks at transitions involving the spin of unpaired electrons rather than the nuclear spin. When species with unpaired electrons are placed in a magnetic field, the M_S components are separated. The relative orientations of the spins with respect to the magnetic field can be of lower energy (parallel alignment) or higher energy (antiparallel alignment). This is called the Zeeman effect and is illustrated on the left of Figure 1.25. The energy difference between levels is given by Equation 1.44, where H, g and μ_B are the same as defined previously. Since H and μ_B are constants, the energy difference is directly proportional to g.

$$\Delta E = g\mu_B H \tag{1.44}$$

The difference in energy corresponds to the microwave part of the electromagnetic spectrum, so absorption of microwave energy induces

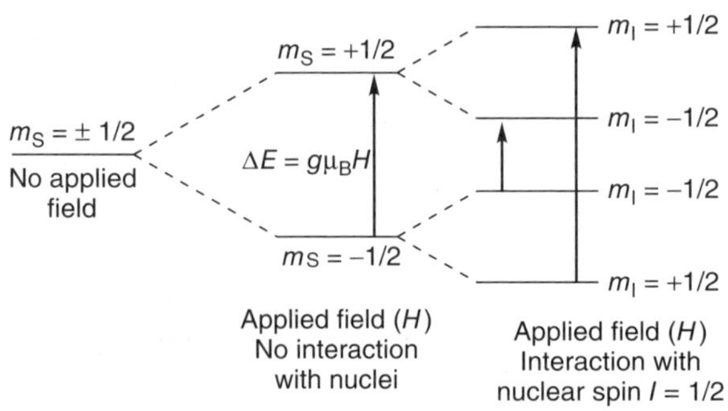

Figure 1.25 Splitting of energy levels by a magnetic field and by interaction with a nuclear spin of $I = 1/2$. Arrows show allowed transitions.

transitions between the magnetic energy levels. For a typical measurement, called a continuous-wave experiment, the microwave frequency is fixed and the magnetic field is swept until resonance occurs. The frequency of the microwave used determines the 'band' of EPR used, with higher frequencies corresponding to increased resolution spectra. The spectra are measured and displayed as the first derivative of an absorption peak, which is a more sensitive method than a simple absorption spectrum. In addition, the energy levels can be split if the electron magnetic moment couples to a nuclear magnetic moment, I. This gives $2I + 1$ absorption lines, as shown in Figure 1.25, for coupling to a nucleus with $I = 1/2$, such as hydrogen. The selection rule for the transition is $\Delta M_S = \pm 1$ and $\Delta M_I = 0$. In addition, information can be obtained about zero-field splitting (ZFS), since the energy separations are a function of the ZFS parameters. For example, if the sample is an axially symmetric paramagnet and the magnetic field is orientated parallel to the unique axis, the transitions are separated by $2|D|/g\mu_B$, where D is the axial ZFC term. By measuring the EPR signal of a single crystal as a function of its orientation with respect to the external field, it is also possible to determine the orientation of the ZFS axis with respect to the crystal. For powdered samples, a weighted average of the three mutually perpendicular g values is found.

1.5.2 Ultraviolet–Visible Spectroscopy

Ultraviolet (UV)–visible spectroscopy (electronic spectroscopy) is a very common technique that records the reflectance or absorption spectra of samples in the UV and visible regions of the electromagnetic spectrum. In these regions, the spectra probe electronic transitions between orbitals, particularly those from the ground state to excited states. The well-known equation governing transitions is the Beer–Lambert law (Equation 1.45), in which A is the dimensionless absorbance of incident light with intensity I_0 and I is the transmitted intensity after absorption. The magnitude of the molar absorptivity (or molar extinction coefficient), ε, reflects the probability of a transition occurring, c is the concentration of the absorbing species and l is the path length of the absorbed light.

$$A = \log_{10}\frac{I_0}{I} = \varepsilon cl \qquad (1.45)$$

With this equation and a calibration curve of A *versus* c in the linear regime, this technique is used commonly to determine the concentrations of samples. However, with respect to magnetic properties, the frequency, or transition energy, of the transitions is important. In particular, for materials containing *d*-block elements, the technique can be used to determine the *d*-orbital splitting parameter, Δ, as used in Equation 1.22. This is done in conjunction with energy-level diagrams for a particular *d*-electron configuration, namely Orgel[60,61] and Tanabe–Sugano diagrams.[62]

1.5.3 Thermal Techniques

The specific heat capacity of a substance is the enthalpy (amount of heat) required to raise the temperature of a unit amount of the substance by one degree (units $\mathrm{J\,K^{-1}\,mol^{-1}}$). When a sample is heated, this energy is converted to kinetic energy (internal vibrations) and other forms of internal energy. The amount that is converted to kinetic energy causes an increase in the sample temperature and a change in entropy, or disorder, within the system. In fact, the temperature is defined as the derivative of the internal energy with respect to entropy. Therefore, other effects that cause large changes in the entropy of a system may be seen on a heat capacity *versus* temperature curve. One such effect can be from magnetic transitions, which move a sample from a disordered paramagnetic state to an ordered state such as ferromagnetic, ferrimagnetic or spin-glass. This appears on a heat capacity curve as a lambda-shaped curve superimposed on a background resulting from contributions from lattice vibrations. The technique is useful for confirming the critical temperatures of magnetic transitions and calculating the entropy changes. For more on thermal techniques, see Chapter 2 of this volume.

1.5.4 Mössbauer Spectroscopy

This technique is a very sensitive method for determining the oxidation state and spin state of some atoms. It is particularly useful for following changes that occur during the spin-transition phenomenon.[63] As mentioned in Section 1.1.7, magnetisation experiments can monitor this effect by, for example, observing the paramagnetic-to-diamagnetic transition in octahedral Fe^{II} complexes. However, it should be clear that there are other mechanisms that can reduce the magnetisation, which cannot be distinguished by bulk magnetic measurements. Particular difficulties arise when partial spin transitions occur, leading to a mix of HS and LS states,[64] or when the transition happens across a large temperature range. An alternative explanation could be introduction of antiferromagnetic coupling in the sample. Mössbauer spectroscopy can confirm whether an effect has its origins in a spin-state change and this technique is quantitative, so that the percentage of metal centres in each spin state can be determined as a function of temperature. The basis of the technique is the Mössbauer effect, which won Mössbauer the 1961 Nobel Prize in Physics. The effect involves the emission and recoilless absorption of γ-rays in solids. The source of the γ-rays is dependent on the element being probed. For iron-based spin-transition materials, the source is nuclear transitions in ^{57}Fe: a radioactive sample of ^{57}Co decays *via* electron capture to an excited state of ^{57}Fe, which itself decays and emits γ-rays, which are absorbed resonantly by the same Fe isotopes in the sample. This technique relies on the fact that the absorption profile detects subtle changes in the environment of the nuclei, arising from, for example, electrical field gradients generated by different electron distributions. The effect is a splitting of nuclear energy levels. This leads to the extraction of parameters including chemical or isomer shift (equivalent to those in NMR spectroscopy), the Zeeman splitting parameter (interaction of the nucleus with a magnetic field) and quadrupole splitting (the effect of electric field gradients). The spectra can be easily correlated to specific spin states of Fe atoms. Other isotopes display an observable Mössbauer effect, such as ^{61}Ni, ^{67}Zn, ^{99}Ru and ^{119}Sn, but most of the research literature focuses on ^{57}Fe because of its interesting magnetic properties and on ^{119}Sn because of its electronic applications. For example, knowledge of the oxidation state of Sn is crucial to its chalcogenide compounds (SnS, SnSe), which have applications in solar cells.[65] For more on Mössbauer spectroscopy, see reference [66].

1.5.5 Measuring Local Magnetic Fields with Muons and Neutrons

The purpose of this chapter was to inform the reader of the basis of the techniques used to measure bulk magnetic properties, *i.e.* the magnetisation or sum of individual magnetic dipoles *per* unit volume, or mass, or mole. There are two other techniques that should not be left out of any discussion on magnetism. These methods look at the local effects of magnetism, as do EPR and Mössbauer spectroscopies. Both have been left to the end of this discussion as they might be termed 'exotic' techniques, since the means of performing them are not found in the standard laboratory. These methods are muon spin rotation (or relaxation, termed μSR)[67,68] and neutron diffraction (for an account of muon and neutron research at the ISIS Neutron and Muon Source, Rutherford Appleton Laboratory, Oxford, UK, see reference [69]).

The muon is a fundamental particle and is produced by accelerated collisions of protons with a suitable target, *e.g.* carbon. The muon carries one unit of elementary charge but can be considered a heavy electron as it has about 200 times the mass of an electron. For the purposes of this discussion, the muon's important property is that it also possesses an intrinsic magnetic moment with a spin of $S = 1/2$. During the experiment, spin-polarised muons are implanted into a sample and stop at interstitial sites as a consequence of their charge. Their magnetic moment interacts with local magnetic fields, causing precession about the field direction, with a rate that depends on the field strength. Since the muon is a short-lived species (2.2 μs), it degrades quickly by expelling detectable positrons. The key to this experiment is that positrons are ejected in a direction related to that in which the muon was pointing at the time of decay. Spatial detection of the positrons can, therefore, reflect the strength of the internal magnetic fields. Variable-temperature measurements, for example, can reveal the dynamics of the local fields, and for this reason the technique is also suited to the investigation of superconductors. A particularly accessible account of this technique is given by Blundell.[67]

Neutron diffraction[68,69] is based on the elastic scattering of neutrons, which occurs such that a neutron's direction is altered but there is no transfer of kinetic energy. This technique is related closely to the more familiar X-ray diffraction, although while X-rays are sensitive to the sample's electrons, neutrons interact with the nucleus and, importantly, magnetic nuclei give additional elastic scatter. In terms of magnetic analysis, the important property is that neutrons also carry a spin ($S = 1/2$), which interacts with magnetic moments that originate from electrons, and thus neutron diffraction can be used to reveal the magnetic structure of a material, ultimately producing spin maps such as those shown schematically in Figures 1.4, 1.6 and 1.9. These are then superimposed onto atomic positions. This technique is particularly useful when complicated arrangements of moments are possible, such as in the spin-ice $Ho_2Ti_2O_7$.[70]

REFERENCES

[1] H. Bethe, *Ann. Physik*, **3**, 122 (1929).
[2] J. H. Van Vleck, *J. Chem. Phys.*, **3**, 807 (1935).
[3] J. H. Van Vleck, *Phys. Rev.*, **41**, 208 (1932).
[4] C. E. Housecroft and A. G. Sharpe, *Inorganic Chemistry*, 3rd edition, Pearson Prentice Hall, 2008.
[5] H. L. Shläfer and G. Gliemann, *Basic Principles of Ligand Field Theory*, Wiley International, New York, 1969.
[6] M.-H. Phan and S.-C. Yu, *J. Magn. Magn. Mater.*, **308**, 325 (2007).
[7] N. Robertson and G. T. Yee, in *Molecular Materials*, D. W. Bruce, D. O'Hare and R. I. Walton (Eds), John Wiley, Chichester, 2010.
[8] O. Kahn, *Molecular Magnetism*, VCH Publisher Inc., New York, 1993.
[9] S. Blundell, *Magnetism in Condensed Matter*, Oxford University Press, Oxford, 2006.
[10] A. F. Orchard, *Magnetochemistry*, Oxford Chemistry Primers, Oxford University Press, Oxford 2003.
[11] F. E. Mabbs and D. J. Machin, *Magnetism and Transition Metal Complexes*, Dover Publications Inc., New York, 2008. Originally published by Chapman and Hall, 1973.
[12] R. J. Carlin, *Magnetochemistry*, Springer, 1968.
[13] W. E. Hatfield, *'Magnetic Measurements'*, in *Solid State Chemistry – Techniques*, P. Day and A. K. Cheetham (Eds), Clarendon Press, Oxford, 1995.
[14] T. I. Quickenden and R. C. Marshal, *J. Chem. Educ.*, **49**(2), 114 (1972).
[15] J. E. Crooks, *J. Chem. Educ.*, **56**, 301 (1979).
[16] G. A. Bain and J. F. Berry, *J. Chem. Educ.*, **85**, 532 (2008).
[17] P. Pascal, *Ann. Chim. Phys.*, **19**, 5 (1910).
[18] J.-S. Zhou, J. B. Goodenough and B. Dabrowski, *Phys. Rev. B*, **67**, 020404 (2003).
[19] C. de la Calle, J. Sánchez-Benitez, F. Barbanson, N. Nemes, M. T. Fernández-Diaz and J. A. Alonso, *J. App. Phys.*, **109**, 123914 (2011).
[20] P. Atkins and J. de Paula, *Atkins'Physical Chemistry*, 9th edition, Oxford University Press, Oxford, 2010, p357.
[21] Data adapted from F. E. Mabbs and D. J. Machin, *Magnetism and Transition Metal Complexes*, Dover Publications Inc., New York, 2008. Originally published by Chapman and Hall, 1973, p. 20.
[22] L. Néel, *Ann. de Physique*, **3**, 137 (1948).
[23] O. Kahn, Y. Pei, M. Verdaguer, J. P. Renard and J. Sletten, *J. Am. Chem. Soc.*, **110**, 782 (1988).
[24] V. Baron, B. Gillon, J. Sletten, C. Mathoniere, E. Codjovi and O. Kahn, *Inorg. Chim. Acta*, **69**, 235 (1995).
[25] S. Turner, O. Kahn and L. Rabardel, *J. Am. Chem. Soc.*, **118**, 6428 (1996).
[26] F. Palacio, *Mol. Cryst. Liq. Cryst.*, **305**, 385 (1997).

[27] I. Koichiro and O. Takehiko, *J. Phys. Soc. Jpn.*, **23**, 765 (1967).
[28] J. Boeckmann, M. Wriedt and C. Näther, *Chem. Eur. J.*, **18**, 5284 (2012).
[29] C. Lacroix, *Physica B*, **404**, 3038 (2009).
[30] A. P. Ramirez, G. P. Espinosa and A. S. Cooper, *Phys. Rev. Lett.*, **64**, 2070 (1990).
[31] A. S. Wills, *Can. J. Phys.*, **79**, 1501 (2001).
[32] M. Onada, H. Imai, Y. Amako and H. Nagasawa, *Phys. Rev. B.*, **56**, 3760 (1997).
[33] C. Y. Huang, *J. Magn. Magn. Mater.*, **51**, 1 (1985).
[34] L. Néel, *Ann. de Phys.*, **3**, 137 (1948).
[35] W. F. Brown, *Phys. Rev.*, **130**, 1677 (1963).
[36] R. E. P. Winpenny and E. J. L. McInnes, in *Molecular Materials*, D. W. Bruce, D. O'Hare and R. I. Walton (Eds), John Wiley, Chichester, 2010.
[37] W. Wernsdorfer, *Int. J. Nanotechnology*, **7**, 497 (2010).
[38] M. Evangelisti and E. K. Brechin, *Dalton Trans.*, **39**, 4672 (2010).
[39] P. Gütlich, A. Hauser and H. Speiring, *Angew. Chem., Int. Ed. Engl.*, **33**, 2024 (1994).
[40] O. Kahn, J. Krober and C. Jay, *Adv. Mater.*, **4**, 718 (1992).
[41] J. M. Holland, J. A. McAllister, Z. Lu, C. A. Kilner, M. Thornton-Pett and M. Halcrow, *Chem. Comm.*, **577** (2001).
[42] M. Halcrow, *Coord. Chem. Rev.*, **249**, 2880 (2004).
[43] R. Pritchard, C. A. Kilner, S. A. Barret and M. A. Halcrow, *Inorg. Chim. Acta.*, **362**, 4365 (2009).
[44] A. Sanderson, *Phys. Educ.*, **3**, 272 (1968).
[45] M. A. Cavanaugh, *J. Chem. Educ.*, **56**, 550 (1979).
[46] C. J. O'Connor, *Prog. Inorg. Chem.*, **29**, 203 (1982).
[47] http://www.imdb.com/title/tt0114558 (last accessed 28 October 2013).
[48] http://www.qdusa.com/products/mpms.html (last accessed 28 October 2013).
[49] J. Bardeen, L. N. Cooper and J. R. Schrieffer, *Phys. Rev. A*, **106**, 162 (1957).
[50] B. D. Josephson, *Phys. Lett.*, **1**, 251 (1962).
[51] R. C. Jaklevic, J. Lambe, A. H. Silver and J. E. Mercereau, *Phys. Rev. Lett.*, **12**, 159 (1964).
[52] P. W. Anderson and J. M. Rowell, *Phys. Rev. Lett.*, **10**, 230 (1963).
[53] H. Hagiwara, O. Sakaguchi, K. Nishi, S. Hashimoto and N. Matsumoto, *Bull. Chem. Soc. Jpn.*, **84**, 936 (2011).
[54] D. W. Bruce, D. O'Hare and R. I. Walton, *Local Structural Characterisation*, John Wiley, Chichester, 2014.
[55] W. Wernsdorfer, K. Hasselbach, A. Benoit, B. D. Mailly, J. Tuaillon, J. P. Perez, V. Dupuis, G. Dupin, G. Girauld and A. Perex, *J. Appl. Phys.*, **78**, 7192 (1995).
[56] S. Foner, *Rev. Sci. Instr.*, **30**, 548 (1959).
[57] R. Engel, D. Halpern and S. Bienenfled, *Anal. Chem.*, **45**(2), 367 (1973).
[58] D. Ostfield and I. A. Cohen, *J. Chem. Educ.*, **49**, 829 (1973).
[59] F. E. Mabbs and D. Collinson, *Electron Paramagnetic Resonance of Transition Metal Compounds*, Elsevier, 1993.
[60] L. E. Orgel, *J. Chem. Soc.*, **4756** (1952).
[61] L. E. Orgel, *J. Chem. Phys.*, **23**, 1819 (1955).
[62] Y. Tanabe and S. Sugano, *J. Phys. Soc. Jpn.*, **9**, 753 (1954).
[63] D. L. Regar, J. D. Elgin, E. A. Foley, M. D. Smith, F. Grandjean and G. U. Long, *Inorg. Chem.*, **48**, 9393 (2009).

[64] C. Grunert, S. Reiman, H. Spiering, J. A. Kitchen, S. Brooker and P. Gütlich, *Angew. Chem., Int. Ed.*, **47**, 2997 (2008).
[65] A. de Kergonmeaux, J. Faure-Vincent, A. Pron, R. de Bettignies, B. Malaman and P. Reiss, *J. Am. Chem. Soc.*, **134**, 11659 (2012).
[66] R. H. Herber, *J. Chem. Educ.*, **42**, 180 (1965).
[67] S. J. Blundell, *Contemp. Phys.*, **40**, 175 (1999).
[68] S. F. J. Cox, *J. Phys. C.*, **20**, 3187 (1987).
[69] http://www.isis.stfc.ac.uk/groups (last accessed 28 October 2013).
[70] T. Fennel, O. A. Petrenko, B. Fak. J. S. Gardner, S. T. Bramwell and B. Ouladdiaf, *Phys. Rev. B*, **72**, 224411 (2005).

2

Thermal Methods

Michel B. Johnson[a] and Mary Anne White[a,b,c]
[a]*Institute for Research in Materials, Dalhousie University, Halifax, NS, Canada*
[b]*Department of Chemistry, Dalhousie University, Halifax, NS, Canada*
[c]*Department of Physics and Atmospheric Science, Dalhousie University, Halifax, NS, Canada*

2.1 INTRODUCTION

Temperature can influence the properties of materials greatly. The effect can be small, as with thermal expansion, which typically gives rise to a change in dimension of a few parts per million for one degree change in temperature, or large, such as a complete reorganisation of the lattice in a solid–solid phase transition.

This chapter presents the methods used to detect and quantify thermally activated changes in materials. The first section addresses three important thermoanalytical tools: **thermogravimetric analysis (TGA)**, **differential thermal analysis (DTA)** and **differential scanning calorimetry (DSC)**. The second presents methods for determining heat capacity, the third methods for determining thermal conductivity and the fourth methods for determining thermal expansion.

Many of the thermal methods described here can be employed using commercial instrumentation, so each is presented with a view to its practical use. Thus the basis for each is given alongside notes on sample considerations, expected accuracy and any precautions that should be taken. The latter point is particularly important as thermal techniques

Multi Length-Scale Characterisation, First Edition. Edited by Duncan W. Bruce, Dermot O'Hare and Richard I. Walton.

tend to produce precise rather than accurate data. Examples illustrating several of the techniques have been selected from the literature.

Thermal properties can be investigated over a wide temperature range, from less than 1 K to greater than 2000 K, revealing details concerning interactions within the materials.

2.2 THERMAL ANALYSIS

Thermal analysis is a powerful tool by which to detect thermal events within a sample, such as mass loss upon heating, reactivity and phase transitions (melting, solid–solid, evaporation, glass transition). Thermal analysis can provide important information in a relatively short span of time using small samples and commercial instruments.[1]

The specific techniques discussed in this section are TGA, DTA and DSC. DTA and DSC are primarily used to characterise energetic events as a function of temperature (either on heating or cooling), while TGA is used to investigate mass loss (or possibly gain) as temperature is increased. All are available commercially and some instruments carry out more than one thermoanalytical measurement simultaneously.

2.2.1 Thermogravimetric Analysis

In TGA, the mass of a sample is monitored as a function of temperature. It is an established technique and can be performed simply by placing a sample on a heating stage and measuring the mass both before and after heating. More commonly, a TGA instrument dedicated to such measurements is employed, and generally more information can be collected, such as the reaction temperature and kinetics associated with structural decomposition, oxidation, corrosion, moisture content, adsorption/desorption and gas evolution.[2]

TGA employs a sample holder (or stage) mounted on a weighing device. An empty sample container is loaded on the stage and the lot is tared. The sample is then added to the container and after equilibration the sample mass is logged as a function of temperature, usually during heating. In some TGA instruments, two pans are used: one contains the sample and the other is an empty reference pan (of the same composition as the first). Conventional TGA methods typically have limits of detection on the order of 1 mg, which, for most samples, is sufficient to acquire reliable data. Some instruments offer μg (or better) scales, allowing detection of very small mass changes.

2.2.1.1 TGA Considerations

Some precautions should be taken in order to obtain meaningful data. Calibration of the electronic balance should not be required as it is very robust. However, it is important that the instrument be calibrated for temperature drift by measuring well-characterised mass-loss standards over the temperature range of interest (at least one standard; more calibration points will give a more accurate temperature correction). A list of common standard materials is available.[3] As with DTA and DSC (see Sections 2.2.2 and 2.2.3), multiple reproducible runs with both standard and sample give the best accuracy in TGA. Repeatability on the same sample should be within 5 °C (or, equivalently, 5 K; for consistency we use °C herein for temperature differences) and reproducibility should be within 10 °C, as recommended by the American Society for Testing and Materials (ASTM).[3]

Some items to consider in TGA experiments are the sample size, the heating/cooling rate and the purge gas. Generally, high-resolution TGA thermograms are obtained when using small samples (a few mg), a slow scan rate and a high thermally conductive purge gas such as helium.[4] Some commercial instruments have a programmed, continuously varied heating rate in order to maximise the resolution.[5] Sample state for TGA is not critical; the sample can be liquid or solid. Special care in ensuring an intimate contact at the sample–pan interface is only required if the TGA is coupled with a DTA (where heat flows are measured; see Section 2.2.2).

Often it is helpful to take the temperature derivative of the TGA curve, in order to highlight subtle features. So-called differential thermogravimetric (DTG) curves show peaks where the rate of mass loss is maximal. An example is given in Figure 2.1.

2.2.1.2 New Technique Developments

The limitations of TGA stem mostly from the measurement of mass (typically at least 1 mg of sample is needed), and in some cases the data are not reproducible when investigating thin films or small surface

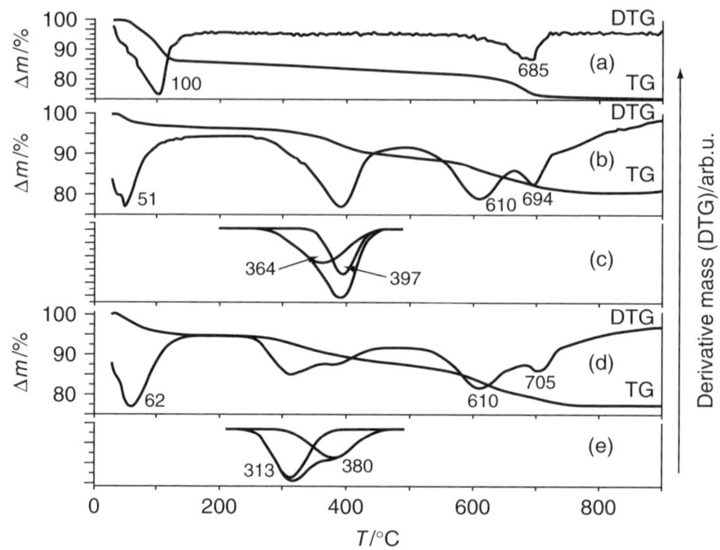

Figure 2.1 TGA (labelled as TG) and DTG curves for several different tetraethylammonium and benzyltrimethylammonium montmorillonites. The DTG curves make it easier to see subtle mass losses. Reprinted with permission from [6] Copyright (2012) Springer Science and Business Media.

modifications to substrates.[2] A newer technique, employing quartz crystal microbalance (QCM) sensor platforms (typically thin piezoelectric AT-cut quartz crystal[7]), allows for the detection of thermal events in microgram sample sizes at elevated temperatures.[2] QCMs are extremely sensitive acoustic devices capable of recording a mass change on the sub-picogram scale on coatings and thin films.[8,9] The operational principle of QCM is based on the resonance frequency of the quartz crystal and the fact that any perturbation to the surface (such as a mass change *via* desorption) alters this frequency in a predictable way. The high sensitivity allows this technique to monitor film thickness on deposition,[10] to detect toxic gases[11] and to carry out moisture assessment.[12]

2.2.1.3 Uses of TGA

Thermal analysis techniques do not indicate the nature of a thermal process *per se*. There are many examples of a mass change with heating; one is chemical decomposition. Such thermal events can be more thoroughly delineated by coupling the TGA to a characterisation instrument such as a gas chromatograph (GC), mass spectrometer (MS), infrared spectrometer or Raman spectrometer.

A straightforward use for TGA is to provide insight into the thermal decomposition kinetics of a material (where decomposition is accompanied by mass loss), *e.g.* in the metal–ligand complexes in reference [13]. ML_2 and $ML_2 \cdot CH_3OH$ complexes (L = the phenate anion derived from 2-(2-hydroxyphenyl)benzimidazole) were decomposed thermally in a TGA instrument; when CH_3OH solvent molecules were incorporated into the crystal, they were released first, followed by the bidentate ligand, so that eventually only the metal oxide remained. Ni and Cu metal centres did not incorporate methanol and the decomposition of the ligand was concerted, while Zn and Co did incorporate methanol and appeared to show non-concerted release of the ligands.[13]

2.2.1.4 *Combined Techniques*

TGA can be coupled to an MS, GC, infrared spectrometer or Raman spectrometer.[5] Coupling of TGA to a characterisation technique enables delineation of the origin of the observed thermal event. For example, if the mass loss in the sample arises from decomposition that gives gaseous products, the latter can be identified and quantified with the other analytical tool.

Consider that the surfaces of virtually all metals and metalloids are covered, in varying degrees, by layers such as oxide or water that play a role in adsorption processes.[14] Determination of the coating is typically carried out by infrared analysis, TGA, titration and/or nuclear magnetic resonance (NMR) spectroscopy. A coupled TGA-MS with a CO_2 sensor makes this determination more efficient, as it is able to distinguish physisorbed water from chemically bound water.[15] Physisorbed water requires less thermal energy to evolve and is removed at much lower temperatures. The desorption information is supported by the mass spectra of the evolved gases as a function of temperature.

2.2.2 Differential Thermal Analysis

In DTA, the thermal response of a sample is measured in comparison to that of a reference material. The sample and reference are heated

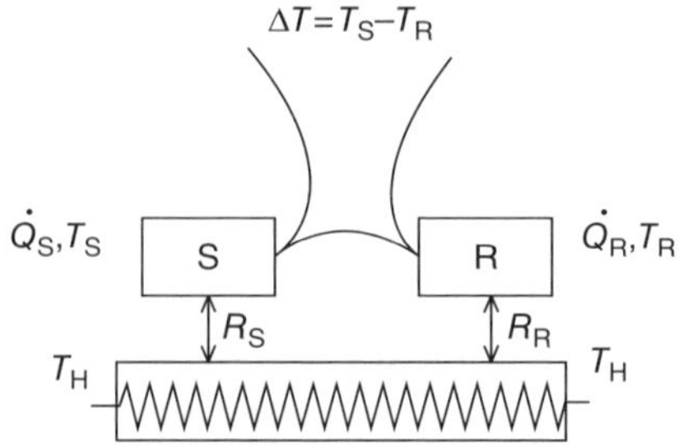

Figure 2.2 Schematic of a DTA apparatus. The sample, S, at temperature T_S and reference, R, at temperature T_R are both heated (or cooled) with corresponding identical powers, and the temperature differential, ΔT, is measured as a function of temperature. The temperature differential can be measured with thermocouples either embedded in the sample and reference or in good thermal contact with them.

(or cooled) with a constant power that gives an approximately linear change in temperature with time. The dependent variable is the temperature differential between the sample and reference, ΔT, which is defined schematically in Figure 2.2. DTA measurements require only the measurement of temperature differences, which can be accomplished rather simply, *e.g.* with a thermocouple. The technique was developed by Roberts-Austen in 1899[16] and first commercialised in the 1950s. Its major advantage is that the user can obtain temperatures of thermal events up to extremely high values (*e.g.* 1740 °C in reference [17]; one company offers a DTA capable of achieving 2400 °C).[18]

DTA experiments require only a small amount of sample (*ca* 10 mg, although larger samples can be accommodated) and can be performed quite readily with commercial instruments. The simplicity of the DTA experiment also means that inexpensive laboratory-built equipment can be used.

2.2.2.1 A More Detailed Look at DTA

In traditional DTA measurements, the thermocouples (temperature sensors for ΔT determination) are placed in direct contact with the sample (or reference) in order to minimise the thermal resistance between the sample/reference and the position at which the temperature is determined. The sample (S at temperature T_S) and the reference (R at temperature T_R) are both heated (or cooled) with the same power and the temperature difference between them ($\Delta T = T_R - T_S$) is monitored as a function of temperature (see Figure 2.2), such that:

$$signal \equiv T_R - T_S = \Delta T. \tag{2.1}$$

Although the thermocouples are embedded directly in the sample and reference, there is still a thermal resistance between heater and sample, R_S, and between heater and reference, R_R, as depicted in Figure 2.2, where the heater is at temperature T_H. Newton's law of energy exchange gives a power to the sample, $\dot{Q}_S$, of:

$$\dot{Q}_S = \frac{1}{R_S}(T_H - T_S) \tag{2.2}$$

and a power to the reference, $\dot{Q}_R$, of:

$$\dot{Q}_R = \frac{1}{R_R}(T_H - T_R). \tag{2.3}$$

Rearrangement and subtraction of Equations 2.2 and 2.3 with the assumption that $R_S = R_R = R$ gives the resultant signal:

$$signal \equiv \Delta T = T_R - T_S = R(\dot{Q}_S - \dot{Q}_R). \tag{2.4}$$

The power to the sample can be related to the heat capacity of the sample, C_S, and the rate of temperature change, $\mathrm{d}T_S/\mathrm{d}t$:

$$\dot{Q}_S = C_S \frac{\mathrm{d}T_S}{\mathrm{d}t} \tag{2.5}$$

and similarly for the reference:

$$\dot{Q}_R = C_R \frac{\mathrm{d}T_R}{\mathrm{d}t}. \tag{2.6}$$

Equation 2.4 can be substituted with Equations 2.5 and 2.6 and the assumption that $\mathrm{d}T_S/\mathrm{d}t = \mathrm{d}T_R/\mathrm{d}t = \mathrm{d}T/\mathrm{d}t$ (*i.e.* the controlled external scanning rate is the same for the sample and the reference):

$$signal \equiv \Delta T = R\frac{\mathrm{d}T}{\mathrm{d}t}(C_S - C_R). \tag{2.7}$$

While Equation 2.7 shows that the signal, ΔT, has a dependence on the difference in the heat capacities of the sample and reference (allowing for a direct calorimetric determination with conventional DTA), the thermal resistance, R, is typically dependent on the sample and reference materials, the instrument geometry and the temperature. As a result, conventional DTA generally provides good results concerning the temperatures of thermal events but is at best a semiquantitative method for specific heat and enthalpy determination.[1]

A modification to classical DTA, introduced by Boersma,[19] makes it more quantitative for enthalpy determination. Figure 2.3 shows Boersma's modified DTA scheme, where the thermocouple measuring the temperature is a differential thermocouple embedded into a stage. In the Boersma method, the sample (at temperature T_S) and the reference (at temperature T_R) are both heated as a function of time. The thermocouple at the mount (see Figure 2.3) measures the difference in temperature between the sample side of the mount, T_{SM}, and the reference side of the mount, T_{RM}, where $\Delta T = T_{RM} - T_{SM}$ is monitored as a function of overall temperature. Because of the placement of the thermocouple, there is a thermal resistance between the thermocouple and the sample, R_S', and between the thermocouple and the reference, R_R', in addition to resistances between the heater and the sample, R_S, and the heater and the reference, R_R, as shown in Figure 2.3. Equation 2.4 can be substituted with Equations 2.5 and 2.6, and $\Delta T = T_{RM} - T_{SM}$ with the assumption that $\mathrm{d}T_S/\mathrm{d}t = \mathrm{d}T_R/\mathrm{d}t = \mathrm{d}T_{SM}/\mathrm{d}t = \mathrm{d}T_{RM}/\mathrm{d}t = \mathrm{d}T/\mathrm{d}t$ and $R_S = R_R = R$, to show that the signal is given by:

$$signal \equiv T_{RM} - T_{SM} = R\frac{\mathrm{d}T}{\mathrm{d}t}(C_S - C_R). \tag{2.8}$$

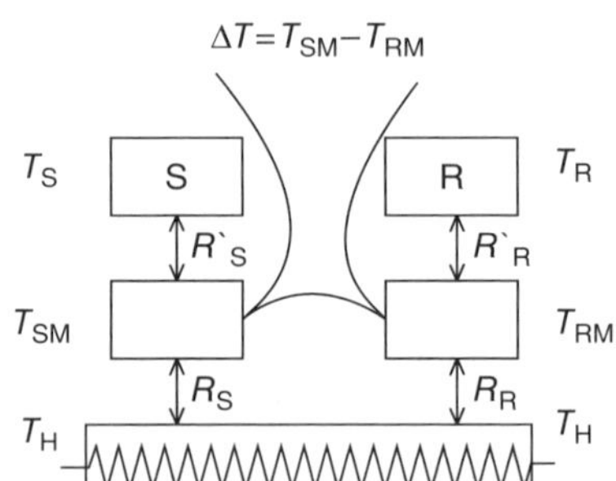

Figure 2.3 Schematic of a Boersma DTA apparatus. The sample, S, at temperature T_S and reference, R, at temperature T_R are both heated (or cooled) as a function of time. The thermocouple junctions used to monitor ΔT are fixed in mounts at temperatures T_{SM} and T_{RM} and the temperature differential is monitored as a function of temperature. In this setup, there are thermal resistances between the mount and the sample and reference (R'_S and R'_R, respectively) and there are further thermal resistances between the heater and the sample and reference stages (R_S and R_R, respectively). The heater is at temperature T_H.

The difference between classic DTA and Boersma DTA is that the *R* in classic DTA is dependent on the sample–heater geometry, whereas in Boersma DTA *R* is a constant of the instrument. As a result of this distinction, Boersma DTA is capable of determining enthalpy changes, provided that calibration has been performed and the appropriate protocols for thermal analysis are followed. A review comparing quantitative DTA with DSC provides more details.[20] It is also important to consider the temperature lag, as classical DTA has none, due to the thermocouple being incorporated in the sample, while in Boersma DTA the thermocouple lies in the sample stage, creating a boundary between the sample and the thermocouple, so the temperature lag is similar to that from power-compensated DSC (see Section 2.2.3.2).

2.2.2.2 *Standards and Uncertainties in DTA*

As with many commercially available thermal analytical instruments, the DTA should be checked regularly for thermometry drift, as measurement of the temperatures of thermal processes is critical to thermal analysis. The calibration material chosen needs to meet several requirements: it should be of high purity, be well characterised and have a thermal process (typically melting point) near the thermal event of the material in question. The best materials are standard reference materials as they are well characterised and can be traced, *e.g.* to the National Institute of Standards and Technology.[21] The choice of material can range from organic to polymeric to metallic materials. In principle, high-purity metals provide excellent standards because their melting points are very sharp, whereas organics and polymers can have broader thermal processes. Equation 2.7 shows that heat capacity and thermal resistance play a role in determining the signal and that the standard should be selected to have a heat capacity (mostly governed by sample mass) and thermal resistance (morphology and thermal conductivity) similar to those of the material under investigation. Choice of calibration standard is critical to an accurate determination of the temperatures of thermal events from the thermogram. It is advisable to measure the melt of more than one standard material in order to determine the temperature dependence of the instrument (*e.g.* the thermometry offset could be 1.0 °C at $T = 420$ °C and 1.6 °C at $T = 500$ °C). Because it is available in high purity and with a known transition temperature and enthalpy change, indium is used frequently as a reference; other calibration materials include tin, zinc and bismuth.[22–24]

The standard calibration should be performed at the beginning of the DTA measurement and again at the end of the session in order to ensure that the data collected are accurate. If there is thermal drift over the course of the measurement, its contribution to the uncertainty can be quantified.

Equation 2.7 indicates that temperature calibration must be performed again after changing the scan rate, as ΔT depends on scan rate. Software in some newer instruments allows multiple calibration corrections for different scan rates, which increases efficiency in scan-rate-dependence studies. For example, the temperature of a thermal event can be measured as a function of scan rate to give its true temperature by extrapolation to a scan rate of $0\,°C\,min^{-1}$.

A further consideration is that the starting temperature should be considerably below the temperature of any anticipated thermal event, in order to be certain that there is a good baseline prior to the temperature region of interest.[25]

It is also worth noting that the purge gas plays a role in heat transfer within the DTA furnace. The gas used should be the same as that used during the DTA calibration, flowing at the same rate. Any deviation from this protocol will give erroneous results.

Samples and reference materials should both be contained in the same type of pan (*e.g.* aluminium) and have the same seal (*e.g.* hermetic), and the same conditions should pertain to the calibration standards.

Sample morphology is another item to consider. Where possible, the sample shape should allow good thermal contact with the pan. This can be achieved by placing one orientation preferentially (flat side to pan) or premelting the sample into the pan (provided that this does not damage the sample). Regarding size: typically small sample sizes (a few mg) are used, as the small thermal mass allows more reliable detection of the beginning of the thermal event and can accommodate faster heating rates.[26] However, small samples limit the sensitivity of detection of minor thermal events, and it is therefore recommended that multiple runs with a range of masses be carried out to establish the thermal event and determine optimal sample sizes for each specific project.

With due care, uncertainty in the temperature of thermal events as determined by DTA can be $1\,°C$ or better.

2.2.2.3 *Measurement of Temperature*

The role of calibration in accurate temperature determination has been discussed above, but determination of the event temperature has not yet

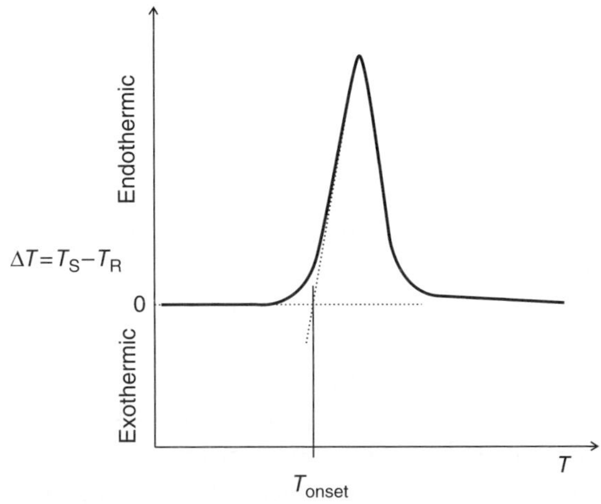

Figure 2.4 The DTA signal, ΔT, as a function of temperature, T, shows a peak associated with a thermal event. The onset temperature, T_{onset}, which is the meaningful temperature to report, is defined as the intersection of the extrapolated baseline and the steepest peak slope, both shown as black dashes.

been presented. The meaningful temperature for a thermal event is the onset temperature, T_{onset}, defined as the intersection between the baseline extension into the peak area and the steepest slope of the peak (see Figure 2.4). The onset temperature is more representative of the true temperature of the thermal event (*e.g.* melting point) than is the peak temperature, due to the thermal lag within the sample. Similarly, results from DTA should always be reported as onset temperatures.

2.2.2.4 Scanning Rate Effects

The DTA signal, ΔT, has a dependence on the scanning rate given by Equation 2.7, because the temperature lag between the sample and the platform increases with increasing scan rate.[27] Selection of a scan rate appropriate to the experiment is critical; faster scan rates generally increase sensitivity to thermal events, whereas slower scan rates (less than 10 °C min^{-1}) normally provide a more accurate value for T_{onset}.[26] Scanning at both fast (*e.g.* 20 °C min^{-1}) and slow (*e.g.* 1 °C min^{-1} or less) rates is beneficial in investigations of a previously unreported material. A scan rate of 10 °C min^{-1} is standard in preliminary experiments, as it provides insight into the thermal events in a material. More appropriate scan rates are then used for subsequent experiments in order to extract more accurate results.

With due care, the temperature accuracy of DTA can be within a few degrees or better, even at very high temperatures.[26]

2.2.2.5 *Example of DTA Use*

DTA was used to investigate the crystallisation of the glass–ceramic Li_2O–SiO_2 in the absence and presence of P_2O_5 at elevated temperatures.[28] In the absence of P_2O_5, the crystallisation was observed to be dependent on the particle size. The larger particle size had a crystallisation temperature of 695 °C and the smaller particle size crystallised at 585 °C (note that these are the published peak temperatures; onset temperatures are lower but were not reported). The addition of P_2O_5 to the large-particle glass had the effect of lowering the crystallisation temperature to 600 °C, while its addition to the small particle size increased the crystallisation temperature slightly to 594 °C.[28]

2.2.3 Differential Scanning Calorimetry

2.2.3.1 *General Principles of DSC: A Simplified View and Examples*

In DSC, the sample and reference are heated (or cooled) at a constant rate that ensures the sample and reference temperatures are identical. The difference in power (sample compared with reference) is recorded as a function of temperature. As the name implies, the measurement is differential (between sample and reference), scanning (temperature) and calorimetric (heat flow within sample). DSC has been extensively used and markedly improved since its inception in 1964.[20]

A schematic diagram of a typical DSC setup is shown in Figure 2.5. The sample (*ca* 10 mg) is typically encased in a metallic pan and heated (or cooled) in the DSC at a constant rate of a few degrees per minute. The reference is typically an empty pan, essentially identical to that used to hold the sample, although for some experiments it is preferable to fill the pan with a material whose mass and specific heat are similar to those of the sample under investigation. The sample and reference are heated (or cooled) such that there is no temperature difference between the sample and the reference – *i.e.* $\Delta T = 0$, as shown in Figure 2.5.

DSC is designed to determine $\Delta\dot{Q}$, the difference between the power to the sample, $\dot{Q}_S$, and the power to the reference, $\dot{Q}_R$ (*i.e.* $\Delta\dot{Q} = \dot{Q}_S - \dot{Q}_R$ as defined here; there is no sign convention for DSC and usually the

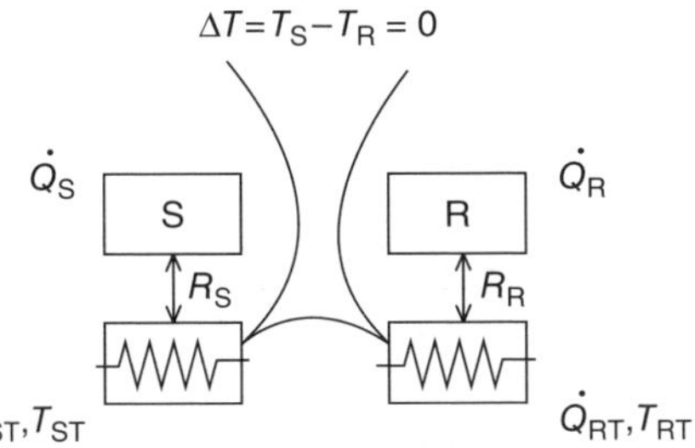

Figure 2.5 Schematic diagram depicting a differential scanning calorimeter. The sample, S, and reference, R, are heated (or cooled) at a constant rate such that their temperatures are kept equal ($\Delta T = 0$) and the difference between the power to the sample, $\dot{Q}_S$, and that to the reference, $\dot{Q}_R$, $\Delta\dot{Q} = \dot{Q}_S - \dot{Q}_R$, is measured as a function of temperature. R_S and R_R represent the thermal resistances between the sample and reference and their respective heater/thermometer assemblies.

exothermic or endothermic direction is shown on the thermogram for clarity). During the experiment, $\Delta\dot{Q}$ is determined as a function of temperature (which is proportional to time, since the ramp rate is constant). When $\Delta\dot{Q} = 0$, the resulting thermogram is flat as the power requirements for both the sample and reference are matched. In the case of an endothermic event in the sample, such as melting, $\Delta\dot{Q} > 0$ and a peak will result, since $\dot{Q}_S > \dot{Q}_R$ during this process. Similarly, for an exothermic event such as crystallisation, $\dot{Q}_S < \dot{Q}_R$, which would result in a peak in the other direction.

A DSC scan showing results for a polymer (PET, polyethylene terephthalate) is presented in Figure 2.6a. In this case, DSC reveals the glass transition temperature, cold crystallisation and eventual melting.

DSC is also used extensively in the characterisation of liquid-crystalline materials as a tool that is complementary to polarised optical microscopy for the determination of phase transitions. An example (TBBA, terephthal-bis-(4-butylaniline)) is shown in Figure 2.6b. In general, transitions from the solid state into the liquid crystal mesophase are strongly first-order in nature, whereas those between different liquid crystal phases are more weakly first-order, as is the transition from the liquid crystal phase into the isotropic liquid state. In addition, transitions between certain pairs of mesophases are found to be second-order, as are glass transitions.

Figure 2.6 shows clearly that valuable information can be recorded with DSC, namely the temperature at which thermal processes occur. The relationship between peak shape and temperature is a matter that requires careful consideration, as DSC is a dynamic experiment.

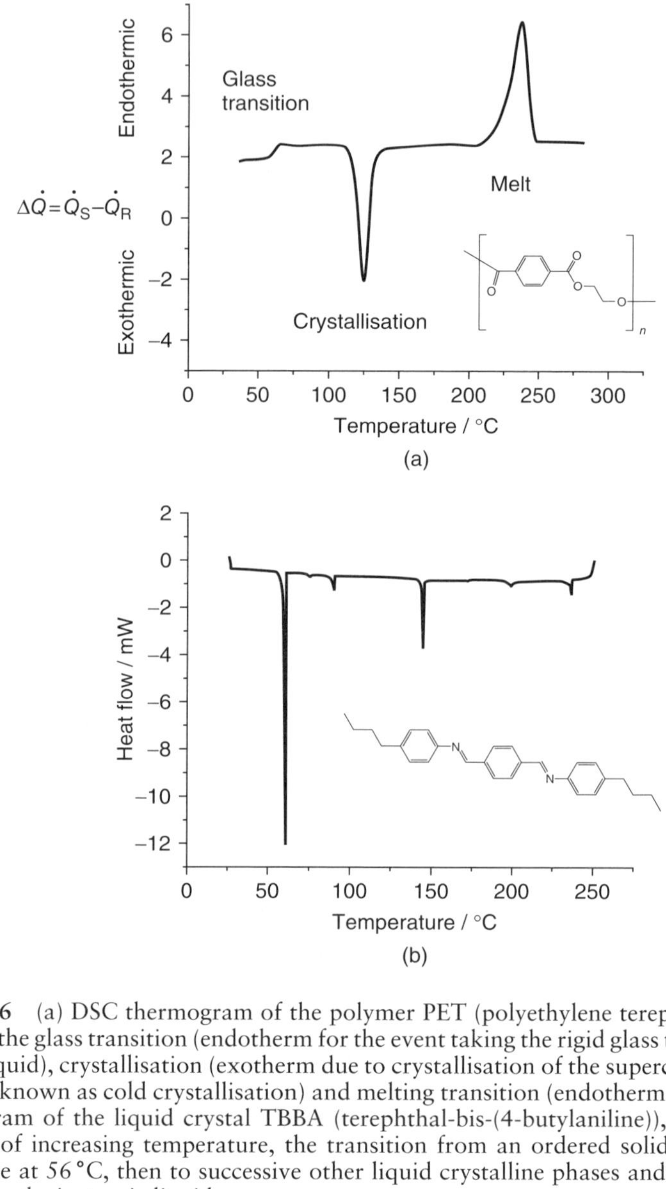

Figure 2.6 (a) DSC thermogram of the polymer PET (polyethylene terephthalate), showing the glass transition (endotherm for the event taking the rigid glass to a supercooled liquid), crystallisation (exotherm due to crystallisation of the supercooled liquid, also known as cold crystallisation) and melting transition (endotherm). (b) DSC thermogram of the liquid crystal TBBA (terephthal-bis-(4-butylaniline)), showing, in order of increasing temperature, the transition from an ordered solid to liquid crystalline at 56 °C, then to successive other liquid crystalline phases and finally, at 233 °C, to the isotropic liquid.

Dimensional analysis of a peak in a DSC scan gives units of energy (power × time), and thus the area of a peak in a DSC thermogram has information pertaining to an enthalpy change, ΔH, associated with a process. Such measurements are the major utility of DSC, but the nature of the process might not be revealed by DSC alone and further experiments could be necessary.

Where first-order transitions are concerned, it is useful to recognise that the transition temperature represents the point of equilibrium between the two states, which have identical Gibbs energies ($\Delta G = 0$), and thus the transition entropy change ΔS can be evaluated ($\Delta G = \Delta H - T\Delta S = 0$). Entropy changes generally indicate the changes in order that accompany a transition. From the DSC trace for TBBA (Figure 2.6b), it is clear that, even qualitatively, ΔH for the first transition (solid–liquid crystal) is much larger than that for the final transition (liquid crystal–isotropic liquid), and hence the corresponding values of ΔS are in the same order. In other words, much more disorder is introduced in the ordered crystal–liquid crystal transition than in the liquid crystal–isotropic liquid transition (see below for further discussion of ΔH determination).

Another common feature of liquid crystals and other materials is that the crystalline form obtained from solution following preparation is not always the same as that obtained on crystallisation from the molten solid. Therefore, where compounds are stable thermally, it is good practice to record two or even three heat–cool DSC cycles in order to identify such possibilities, determining the total enthalpy change associated with each heating and cooling ramp as an extra check.

2.2.3.2 A More Detailed View of DSC

The principles of DSC have been oversimplified thus far in one respect: the thermal resistance between the sample and the thermometer, R_S, and that between the reference and the thermometer, R_R, have not been considered.[29] These resistances arise from factors relating to surface contact at the interface between the sample and the pan and between the pan and the heating element. Thermal resistance can be minimised by choosing a pan with an inherently flat bottom (and maintaining the shape during encapsulation) or by using sample morphologies that increase the sample–pan contact. Such measures will reduce thermal resistance but it cannot be eliminated fully.

In power-compensated DSC, the signal is the difference in power input to the sample and its sample platform ($\dot{Q}_S + \dot{Q}_{SP}$) compared to that of the reference and the reference platform, ($\dot{Q}_R + \dot{Q}_{RP}$):

$$signal \equiv (\dot{Q}_S + \dot{Q}_{SP}) - (\dot{Q}_R + \dot{Q}_{RP}). \quad (2.9)$$

When the sample and reference platforms are well matched, such that $\dot{Q}_{SP} = \dot{Q}_{RP}$, the signal is:

$$signal \equiv \dot{Q}_S - \dot{Q}_R, \tag{2.10}$$

as before. The power to the sample, S, is related to its heat capacity, C_S, and the drift rate at the sample–platform thermocouple, dT_{SP}/dt :

$$\dot{Q}_S = C_S \frac{dT_{SP}}{dt}. \tag{2.11}$$

Similarly, the power to the reference, R, is related to its heat capacity, C_R, and the drift rate at the reference–platform thermocouple, dT_{RP}/dt :

$$\dot{Q}_R = C_R \frac{dT_{RP}}{dt}. \tag{2.12}$$

This system can be simplified somewhat by assuming that the temperature drifts at the platform and sample are similar to those at the platform and reference:

$$\frac{dT_S}{dt} = \frac{dT_{SP}}{dt} = \frac{dT_R}{dt} = \frac{dT_{RP}}{dt} = \frac{dT}{dt}. \tag{2.13}$$

Hereafter the drift is denoted simply by dT/dt. Equations 2.11–2.13 can be substituted into Equation 2.10 to give:

$$signal \equiv (C_S - C_R)\frac{dT}{dt} \tag{2.14}$$

where the signal is proportional to the difference between the heat capacities of the sample and the reference, and independent of the sample–platform and reference–platform thermal resistances.

While the temperature drifts may be similar, the temperature lag can vary between the sample–platform and reference–platform, and this becomes a concern for the observed temperature during thermal processes. Newton's law of energy exchange indicates that the amount of heat flowing to the sample from the platform is proportional to the temperature difference between the two. From this, the power to the sample is given by:

$$\dot{Q}_S = \frac{1}{R_S}(T_S - T_{SP}). \tag{2.15}$$

Substituting Equations 2.11 and 2.13 into Equation 2.15 leads to the relationship between the temperature at the platform and the sample temperature:

$$(T_S - T_{SP}) = R_S C_S \frac{dT}{dt} \tag{2.16}$$

which demonstrates the temperature lag at the sample. At a scan rate of 1 °C min^{-1}, a typical thermal resistance of 0.06 K s^{-1} mJ^{-1} and a heat capacity of 50 mJ K^{-1}, this leads to a temperature lag of 0.05 °C. It is evident from Equation 2.16 that the temperature lag becomes greater as the scan rate is increased, and lag should be considered when conducting a scan. For very accurate temperature determination with DSC, onset temperatures (T_{onset}, using the same method as discussed for DTA) should be determined as a function of scan rate and extrapolated to a scan rate of 0 °C min^{-1}.

2.2.3.3 *Modulated DSC*

Modulated DSC (MDSC) is related closely to conventional DSC, except that the heating rate in MDSC follows a sinusoidal form rather than the conventional linear ramp.[30] This change results in three heating-related variables that improve the DSC signal and is explained by:

$$\frac{dQ}{dt} = -\frac{dT}{dt}\left[C_S + f''(t,T)\right] + f(t,T) \tag{2.17}$$

where dQ/dt is the heat flow out of the sample, dT/dt is the heating rate, $f'(t,T)$ is a thermodynamic heat flow component and $f(t,T)$ is a kinetically limited heat flow. The signal obtained with MDSC (oscillating signal) is processed by Fourier transformation to give a typical DSC thermogram.

MDSC provides a method for measuring reversing and non-reversing heat flows and an improved method (over conventional DSC) for measuring the heat capacity of a material. Heat capacity can be determined by MDSC in one run, whereas the conventional DSC method requires three temperature scans: empty cell, reference material (*e.g.* sapphire) and sample.[30] MDSC's heat-capacity-determination accuracy is also better than that of conventional DSC; with due care, accuracies better than 5% can be obtained.[30]

2.2.3.4 *DSC Methods/Precautions*

The comments concerning the need for calibration standards in DTA also apply to DSC measurements. Some of the keys to successful DSC measurements are a suitable sample morphology (in order to get good thermal contact between the sample and the pan), temperature ramp, similar thermal conductivity between the sample and standard materials and identical operating conditions (sample pan type, sealing method, purge gas, flow rate, heating/cooling rate) in the calibration and experimental runs. The precision and accuracy of DSC temperature measurements are similar to those in DTA. To reiterate, the scanning rate has a large effect on the heat flow in a sample and therefore on temperature accuracy. Scanning at a high rate increases the sensitivity of a thermal event that would be washed into the baseline at lower rates. However, a slow rate increases peak resolution. Different scan rates should be explored in probing the sample's response in order to optimise results.

2.2.3.5 Precision and Accuracy

Precision and accuracy in DSC measurements depend very much on the ability to reproduce the experiment – especially the thermal resistance (R_S in Figure 2.5). This is particularly important for samples with low thermal conductance. With due diligence, temperature precision can be within 0.2 °C, and systematic errors can lead to uncertainty within 0.2 °C. The operator must be wary of DSC results that are precise but not accurate. The best way to improve accuracy is to carry out several well-calibrated experiments for multiple samples or even for the same sample multiple times (after re-setting the pan on the platform).

2.2.3.6 Determination of Enthalpy Changes

Because DSC determines power, the area under a DSC peak is proportional to the enthalpy change, ΔH, associated with the thermal process in the sample. However, DSC is not adiabatic and there are unaccounted heat exchanges with the environment. The geometry is invariant from one experiment to the next, so ΔH can be calibrated with known standards. The ability to determine ΔH is the main advantage of DSC over conventional DTA.

As for temperature calibration, enthalpy calibration requires a well-characterised standard that has a thermal process near the temperature under investigation. Metals are commonly used for enthalpy correction (also called cell-constant correction) due to their sharp melting points and established enthalpy changes.[22–24] For low temperatures, a solid–solid transition (such as that of ammonium phosphate, which has a transformation at 151.8 K[22]) can be used. It is worth noting that impurities do not have such a detrimental effect on ΔH as they do on temperature. However, for the sake of efficiency, the material used for temperature correction is usually also used for enthalpy correction. The following precautions apply: the reference should have similar morphology and thermal conductivity to the sample and should have an event in the appropriate temperature range; the same heating/cooling rates should be used for the calibration and experiment; calibration checks should be made before and after the experiment; and the calibration and experiment should be made with the same purge gas and flow rate and the same pan type and sealing method.

Enthalpy is determined in a DSC experiment by measuring the area under the peak of a thermal event, which is proportional to ΔH. This determination can be subject to a large uncertainty, due to the need to know the baseline. In a DSC thermogram, $\Delta\dot{Q}$ is plotted as a function of temperature. A perfect baseline would have $\Delta\dot{Q} = 0$, but this is not normally found in practice as it implies a perfect match between the reference and sample heat capacities at all temperatures except during the thermal event. The best case scenario is where the baseline is flat (but displaced from zero due to sample–reference heat capacity mismatch), as shown in Figure 2.7. The important task here is to use a proper baseline fit (*e.g.* linear, stepped or sigmoidal, as shown in Figure 2.7) in order to determine ΔH from the resultant peak area. The software supplied with modern DSC instruments can be used to calculate the baseline and hence ΔH.

Although low scan rates give more accurate temperatures for the onset of thermal events, if the scan rate is too low, ΔH can be underestimated, because some of the enthalpy change is swept into the baseline.[31] Furthermore, if two (or more) thermal events occur very close together in temperature, the resolution of these events (and hence the ΔH values for the individual events) can be adversely affected by too high a scan rate.

As with temperature determination, enthalpy change determination can be very precise if enough attention is paid to the details that give reproducible thermograms (sample morphology, size, pan type and seal, purge gas and heating rate similar to calibration conditions). With such precautions, the uncertainty in ΔH can be within a few per cent if the thermal feature is large and sharp. For smaller thermal events, the uncertainty in ΔH can be quite large (>10%). If there is no independent knowledge of the baseline (*e.g.* data for the step in heat capacity over the event) then uncertainty in the baseline should be accounted for in the propagation of error. Improper drawing of the baseline can easily lead to

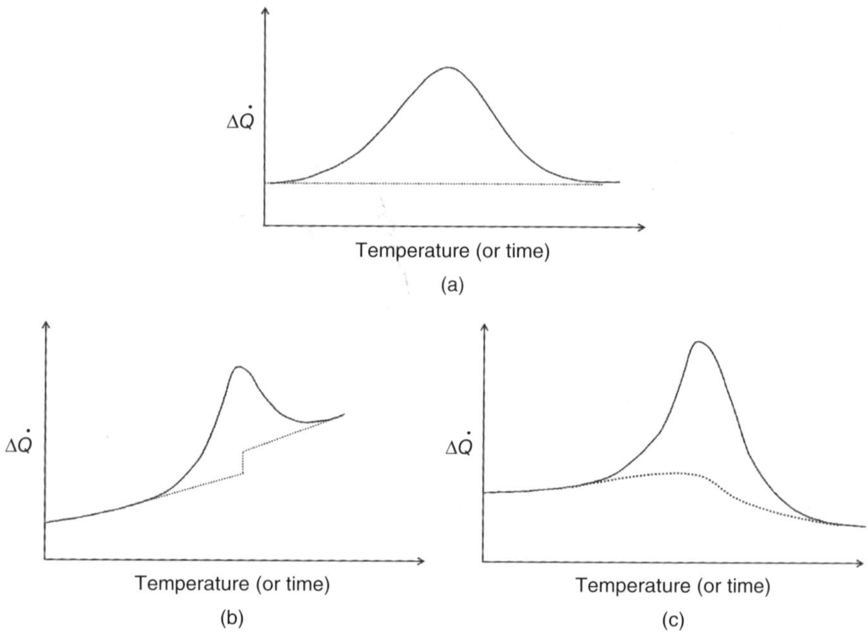

Figure 2.7 Three common types of baseline in stylised DSC thermograms: (a) linear, for a flat (or linearly sloping) baseline; (b) stepped, for a discontinuity between the non-event slope before and after the thermal event; and (c) sigmoidal, for a shifted baseline.

systematic errors, which can be especially large in the case of broad thermal events. There is no general rule for uncertainty in ΔH determination from DSC, and experiments with several samples of different masses, run at several different scan rates, are required in order to best assess it.

2.2.3.7 *Phase-Transition Resolution Example*

In 1984, the solid–solid phase transitions in bis(*n* -octadecylammonium) tetrachlorocadmate(II) ($(n\text{-}C_{18}H_{37}NH_3)_2[CdCl_4]$) was investigated by both (high-accuracy) adiabatic calorimetry and DSC.[31] Adiabatic calorimetry showed four distinct solid–solid transitions between 340 and 370 K, observed as discontinuities in the heat capacity. A state-of-the-art (for 1984) DSC with varying scan rates of 10.0 to 0.5 °C min^{-1} failed to resolve all four peaks: only the lowest- and highest-temperature peaks were clear, with a high baseline between them.[31] Recently (in 2012), we revisited this subject using a new DSC and observed a marked

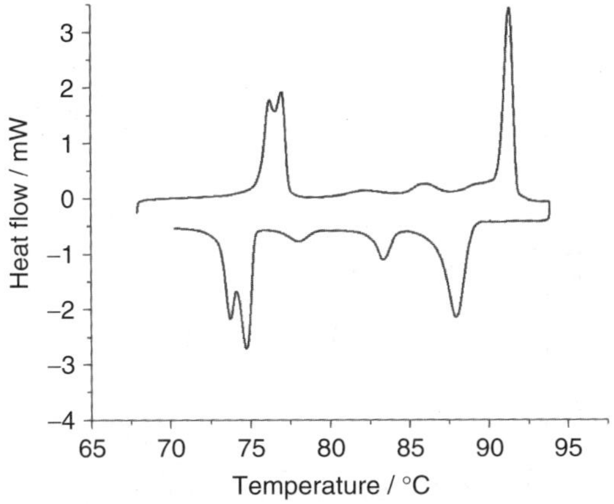

Figure 2.8 DSC thermogram of (n-$C_{18}H_{37}NH_3)_2[CdCl_4]$) as measured by the authors in 2012. The upper trace indicates data on heating and the lower one data on cooling. The sample was measured with the following parameters: temperature ramp 0.75 °C min^{-1}, helium atmosphere, sample mass 7.03 mg, endotherm up.

improvement in sensitivity, resolving all four main phase transitions and showing additional structure in the lowest-temperature transition (Figure 2.8).

2.2.4 Example of a Coupled Thermoanalytical Technique

Many commercial TGA instruments have combined TGA/DSC/DTA (either classical DTA or Boersma DTA) in one unit for sequential or concurrent experiments.

The investigation of aluminium nanoparticle synthesis is of interest because the formation of surface Al_2O_3 ($\Delta H = 31\,\mathrm{kJ\,g^{-1}}$) can give a very high heat of reaction and the exothermicity can become a concern due to the high surface area-to-volume ratio.[32] Passivation of the surface with an organic agent is an effective method for limiting oxidation; for example, the thermal decomposition of alane with titanium isopropoxide is used to make aluminium–oleic acid core–shell nanoparticles.[32] A coupled TGA/DSC experiment in which the DSC scan is recorded concurrently with the mass change has been used to investigate the effect of gas (*e.g.* inert *versus* oxidiser). Figure 2.9 shows the combined thermogram from such a measurement; the knowledge of exothermicity concurrent with mass loss is very useful in its interpretation.[32]

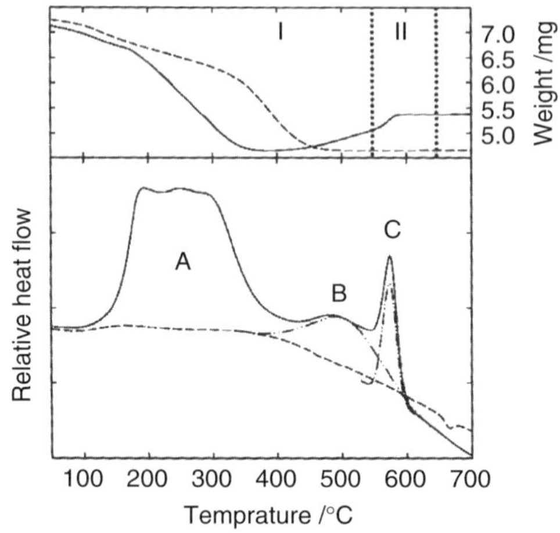

Figure 2.9 A coupled TGA/DSC technique was used to probe the decomposition mechanism of aluminium–oleic acid core–shell nanoparticles measured in air (solid line) and in argon (dashed line). Event A is the reaction of the oelic acid; this occurs only in the air atmosphere, resulting from the combustion of the organic material. Events B and C, also present only in air, are caused by the initial oxide layer and oxide-coated aluminium particles, respectively.[32] Reprinted with permission from [32] Copyright (2009) American Chemical Society.

2.2.5 Concluding Comments Concerning Thermal Analysis

Thermoanalytical techniques can use small samples (*ca* mg) and commercial instruments to determine many types of thermal event in a condensed matter. Accurate determination of temperatures (TGA, DTA and DSC) and enthalpy changes (DCS and Boersma DTA) requires appropriate calibration and care during the experiment and data analysis. It is easier to get precise data than accurate data, but high accuracy is possible. Thermal analysis results can be combined with information from other techniques to reveal the processes that give rise to the thermal events. Other thermoanalytical techniques include dynamic mechanical analysis, which can be used to assess the temperature-dependence of viscoelastic behaviour, microthermal analysis, which gives results similar to DSC but with high spatial resolution, micromechanical calorimetry, which gives the rate of change of mass for very small samples as a function of temperature, thermal nanoprobe methods, which combine thermal analysis and scanning probe techniques, and nanocalorimetry, which provides very sensitive enthalpy measurements.[1,33–35]

2.3 HEAT CAPACITY

2.3.1 Background

Heat capacity, C, is a fundamental physical property that can provide insight into materials. It is representative of the change in energy per change in temperature, described at constant volume, C_V, as:

$$C_V = \left(\frac{dU}{dT}\right)_V \tag{2.18}$$

where U is the internal energy, or at constant pressure, C_P, as:

$$C_P = \left(\frac{dH}{dT}\right)_P \tag{2.19}$$

where H is enthalpy. In general, for solids C_P is determined experimentally but C_V is easier to model from a theoretical standpoint. Heat capacity can also be expressed per unit mass (specific heat capacity) or per mole (molar heat capacity).

For solids, both C_V and C_P vary with temperature. The difference between C_V and C_P reflects the difference in the amount of energy required to expand the lattice with temperature, which also varies with temperature. For a cubic lattice (or any isotropic matter):[36]

$$C_P - C_V = \frac{\alpha_V{}^2 VT}{\beta_T} \tag{2.20}$$

where α_V is the volume coefficient of thermal expansion (see Section 2.5.1), V is the volume and β_T is the isothermal compressibility $\left(-\frac{1}{V}\left(\frac{\partial V}{\partial P}\right)_T\right)$. (For a non-cubic solid, a tensor form of Equation 2.20 must be used to calculate $C_P - C_V$.)

The temperature dependence of the heat capacity of a solid reflects the thermal excitation of vibrational modes.[37] For n atoms per unit cell, there will be $3n$ degrees of freedom. From equipartition theory, we can expect a high-temperature limit of $C_V = 3nR$, where R is the gas constant and the heat capacity is expressed per unit cell[36] (with Z formula units per unit cell, the high-temperature limit of the molar heat capacity would be $3nR/Z$). At lower temperatures, the heat capacity drops such that $C_P \rightarrow 0$ *as* $T \rightarrow 0\,K$. Of the $3n$ degrees of freedom in the unit cell, $(3n - 3)$ will correspond to optic modes, each well described by the

Einstein model:[36]

$$C_V^E = R\left(\frac{h\nu}{kT}\right)^2 \frac{e^{h\nu/kT}}{(e^{h\nu/kT}-1)^2} \tag{2.21}$$

where h is Planck's constant, ν is the frequency (in s^{-1}) for that particular vibration and k is Boltzmann's constant. (Note that the frequency is sometimes expressed as an Einstein characteristic temperature, $\theta_E = h\nu/k$.)

The remaining three modes are acoustic and are better described by the Debye function:[36]

$$C_V^D = 9R\left(\frac{kT}{h\nu_D}\right)^3 \int_0^{h\nu_D/kT} \frac{\left(\frac{h\nu}{kT}\right)^4 e^{h\nu/kT}}{(e^{h\nu/kT}-1)^2} \mathrm{d}\left(\frac{h\nu}{kT}\right) \tag{2.22}$$

where ν_D is the Debye frequency (sometimes expressed in terms of the Debye characteristic temperature, $\theta_D = h\nu_D/k$). The main difference between the Einstein and the Debye heat capacities is that the former is essentially dispersionless (same frequency throughout the full Brillouin zone), whereas the latter has a frequency distribution (all frequencies up to ν_D, as manifest in the integral in Equation 2.22).

Analysis of experimental heat capacities in terms of Einstein and Debye frequencies, which can be determined independently by vibrational spectroscopies such as Raman, infrared and inelastic neutron scattering, can reveal information concerning the dynamics of the lattice.

Although the techniques presented here are primarily for bulk materials, recent developments have allowed highly accurate heat-capacity determinations for thin films.[38]

2.3.2 Adiabatic Calorimetry

One of the most direct ways to determine C_P for a solid or liquid is through consideration of the operational form of Equation 2.19:

$$C_P = \left(\frac{Q}{\Delta T}\right)_P \tag{2.23}$$

where heat, Q, is input under adiabatic conditions and the temperature rise, ΔT, is determined. In typical adiabatic calorimetry experiments, ΔT is approximately 5% of the absolute temperature, or less. Adiabatic conditions imply no extraneous heat exchange with the surroundings, so the sample must be held in a vacuum and suspended by a mechanism that has negligible thermal conductivity. In addition, the sample should be surrounded by a temperature-matched heat shield in order to reduce radiative losses.

Adiabatic calorimetry is an absolute method, directly giving the heat capacity of the sample and its containment vessel; the heat capacity of the latter is determined in a separate experiment. Since it is not a relative measurement, care must be taken to determine both heat input and temperature accurately. The latter requires calibration to standard scales, such as IPTS-90,[39] as precision within ±0.0001 °C is necessary. Adiabatic calorimeters should be checked by measurement of standard materials from NIST, such as benzoic acid, sapphire and high-purity copper. With due care, accuracy in values of the heat capacity can be better than 0.1% using adiabatic calorimetry over the temperature range from <1 to about 800 K.[1]

The sample size for adiabatic calorimetry is determined by the need for the heat capacity of the sample to be a significant component of the total heat capacity of the containment vessel plus sample. Ideally, the sample should contribute more than 50% to the total heat capacity, and, given the need for a relatively large thermometer, most adiabatic calorimeters require at least 1 g of sample.[40] Although adiabatic calorimetry can give the most accurate determination of heat capacity,[41] it is time-consuming (taking days to achieve a series of measurements) and adiabatic calorimeters are not widely available commercially.

2.3.3 Relaxation Calorimetry

The commercial availability of relaxation calorimeters and their ability to achieve accurate determination of heat capacities on samples as small as 1 mg, with measurement of heat capacities over a wide range of temperatures in a matter of hours, has led to an upsurge in interest in heat capacity measurements using this technique.

The basic concept of relaxation calorimetry is that a sample is placed on a platform that is then heated, and the sample temperature is monitored.[42] A schematic of a simple relaxation calorimetry experiment is shown in Figure 2.10. In this case, the sample is in excellent thermal contact with the platform and the platform is weakly linked to the heat sink. Constant power, $\dot{Q}$, is applied to the platform for a period of time, during which the temperature rises as shown schematically in

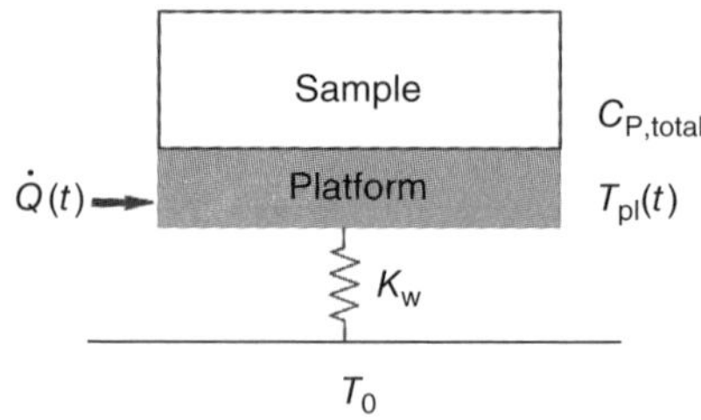

Figure 2.10 Simplified view of relaxation calorimetry, showing a sample in excellent thermal contact with a heat platform that is weakly linked to a heat sink.

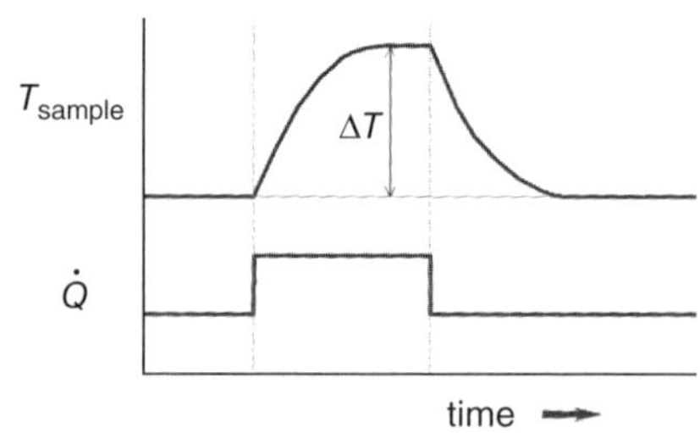

Figure 2.11 Simplified view of relaxation calorimetry showing that when the power is turned on, the sample temperature increases, and when the heat is switched off, the temperature decays.

Figure 2.11.[1] In this configuration, the sample temperature is the same as the platform temperature. With T_0 as the temperature of the heat sink, the sample rises to a maximum temperature of $T_0 + \Delta T$ and during the heating the temperature of the platform, T, is described by:[1]

$$C_{P,total}\left(\frac{dT}{dt}\right) = -K_w(T - T_0) + \dot{Q}(t) \qquad (2.24)$$

where K_w is the thermal conductance of the wires connecting the platform to the heat sink and $C_{P,total}$ is the total heat capacity (platform + sample). When the heater power is switched off, the temperature relaxes, as:[1]

$$T = T_0 + \Delta T e^{-t/\tau} \qquad (2.25)$$

where the relaxation time, τ, is given by:

$$\tau = C_{P,total}/K_w. \qquad (2.26)$$

The total heat capacity can be determined from Equations 2.24 and 2.25. A separate measurement of the platform (which usually includes grease

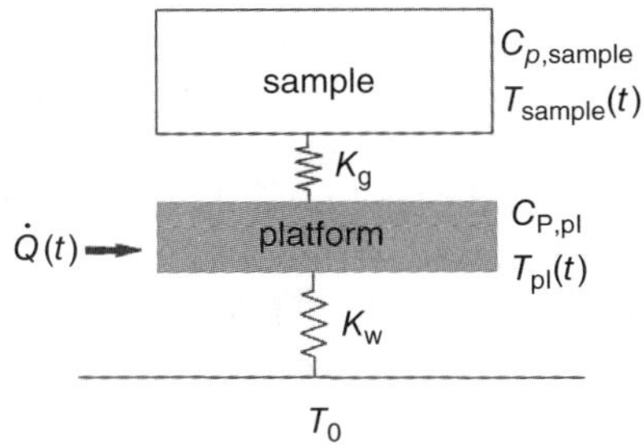

Figure 2.12 A more exacting view of relaxation calorimetry than that in Figure 2.10, showing a thermal resistance between the sample and the platform, which must be taken into account in determining the heat capacity of the sample.

for thermal contact) can be used to determine the heat capacity of the sample.

Although the above description gives the essence of relaxation calorimetry, the thermal contact between the sample and the platform is usually far from perfect. A more realistic schematic of a relaxation calorimetric setup is shown in Figure 2.12. Here, the temperatures of the sample, T_{sample}, and the platform, T_{pl}, are not identical. The thermometer is in contact with the platform and therefore determines T_{pl}, which can be described by two exponential terms:[43]

$$T_{\text{pl}} = T_0 + A_1 e^{-t/\tau_1} + A_2 e^{-t/\tau_2}, \tag{2.27}$$

where A_1 and A_2 are constants and τ_1 and τ_2 are the long and short relaxation times for the sample and platform, respectively (this analysis is often referred to as the two-tau (two-τ) model). The heat balance equation for the platform is:[43]

$$C_{\text{pl}} \frac{dT_{\text{pl}}}{dt} = \dot{Q}(t) - K_w(T_{\text{pl}}(t) - T_0) + K_g(T_{\text{sample}}(t) - T_{\text{pl}}(t)) \tag{2.28}$$

where K_g is the conductance of the grease between the sample and the platform. The heat balance equation for the sample is:[43]

$$C_{\text{sample}} \frac{dT_{\text{sample}}}{dt} = K_g(T_{\text{sample}}(t) - T_{\text{pl}}(t)). \tag{2.29}$$

As in the simple case, C_{pl} must be determined in a separate experiment. Then, from Equations 2.28 and 2.29 and the curve-fitting procedure of Hwang *et al.*[43] C_{sample} can be determined.

As with thermal analysis, it is relatively easy to obtain precise data with relaxation calorimetry, but special care is required to determine heat capacity accurately. Lashley *et al.* made a thorough assessment of the requirements in 2003[44] and concluded that, with due care, heat capacity determinations can be made within 1% accuracy in the temperature range from 100 to 300 K and within 3% for *ca* 4 K. Because the heat must penetrate the sample, the influence on accuracy of sample geometry can be substantial. Particularly for samples with low thermal conductivity, several of different masses should be investigated in order to determine the ideal sample mass (sufficiently large to contribute at least 30% to the total heat capacity but not so large as to show mass-dependent effective heat capacities due to thermal conductivity lag).[45] The sample coupling, defined as $100 \times K_g/(K_g + K_w)$, must be high for accurate measurement.

Relaxation calorimetry generally has the sample sitting on an open platform. The absence of a closed cell can be problematic for volatile samples, but procedures that use hermetically sealed DSC pans allow accurate measurements to be made.[46] DSC pans can be used to contain powders that cannot be measured directly on the platform[47] and powders can also be consolidated with sapphire or epoxy containment for measurement on the platform.[48] The overarching point is that care must be taken and many samples measured and analysed over a wide range of temperatures in order to be assured of high accuracy.

Because of the latent heat associated with first-order phase transitions, special methods are required for the accurate determination of heat capacity with relaxation calorimetry near phase transitions.[44,49] With due care, accuracy can be comparable to that in adiabatic calorimetry, based on much smaller samples sizes (*e.g.* 30 mg compared with several grams).[44]

Relaxation calorimetry has been especially useful in advancing the understanding of magnetic materials, because heat capacity can be measured directly in a magnetic field. An example is shown in Figure 2.13.[50] The accuracy of field-dependent heat-capacity measurements has been addressed by Lashley *et al.*[44] and by round-robin experiments.[51]

2.3.4 Other Heat-Capacity Methods

As described above, the most common methods for determining heat capacity accurately are adiabatic calorimetry and relaxation calorimetry;

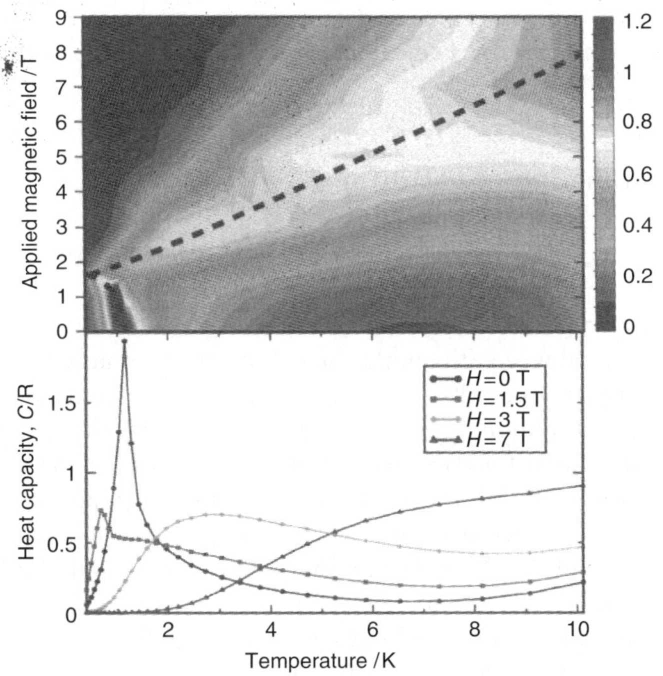

Figure 2.13 Magnetic field dependence of the heat capacity of the frustrated XY pyrochlore magnet, $Er_2Ti_2O_7$, showing a phase transition at $T = 1.5$ K and zero field that first moves to lower temperature at applied field and then disappears above a critical field of about 1.5 T. The upper diagram shows the heat capacity as a contour plot and the dashed line tracks the peak as a function of temperature and magnetic field.[50] Reprinted with permission from [50] Copyright (2008) American Physical Society.

DSC can also be used but usually achieves a lower accuracy, although its precision can be high.

Of the other methods, the most common is AC calorimetry. In this method, the sample is subjected to an AC power (heat) flux and the heat capacity is determined from the resultant temperature oscillations. Very high temperature resolution is possible (*e.g.* 1–10 mK), which makes AC calorimetry very suitable for critical exponent analysis near a phase transition. As with relaxation calorimetry, the sample size can be quite small (a few mg).[1]

Another way of determining heat capacity is the use of laser flash techniques. These are especially suitable above room temperature, but require quantitative information concerning the power input, which is rarely available. However, laser flash experiments can give information on thermal conductivity. They are addressed in Section 2.4.6.[52]

2.3.5 Estimation of Heat Capacity

In some instances, an approximate heat capacity can be sufficiently accurate for the requirement at hand.

The simplest possible estimate of the heat capacity of a solid is the Dulong–Petit value, *i.e.* the molar heat capacity for many non-metallic solids at room temperature is $\approx 3R$.[36] This value is reasonably close to experiment for monatomics, because all the vibrational degrees of freedom are nearly fully excited. However, it does not hold well for more complex materials since that assumption no longer is true. For example, the room-temperature heat capacity of boron carbide (B_4C) is $53.37\,J\,K^{-1}\,mol^{-1}$, [53] while a Dulong–Petit value would be $5 \times 3R = 124.7\,J\,K^{-1}\,mol^{-1}$. As a very hard material, boron carbide's internal vibrations are not fully excited until $T > 1000\,K$.[53]

The Neumann–Kopp law can provide a reasonable estimate of the heat capacities of more complex solids. For example, for MX_2, the heat capacity is often closely approximated as $C_{P,m}(MX_2) \approx C_{P,m}(M) + 2C_{P,m}(X)$, where $C_{P,m}$ is the experimental molar heat capacity at constant pressure over the temperature range under which M, X and MX_2 are all solids.[36] The theoretical basis for the Neumann–Kopp law is that the vibrational frequency of a given atom in condensed matter is almost fully determined by the atomic mass. An example where this works well is the fully stuffed semiconductor clathrate, $Sr_8Zn_8Ge_{38}$, in which the heat capacity is within a few per cent of $8C_{P,m}(Sr) + 8C_{P,m}(Zn) + 38C_{P,m}(Ge)$ over the temperature range from 20 to 300 K.[54]

The Neumann–Kopp law has its limitation. For example, it does not work for oxides as oxygen is not a solid until very low temperatures. Another approach is the so-called constituent additivity model, wherein *e.g.* $C_{P,m}(A_xB_yO_z) \approx C_{P,m}(A_xO_a) + C_{P,m}(B_yO_{z-a})$.[55] The constituent additivity approach works well for giving estimates of the heat capacities of oxides within a few per cent of the experimental data over a wide temperature range, provided that: (i) the experimental data for the constituent solids are reliable, (ii) the coordination within the

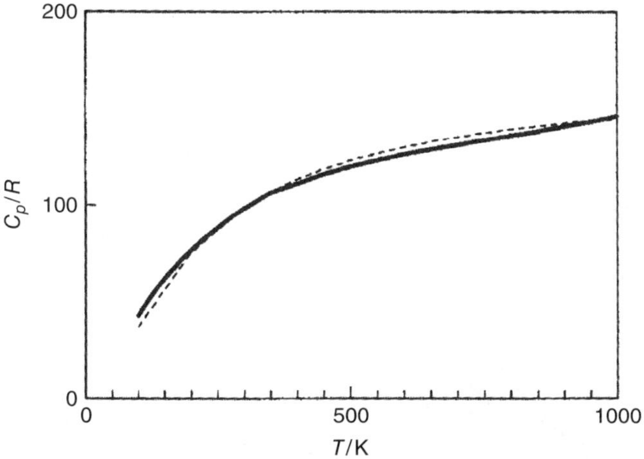

Figure 2.14 The constituent additivity method works well in describing the heat capacity of sodalite over a wide temperature range. The experimental data for $Na_8Al_6Si_6O_{24}Cl_2$ (solid line) are within a few per cent of data calculated from appropriately weighted heat capacities of $Na_2O, Al_2O_3, \alpha\text{-}SiO_2$ and NaCl (dashed line).[55] Reprinted with permission from [55] Copyright (2001) American Chemical Society.

constituent solids is similar to that in the complex solids and (iii) there are no solid–solid phase transitions.[55] As an example of this approach, sodalite (a mineral of idealised formula $Na_8Al_6Si_6O_{24}Cl_2$) is shown in Figure 2.14.[55] A similar method has been shown to be accurate for the extrapolation of heat capacities of minerals to high temperatures where experimental data are not otherwise available.[56]

If highly accurate heat-capacity data are not required, the Neumann–Kopp law and/or constituent additivity can be very useful, especially since the heat capacities of many elements and simple compounds are known accurately over a wide temperature range.[53] For example, if knowledge of heat capacity is required for the conversion of thermal diffusivity to thermal conductivity (see Section 2.4.6), an uncertainty of 10% is tolerable and Neumann–Kopp or constituent additivity approaches are reasonable, and certainly much more accurate than Dulong–Petit values.

When the Neumann–Kopp law and/or constituent additivity do not represent the heat capacity of a solid accurately, this can reveal information concerning the material. For example, the heat capacity of the open

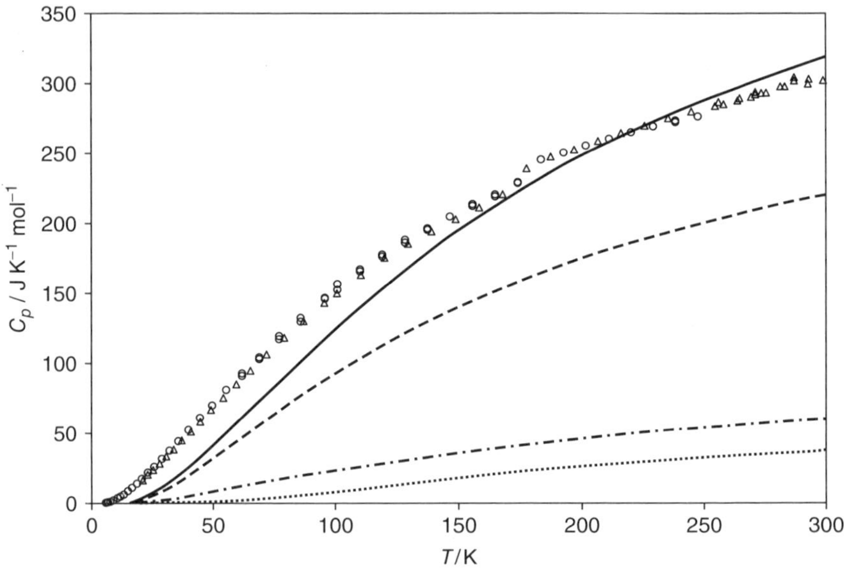

Figure 2.15 The heat capacity of the open-framework material $HfMgMo_3O_{12}$ is not well represented by the constituent additivity model. ○ and Δ represent the experimental data, the solid line shows the heat capacity calculated from $C_{P,m}(HfO_2) + C_{P,m}(MgO) + 3C_{P,m}(MoO_3)$ and the broken lines show the contributions of MoO_3 (largest), HfO_2 (intermediate) and MgO (smallest).[57] Reprinted with permission from [57] Copyright (2012) Elsevier.

framework material, $HfMgMo_3O_{12}$, is underestimated significantly by the constituent additivity method, as shown in Figure 2.15.[57] In this case, the anomalously high heat capacity, especially at low temperatures, is indicative of the unusually low frequencies of the lattice vibrations as a result of the open framework structure.[57]

Note also that for mixtures, such as composites, the rule of mixtures often works well; here the property $\boldsymbol{\Phi}$ is given by:[36]

$$\boldsymbol{\Phi} = \sum_{i} f_i \boldsymbol{\Phi}_i \tag{2.30}$$

where f_i is the fraction of component i and $\boldsymbol{\Phi}_i$ is the property of component i (*e.g.* the heat capacity of component i). Because the primary determinant of the heat capacity of condensed matter is composition, especially the masses of the atoms, the rule of mixtures can give estimates of the heat capacities of complex systems within about 10%.

2.4 THERMAL CONDUCTIVITY

2.4.1 Background

Heat transport within a system is governed by the movement of 'particles' such as molecules (in a gas) or lattice waves and/or electrons (in a solid), which try to restore the thermal equilibrium in the system when it is exposed to a temperature gradient.[58]

For a classical, isotropic material, the heat conduction is defined by Fourier's law:

$$\dot{Q} = -\kappa \nabla T \tag{2.31}$$

where $\dot{Q}$ is the heat flux, ∇T is the temperature gradient and κ is the scalar thermal conductivity. The isotropic scalar equation form is applicable to many structures, including cubic crystals, polycrystalline materials and amorphous solids (for anisotropic materials, the thermal conductivity depends on the direction of the heat flux).

The thermal conductivity, κ, of a solid material can be dominated by a single energy carrier, but all possible carriers contribute in an additive fashion:

$$\kappa = \kappa_L + \kappa_e + \kappa_r + \kappa_g + \ldots \tag{2.32}$$

where the thermal conductivities of the lattice (κ_L), electrons (κ_e), photon radiation (κ_r), gas molecules (κ_g) *etc.* make up the total thermal conductivity. For example, in semiconductors the main contributors are κ_L and κ_e, whereas for high-temperature thermal barrier coatings, κ_L and κ_r dominate.[59–62] For metals, both κ_L and κ_e contribute, and the total thermal conductivity is given by the Wiedemann–Franz law:[36]

$$\frac{\kappa}{\sigma T} = L_0 \tag{2.33}$$

where σ is the electrical conductivity and L_0, the Lorentz number, is $2.45 \times 10^{-8}\ \mathrm{S^{-1}\,K^{-2}}$. The ratio on the left side of Equation 2.33 can be used to assess the relative contributions of lattice vibrations and charge carriers in heat transport in a material. If the effective Lorentz number, $L_{eff} = \kappa/(\sigma T)$, is significantly greater than L_0 then heat transport by charge carriers is negligible with respect to transport by lattice vibrations.

Theoretical understanding of the thermal conductivity of matter was advanced by Debye in his consideration of the contributors to heat flux.

For a gas, he derived that:[36,63]

$$\kappa = \frac{1}{3}C\bar{v}\lambda \tag{2.34}$$

where C is the heat capacity per unit volume, $\bar{v}$ is the mean velocity of the energy carriers and λ is their mean free path (the average distance between collisions). Equation 2.34 also applies to solids, where, for insulators, $\bar{v}$ is the mean speed of the heat-carrying, acoustic phonons (*i.e.* velocity of sound) and λ is their mean free path (the average distance between phonon collisions). The contribution of the optic branches of the dispersion curves to the heat capacity per unit volume, C, can be approximated by the Einstein model of isolated atomic vibrations (see above),[64] but the optic phonons are ineffective carriers of thermal energy due to their near-zero speed. Although in certain circumstances the optic modes can interact with acoustic phonons to lower the thermal conductivity, *e.g.* in materials with low-frequency optic modes such as clathrates[65] or framework lattices,[66] in general most of the thermal energy in insulators is carried by the acoustic phonons.[59] A typical, simple, crystalline solid shows a peak in thermal conductivity for the following reasons: at high temperatures where C is approximately constant, the temperature dependence of κ is dominated by the reduction in mean free path as temperature is increased due to the increased probability of phonon–phonon scattering; at very low temperatures, where the population of phonons is low and the mean free path is determined by the average distance between defects, impurities or grain boundaries, κ parallels C and decreases as the temperature is lowered. Thermal conductivity mechanisms are detailed in references [36] and [59].

The thermal conductivity of a solid can be very dependent on purity. For a semiconductor, impurities can increase the thermal conductivity by opening electronic channels for heat conduction. However, more generally impurities in a crystalline lattice lead to disruption of the phonon mean free path and a reduction in κ.

Thermal conductivity values can range widely with material and temperature, from $>2000\,\mathrm{W\,m^{-1}\,K^{-1}}$ for diamond to $<0.2\,\mathrm{W\,m^{-1}\,K^{-1}}$ for aerogels.[36] Extensive lists are available for the thermal conductivities of such materials as heat insulators, metals, non-metallic solids and gases.[67,68]

Many different methods can be used to determine thermal conductivity. An extensive reference concerning thermal conductivity measurements is provided by Slack[69] and several of the most important methods are presented there.

2.4.2 Steady-State Method

The most straightforward way to measure thermal conductivity is by direct measurement of the temperature gradient, dT/dz, when a sample of cross-sectional area A is heated along the z -direction with power $\dot{Q}$:

$$\kappa = \frac{\dot{Q}}{A}\left(\frac{dz}{dT}\right). \tag{2.35}$$

A schematic of such a set-up is shown in Figure 2.16.[70]

Although Equation 2.35 gives κ directly, in practice κ is determined by measuring the temperature difference between two points, ΔT (typically the temperature drop across the thermocouple shown in Figure 2.16), as a function of heater power at a number of power values, giving the thermal conductance, K:

$$K = \frac{\dot{Q}}{\Delta T}. \tag{2.36}$$

By factoring in the sample geometry, thermal conductivity, κ, is obtained from:

$$\kappa = K\frac{L}{A} \tag{2.37}$$

where L is the distance across which ΔT is determined (Figure 2.16).

However, Equations 2.36 and 2.37 do not account for radiative heater-power losses, which can be substantial, especially at high temperatures. The radiative heat loss can be accounted for through:

$$\dot{Q} = I^2R - P_{\text{rad}} \tag{2.38}$$

where I is the heater current, R is the heater resistance and the radiative loss is given by:

$$P_{\text{rad}} = \sigma_{\text{T}}\frac{S}{2}\varepsilon(T_{\text{sam}}^4 - T_{\text{surr}}^4) \tag{2.39}$$

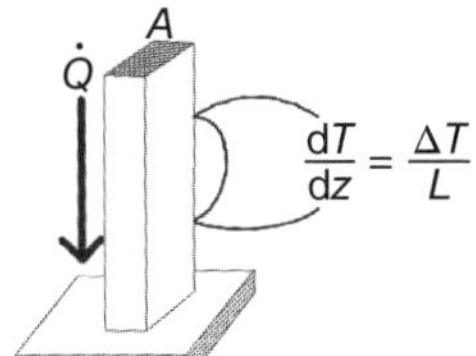

Figure 2.16 Set-up for steady-state determination of thermal conductivity, where the sample is shown mounted vertically on a heat-sink base.

where σ_T is the Stefan–Boltzmann constant (5.67×10^{-8} W m^{-2} K^{-4}), S is the sample surface area, ε is the sample emissivity, T_{sam} is the sample temperature and T_{surr} is the temperature of the surroundings. The need for radiative correction can be greatly reduced by the presence of a radiation shield around the sample, such that $T_{sam} \approx T_{surr}$. Other corrections, such as any attachments to the sample, *e.g.* heater–thermometer assemblies, must also be taken into account.

Steady-state thermal conductivity measurements can be carried out with laboratory-built or commercial equipment and the temperature range can be from <1 to 300 K or slightly higher. The upper limit ensures that sample radiation is small compared to the heat transported through the sample;[71] note the T^4 temperature dependence for radiative losses in Equation 2.39.

2.4.2.1 *Sample Considerations*

Samples for steady-state thermal conductivity measurement are typically solid and can include metals, composites and consolidated powders. Accurate and regular dimensions, especially regular cross-sectional area, are required for a high accuracy of the determination. While metal and composite samples can be shaped easily for measurement, powdered samples often need to be consolidated. Methods include cold pressing, hot pressing, cold pressing/sintering and spark plasma sintering, but due care must be taken to ensure that the sample has not been changed by the consolidation process.

Generally, the required sample dimensions will depend on the thermal conductivity of the material. For low–thermal-conductivity materials (<5 W m^{-1} K^{-1}), a sample thickness on the order of a few mm will produce a measureable temperature gradient. For the thermal transport option of a commercial physical property measurement system, typical sample cylinder dimensions are 3 mm length and 16 mm^2 cross-sectional area. For high-κ materials (>100 W m^{-1} K^{-1}), a longer sample (on the order of 1 cm) is required in order to obtain a measureable temperature gradient.[72]

2.4.2.2 *Accuracy*

Excellent heat sinking to the surroundings, excellent contacts between the sample and the heater and excellent thermometry are all necessary;

these factors are discussed at length by Kopp and Slack.[73] It is recommended strongly that the technique (mounting heater, thermometry and heat sink) is tested against standard materials with known κ in approximately the same range of κ. Standard materials are numerous and available commercially; some notable examples are Pyrex, copper, nickel and aluminium. Furthermore, the procedure should be checked for reproducibility with specimens of different geometries. Samples should be of high purity, since perturbations in the lattice can disrupt the phononic pathway and lead to lower measured values of κ.

Even with great care, highly accurate, steady-state thermal conductivity measurements are difficult to achieve as there are many experimental variables that can give rise to a variation in κ for nominally identical samples.[74] Intrinsic differences in κ can result from subtle chemical and microstructural differences, such as grain size, porosity, density, defects and impurities. Generally, with good techniques, the accuracy of κ from steady-state determination can be within 10%. This uncertainty stems in part from thermal contact resistances, which can be reduced from typical values of 10^{-3}–10^{-4} m^2 K W^{-1} to 10^{-5} m^2 K W^{-1} by polishing sample surfaces, clamping during adhesion and using adhesive grease or foil.[75] As a check on the accuracy of the measurement of κ, the contact resistance should be less than the sample's area-specific thermal resistance, *i.e.* L/κ, where L is the sample length (or thickness).[59] The other major contributor to uncertainty is geometric; this can be minimised with uniformly shaped samples.

2.4.3 Guarded Hot-Plate Method

The guarded hot-plate method can be considered a special case of a steady-state thermal conductivity apparatus, especially suited to large amounts of sample.

In this technique, the sample is placed between two plates, one hot and one cold. Uniform and excellent contact between the sample and the plates is essential for accurate determination of κ. Thermal contact fluids such as silicone oil can be used, provided that there is no interaction with the sample.

The presence of a guard ring around the sample reduces the edge losses and ideally the ring should have the same temperature profile as the sample at its edge. In practical terms, a tight-fitting guard of thermal conductivity similar to that of the sample is sufficient. A wide range of thermal conductivities can be determined over temperatures from *e.g.* −75 to 200 °C, with uncertainties of a few per cent for a well-controlled experiment and uniform sample.[76] Other designs show measurements to much lower temperatures[77] with similar accuracy.

The guarded hot plate has the major advantages that it is available in relatively inexpensive commercial instruments and that there are ASTM standard methods available to guide measurements.[78,79] The major disadvantage is the need for rather large samples (*e.g.* the plate might be several cm in diameter) with extremely uniform thickness (tolerance better than 0.1 mm).

2.4.4 Parallel Thermal Conductance Method

The thermal conductivities of large bulk structures can be determined readily by steady-state measurements, because such samples can support the heater and thermometry required to conduct the measurements. There are, however, special cases where a material is small and fragile or flexible and the mounting of heater/thermometry assemblies is not possible. A technique has been developed for measurements on needle-like samples: the parallel thermal conductance (PTC) method.[80] This is relatively straightforward compared to the thermal potentiometeric method, for which measurements are laborious and tedious.[81]

In a typical steady-state measurement, thermocouples attached to the sample provide temperature measurements and a heater provides the temperature gradient. For small samples that do not have much strength and/or stiffness, the most challenging aspect is that the sample cannot support the heater and thermometry required for conventional methods. The PTC method addresses this challenge by having the small sample supported on a stage (Figure 2.17) and it operates as a variant of the steady-state technique (power *versus* ΔT). In its essence, the PTC stage comprises a cold side heat-sinked to the surroundings, a hot side isolated from the system by a strip of low-κ material (*e.g.* acetate transparency film) as a support, a differential thermocouple with junctions in contact with the hot and cold sides to measure the temperature gradient, ΔT, directly and a heater attached to the hot side to provide the heat flux. The sample spans the gap between the hot and cold sides, as shown schematically in Figure 2.17.[80]

With the PTC technique, the thermal conductance of the holder (K_{holder}) must be measured throughout the temperature range as the

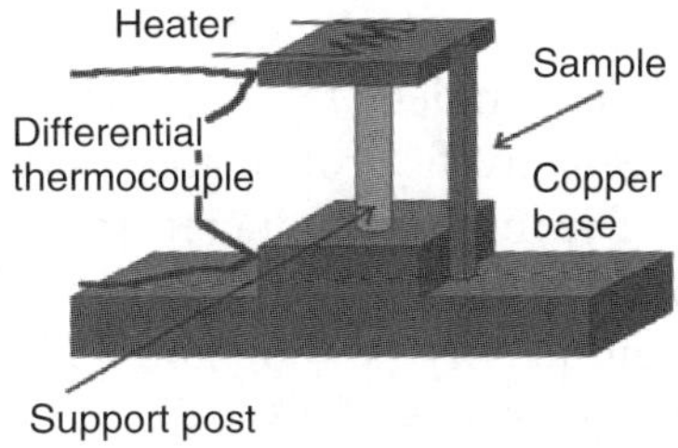

Figure 2.17 Schematic of the sample holder for the PTC method.[80] Reprinted with permission from [80] Copyright (2001) American Institute of Physics.

first of three steps. The thermal conductance is calculated by the slope of power *versus* ΔT at the set point (*i.e.* cryostat temperature), where the power is set such that ΔT is on the order of 0, 1, 2 and 3% of the cryostat temperature. The second step is the measurement of the stage with a sample across the hot/cold platforms (giving K_{tot}) and the third is a measurement with the sample disconnected from the cold side, in order to determine the sample radiation, K_{rad}. The sample conductance is calculated by:

$$K_{sample} = K_{tot} - \left(\frac{K_{rad} - K_{holder}}{2}\right) - K_{holder} \tag{2.40}$$

where the factor of 1/2 is due to the assumption that only half of the sample will be hot and hence radiating. In order to obtain the best PTC technique efficiency, the use of thermally conductive materials in the construction of the base is recommended. For electrical leads, very small-diameter (20 μm) alloys are recommended, due to their low thermal conductivities.[80]

The advantage of the PTC technique is that the thermal conductivities of very small samples and/or samples that lack rigidity can be measured. However, this technique does have some drawbacks. The measurements can be time-intensive and reaching equilibrium after changing the cryostat temperature takes time (up to hours). With due care, PTC measurements can be very accurate. PTC depends critically on the accuracy of the sample dimensions, but overall uncertainty in κ can be within 10%.[80]

The technique was developed and validated by the Tritt group,[80] who measured materials (various steels, Pyroceram and a carbon fibre material) and compared the results with previous measurements on the same materials by other techniques. More recently, this technique has been used for the measurement of macroscopic structures composed of pure carbon nanotubes (CNTs), specifically CNT yarns[82] and sheets.[83] For the CNT yarns, multiple samples (15–18) were loaded onto the PTC stage and the method effectively probed the thermal conductivity of a total length of ~8 cm of yarn, thereby increasing the sample signal relative to the background. The results showed that smaller-diameter yarns had higher thermal conductivity ($\kappa = 60$ and $25\,\mathrm{W\,m^{-1}\,K^{-1}}$ for 10 and $34\,\mu\mathrm{m}$ yarns, respectively, at $T = 300\,\mathrm{K}$), likely due to tighter winding and a lower number of defects during spinning.[82] The CNT sheet had a κ of *ca* $2.5\,\mathrm{W\,m^{-1}\,K^{-1}}$ at $T = 300\,\mathrm{K}$.[83] This much lower value is due mostly to the very low density of the material and the reduced alignment of the CNTs composing the sheet.

2.4.5 Power-Pulse Method

The power-pulse method is very similar to the steady-state technique, except that the temperature of the sample/mount ensemble changes at a constant rate, *i.e.* it is not isothermal. Furthermore, as the name implies, in the power-pulse method the heater current is pulsed (square wave) to produce the heat flux.[84] The benefit of this so-called Maldano technique over steady-state methods is a greater density of data points at a given time, as the need for isothermal conditions is eliminated.

The power-pulse technique is employed in a commercial instrument,[72] in which ΔT across a sample is measured as a function of time during the heat pulse and the data are fit with a non-linear, least-squares fitting algorithm to the following form:[85]

$$\Delta T = \Delta T_{\infty}\left[1 - \frac{\left(\tau_1 e^{\frac{-t}{\tau_1}}\right) - \left(\tau_2 e^{\frac{-t}{\tau_2}}\right)}{\tau_1 - \tau_2}\right] \tag{2.41}$$

where the asymptotic temperature, ΔT_{∞}, represents the hypothetical steady-state rise in temperature (*i.e.* if the heater remained on) and τ_1 and τ_2 are relaxation time constants. The value ΔT_{∞} can be used to determine κ through Equation 2.36, with ΔT_{∞} replacing ΔT.

Sample considerations, preparation, mounting, instrument calibration and calibration testing with standard materials are the same as for steady-state measurements. If the same care is given to sample preparation and mounting as in steady-state measurements, the same accuracy can be obtained. To obtain the best possible fits to Equation 2.41 using this method, the cryostat temperature scan rate should be small, on the order of $1\,\mathrm{K\,min^{-1}}$ or less.

Measurements of κ using commercial instrumentation and the Maldonado technique have been reported. A recent illustrative example involves κ investigations of the ceramic $HfMgMo_3O_{12}$. [57] In this study, methanol was used to facilitate consolidation of the powdered samples into a pellet and the samples were sintered to remove the solvent and provide mechanical strength. Each pellet (*ca* $6\,\mathrm{mm}^2$ in area and 1.5 mm in length) was adhered between two copper plates, with the heater/hot thermometer on one end and the cold thermometer/heat sink on the other. Measurements were conducted with a physical property measurement system, automated through a programmable software interface. The experiments were carried out such that the heater power was modulated to yield a defined ΔT, typically 3% of the cryostat temperature (*e.g.* $\Delta T = 9\,\mathrm{K}$ for $T = 300\,\mathrm{K}$), where ΔT changed as the cryostat temperature changed.[85] The samples were porous and the ideal thermal conductivity of the fully dense material, κ_{dense}, was determined from the experimental thermal conductivity, κ_{exp}:[86]

$$\frac{\kappa_{\mathrm{exp}}}{\kappa_{\mathrm{dense}}} = 1 - \frac{4}{3}\phi \tag{2.42}$$

where ϕ is the sample porosity. The thermal conductivity of $HfMgMo_3O_{12}$ was found to be low, near $1\,\mathrm{W\,m^{-1}\,K^{-1}}$ at room temperature, due to the presence of low-frequency optic modes associated with easy twisting of polyhedra in the open-framework structure.[57]

2.4.6 Laser-Flash Diffusivity

Laser-flash diffusivity (LFD) is a technique devised[52] to determine the thermal properties of both thin films and bulk samples.[87] It involves irradiating the surface of a sample with a short laser pulse ($\leq$1 ms) while the temperature is monitored by infrared detection on the sample's other side. Analysis of the time evolution of the temperature at the back side of the sample allows determination of the thermal diffusivity, a. Thermal conductivity and diffusivity are related through the specific heat, C, and density, ρ, by:

$$a = \frac{\kappa}{\rho C}. \tag{2.43}$$

Specifically, a is determined by the so-called $t_{1/2}$ method, where $t_{1/2}$ is the time at which $V = 1/2$. V is the dimensionless temperature at the back of the sample, defined as:

$$V = \frac{T}{T_{max}} \tag{2.44}$$

where T_{max} is the maximum temperature on the back of the sample after the laser flash, and:

$$a = \frac{\ln 4L^2}{\pi^2 t_{1/2}} \tag{2.45}$$

where L is the sample thickness.[52]

The LFD method is used above room temperature so that there is sufficient black-body radiation for infrared detection. In fact, it can be used up to quite high temperatures – more than 2000 K. At high temperatures, the heat capacity of a material is essentially a constant and thus the thermal diffusivity is proportional to the thermal conductivity. Therefore, in principle, LFD can be used to determine κ.

However, this technique's utility is very dependent on meeting strict sample preparation requirements. Thin, flat discs or plates are required to prevent flash-through to the infrared detector. Also, the sample surface must be highly emissive in order to maximise the thermal energy transmitted from the surface and the observed signal at the detector. Often there is a need for a thin coat of graphite on the surface and if this is not properly adhered, there can be significant errors in the measurements. Furthermore, the sample properties must be such that the characteristic thermal diffusion time is much less than the flash duration.[52] Although the sample should normally be homogeneous, LFD can be used to measure diffusivity in inhomogeneous materials by varying the beam focus.[88,89]

Commercial LFD instruments are available for thermal diffusivity measurements over the temperature range 77 to ~2300 K.[90–92] Such instruments typically have facile experimental set-ups and the measurements are automated; sample considerations as mentioned above are required for high accuracy.

A major limitation of using the LFD method to determine κ is that the heat capacity and density must be known over the temperature range investigated (*i.e.* from separate experiments or from reliable estimates) in order to obtain the sample thermal conductivity. Also, a large sample (typically 5 cm diameter) is required for most LFD measurements.

If the LFD method is used to determine thermal diffusivity, it is strongly recommended that the same sample be used for the separate measurements of heat capacity and density, in order not to unnecessarily add to uncertainty in the thermal conductivity determination. Uncertainties for diffusivity measurements can be within 4%,[93] but when propagating the uncertainty for C_P and ρ, the uncertainty in κ can be 10% or more.

A recent example involves the use of the LFD method in studying hafnium–zirconium tantalum oxides, $(Hf_{1-x}Zr_x)_6Ta_2O_{17}$ ($x = 0$, 0.1, 0.3 and 0.5), up to temperatures of 1000 °C in an argon atmosphere.[94] A commercial instrument was used for these measurements and a noticeable improvement was made with respect to sample size. Samples 13 mm in diameter and 1 mm in thickness were used and these were coated with graphite to prevent transmission of the laser beam. The study looked at the effect on the thermal conductivity of an increasing zirconium content. It was found that when $x = 0.1$ there was little change in κ. When $x = 0.3, \kappa$ decreased by less than 10%, but when $x = 0.5$, a substantial decrease (*ca* 25%) in κ occurred. The increased Zr content increased the impurity scattering of the phonons compared with pristine hafnium tantalum oxide and perturbed the phonon mean free path, leading to a reduction in κ, as was observed.[94]

2.4.7 Hot-Wire Method

Another method that can be used to measure the thermal conductivity is the hot-wire technique, also referred to as the hot-strip method. This was developed to measure the κ of liquids,[95] gases[96] and solids.[97] Hot-wire measurements involve placing a metal strip between two sample substrates; the metal strip acts as the heat source. The temperature distribution in the sample is monitored (*e.g.* by optical analysis[97]). Alternately, the thin metal strip can act as both the heat source and the temperature sensor;[98] a constant current is applied to the strip so that the power is nearly constant and the voltage across the strip is monitored. The hot-wire technique can provide the thermal conductivity and thermal diffusivity of a sample within a few minutes of experimentation. The theory behind the extraction of the thermal property values using this method is described in detail in references [98] and [99].

Because the metal strip is sandwiched between two solid samples, the samples should be kept flat and parallel to one another in order to ensure the best possible thermal contact with the strip. The ideal situation is to have the strip as a plane source. The sample size can be as large as

$80 \times 50 \times 15\,\text{mm}^3$ and the heater should be the length of the sample, with an optimal width between 3 and 5 mm.[98] Typically, strip thickness should be <1 mm and the heat pulse duration less than one second.[98] The thin strip can be either an evaporated thin film or an applied thin strip. Good contact greatly decreases the errors in measurement. For evaporated thin films, thermal contact problems are eliminated as the metal is bonded directly to the substrate. If a thin strip is applied, it is typically held with a fluid (*e.g.* low-viscosity oil) and thermal contact can become a concern.[98] Uncertainty in hot-wire measurements can be as large as 10%, but with due care in sample preparation (evaporated thin film, flat, oxide-free surface) the accuracy can be better than 1%.[98]

Hot-wire measurements have a major advantage in that materials can be investigated under pressure, even at low temperatures. For example, the pressure collapse of ice in a polymer hydrogel has been studied using hot-wire thermal conductivity measurements.[100]

The hot-wire method has been reinvestigated recently in order to advance understanding of modern, small-scale materials.[101] Conventional hot-strip methods require large sample sizes and would not work for thin-film samples, but an ultra-fast hot-strip technique can work for such samples.[101–103] In this method, Al lines ($0.6 \times 500\,\mu\text{m}^2$) are patterned onto a thin film surface as the heater and thermometer. Short heat pulses (10 μs) at 1 kHz frequency give a voltage change, representing the temperature, which is monitored during the experiment. The thermal impedance, Z, is fit to a heat diffusion model over a short time window (100 ns to 10 μs) in order to estimate κ: [101]

$$Z = \frac{16 R_0 \Delta V}{\beta U^3} \tag{2.46}$$

where R_0 is the strip resistance without current, ΔV is the change in voltage during measurement (output voltage), β is the temperature coefficient of the strip and U is the input voltage. Using such short pulses leads to improved signal-to-noise ratios (SNRs),[102] and recent work has shown that the limit of detection for thermal resistance can be $10^{-9}\,\text{K}\,\text{m}^2\,\text{W}^{-1}$.[101] Such a technique should best be used with high-κ materials in order to have appreciable ΔV.

2.4.8 3ω Method

Measurements of thermal conductivity can also be made by the so-called 3ω method.[71,104] This method is closely related to the hot-strip method in that the measurements of κ involve monitoring the resistance of a metal heater/thermometer as power is applied to it. The difference between the 3ω method and hot-strip method is that the former uses lock-in amplifier technology and its measurement is in the frequency domain rather than the time domain. In the 3ω method, the amplitude and phase of the resistance changes of the metal film are monitored as a function of frequency. Typically, the heater and thermometer are incorporated in one unit, which is evaporated onto the sample surface. The film width used for 3ω measurements must be small compared to the wavelength of the diffusive thermal wave. This configuration leads to a cylindrical heat flow pattern, from which κ can be directly determined:

$$\Delta T = \frac{P}{\pi l \kappa}\left[\frac{1}{2}\ln\left(\frac{\kappa}{r^2 \rho C}\right) + \ln 2 - 0.5772 - \frac{1}{2}\ln(2\omega) - \frac{i\pi}{4}\right] \quad (2.47)$$

where P/l is the power amplitude per unit length, ω is the angular frequency of the heater current, ρ is the density, C is the specific heat and r is the distance from the line source of heat.[104] The slope of the in-phase temperature oscillations plotted against $\ln \omega$ gives κ directly. The temperature (voltage) oscillations are measured at the third harmonic, which is the origin of the name: 3ω.

The temperature range for this method can be from cryogenic (*e.g.* 30 K) up to ~ 750 K. Black-body radiation is often a concern in κ measurements, especially when experimenting at high temperatures, but it does not affect 3ω measurements for temperatures as high as 1000 K.[71,104] This is due to the fact that only a very small area of the sample is heated (because of the small heater/thermometer strip) and only for a short period of time. As a result, the thermal penetration is small and so the hot area is small, leading to a small radiative area.

Sample preparation for 3ω measurements typically involves deposition of a metal through a mask onto the sample (the only option for water-soluble samples) or photolithography of a metal film deposited onto the sample (photolithography produces more accurate results). The heater metal chosen should be compatible with the temperature range under investigation; typically, nichrome is deposited onto the sample, with Ag deposited on top of it (nichrome promotes adhesion and Ag is a good thermometer at low T).[104] The sample surface should be smooth and flat and is typically on the order of $1 \times 0.5\ \text{cm}^2$ due to electrical connection constraints.

There are three main advantages to the 3ω technique. First, black-body radiation is virtually eliminated due to the very small size of the heater and thus small area of heating (within the thermal penetration depth of the heater–thermometer) of the sample, virtually removing this source of uncertainty. Second, the sample size can be very small; the constraint for such measurements is that the sample must accommodate the metal pattern evaporated onto the surface and the sample thickness should be five times the width of the patterned line, in order to prevent reflections where a sample is evaporated onto a substrate.[71] Third, the equilibration time is only a few periods of the temperature oscillation of the metal line (at most a few seconds); this allows for collection of κ data with slow heating/cooling rates.

Recent advances in the 3ω technique involve using a metal-coated glass fibre as the heater–thermometer: so-called metal-coated 3ω.[105] The metal-coated 3ω method can be used to measure the thermal conductivity of fluids, aerogels, gases and powdered samples. While conventional 3ω can be used with fluids, it requires the use of a substrate to pattern the heater and the submerging of the lot. The disadvantage here is that the thermal effusivity of the substrate is much larger than that of the fluid. Metal-coated 3ω is essentially a combination of the hot-wire and 3ω techniques, where a micron-scale optical fibre (20 μm diameter) is coated with 10 nm Ti (adhesion layer) and 50–100 nm Au or Pt. These thin coatings have a high resistance and therefore give an increased 3ω signal and thus an increase in the SNR for a given ΔT. The optical fibre used is SiO_2; this material has a low thermal mass, which does not interfere with the measurements. As with 3ω, black-body radiation is only a concern at very high temperatures. Other benefits include rapid measurements and the ability to measure millimetre-sized samples even if they have low thermal effusivity. The accuracy of metal-coated 3ω depends on the thermal conductivity range: uncertainty is <10% for low-κ materials (*e.g.* Ar gas) and can be better than 5% for higher-κ materials such as water.[105]

2.5 THERMAL EXPANSION

2.5.1 Terminology and Atomic Origins

Thermal expansion is a manifestation of the change in the interactions between atoms as the temperature changes. The value and even the sign of thermal expansion can provide significant insights into the structure and dynamics of a solid.

The coefficient of thermal expansion, known as CTE and generally represented by the symbol α, can be expressed in terms of volume, V, as α_V:

$$\alpha_V = \frac{1}{V}\left(\frac{dV}{dT}\right)_P \tag{2.48}$$

where T is temperature and P is pressure. The CTE can also be expressed in terms of a length, ℓ, *i.e.*:

$$\alpha_\ell = \frac{1}{\ell}\left(\frac{d\ell}{dT}\right)_P \tag{2.49}$$

where ℓ might be a unit cell length (*e.g.* giving α_a) or a length in a polycrystalline sample (giving the averaged linear CTE). For a cubic solid, $\alpha_V = 3\alpha_a$, but this is not true in general.

Typical values of α_V (or α_ℓ) for inorganic solids at room temperature are *ca* $10^{-5}\ K^{-1}$. However, the actual values depend on the material and the temperature.

The atomic origins of thermal expansion can be understood in terms of Figure 2.18. A harmonic interaction potential is perfectly symmetric and leads to the same interatomic separation regardless of the thermal energy available; hence, $\alpha_\ell = 0$. However, a true potential is stiffer on compression and less stiff on expansion than a harmonic potential. The asymmetry leads to an extended interatomic separation when there is more available thermal energy, *i.e.* thermal expansion. The asymmetry of the potential is higher at higher temperatures and the CTE generally increases as the temperature increases. Conversely, all potentials become

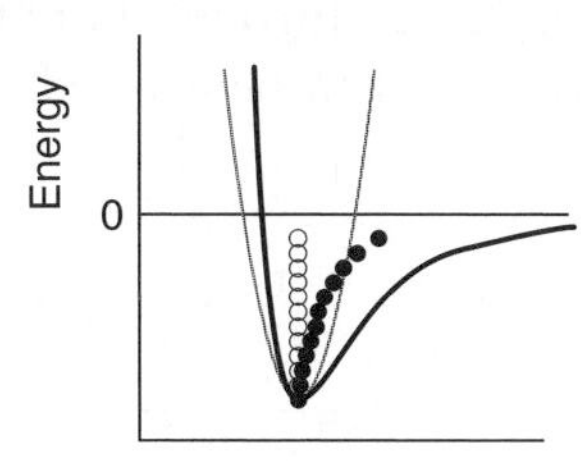

Figure 2.18 The harmonic potential (parabola shown as a faint curve) has the same thermally averaged interatomic spacing, $\langle r \rangle$, at all temperatures, as shown by ○, and therefore the CTE, α, is zero. The true, anharmonic potential, shown by the heavy curve, is asymmetric and therefore the thermally averaged interatomic spacing, $\langle r \rangle$, increases with increasing temperature, as shown by •, which moves to larger r as the temperature increases, giving $\alpha > 0$.

harmonic in the limit of $T \rightarrow 0\,\mathrm{K}$, and α_V and $\alpha_\ell \rightarrow 0$ as $T \rightarrow 0\,\mathrm{K}$, for virtually all materials.[36]

Although thermal expansion is usually positive, some materials exhibit negative thermal expansion (NTE, also known as thermomiotic behaviour[106]). The archetypal example is ZrW_2O_8, which shows negative thermal expansion (average value of $\alpha_a = -9 \times 10^{-6}\,\mathrm{K}^{-1}$ from 0.3 to 1050 K).[107] The interested reader is referred to a recent comprehensive review of thermomiotic materials.[106]

2.5.2 Diffraction Methods

If a material is crystalline, diffraction can be used to determine CTEs as a function of both temperature and crystal axis. Usually, this is X-ray diffraction (XRD), but neutron diffraction can also be used.

The principle of diffraction is based on interference effects. For a beam of X-rays impinging on a layered structure at an angle θ, the reflections from subsequent layers will only give constructive interference, *i.e.* be in phase, if the additional path travelled by the beam reflecting from the lower layer (see Figure 2.19) is an integral number of wavelengths ($n\lambda$, where n is an integer and λ is the wavelength of the X-ray). The constructive interference condition is succinctly expressed by Bragg's law:

$$n\lambda = 2d \sin\theta \tag{2.50}$$

where d is the distance between layers.

Determination of the lattice parameters as a function of temperature directly yields the CTEs $\alpha_a, \alpha_b, \alpha_c$ through Equation 2.49. Furthermore, if the space group is known, α_V can be derived.

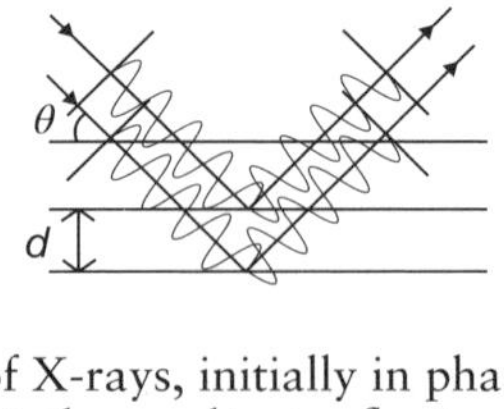

Figure 2.19 When a beam of X-rays, initially in phase, hits a layered sample, with repeat distance d, at an angle θ, the resultant reflecting beams are only in phase if the additional path distance travelled by the lower beam ($2d\sin\theta$) is an integral number of wavelengths, $n\lambda$. If this condition is satisfied, the intensity observed on reflection is at a maximum.

The change in unit cell parameters with temperature can provide the CTE, which can give insights into interatomic interactions, as discussed above. Furthermore, discontinuous changes in the unit cell parameter with temperature (leading to steps in the CTE) can delineate phase transitions.

The CTE given by diffraction techniques is the intrinsic CTE, as it tracks the change in spacing between the atoms as a function of temperature. As shown below, the intrinsic CTE can differ from the CTE for a bulk sample.

Determination of the unit cell parameters as a function of temperature using diffraction can be carried out with powder samples or with single crystals. For powder methods, although the space group must be known, the detailed crystal structure is not required (the Rietveld method[108] is commonly used but the Le Bail method[109] can be used even if atomic positions are unknown). In general, unit cell parameters can be determined by XRD for samples on the order of 10s of mg.

CTE determination from diffraction techniques requires calibration involving adjusting the set parameters to the diffraction pattern collected from standard specimens, such as a LaB_6 or an Si standard (both available from NIST). With due care and a high-resolution apparatus, such as that found at a synchrotron source, unit cell dimensions can be determined to within 10^{-6} Å.[110] Using lab-based equipment with good calibration, unit cell volumes can be determined within about 0.03%. In addition to accurate distance determination, properly calibrated thermometry is required in order to obtain accurate CTEs. It is not possible to give a general statement concerning the accuracy of CTEs from diffraction, because all uncertainties in lattice parameters and in temperature must be propagated. Also, since the CTE is itself temperature dependent, it is necessary to quote the temperature range over which it is determined. Since the CTE relates to changes in lattice parameters, the relative uncertainty is higher when the CTE is closer to zero.

As an example, the lattice parameters of diamond and the corresponding linear CTE, as a function of dopant, are shown in Figure 2.20.[111] Note that $\alpha_a \to 0$ as $T \to 0\,K$ and, because diamond has a very stiff lattice, the CTE at room temperature is relatively low (*ca* $10^{-6}\,K^{-1}$).

2.5.3 Dilatometry

Dilatometry is a form of thermomechanical analysis in which the length of a macroscale specimen is determined as a function of temperature.

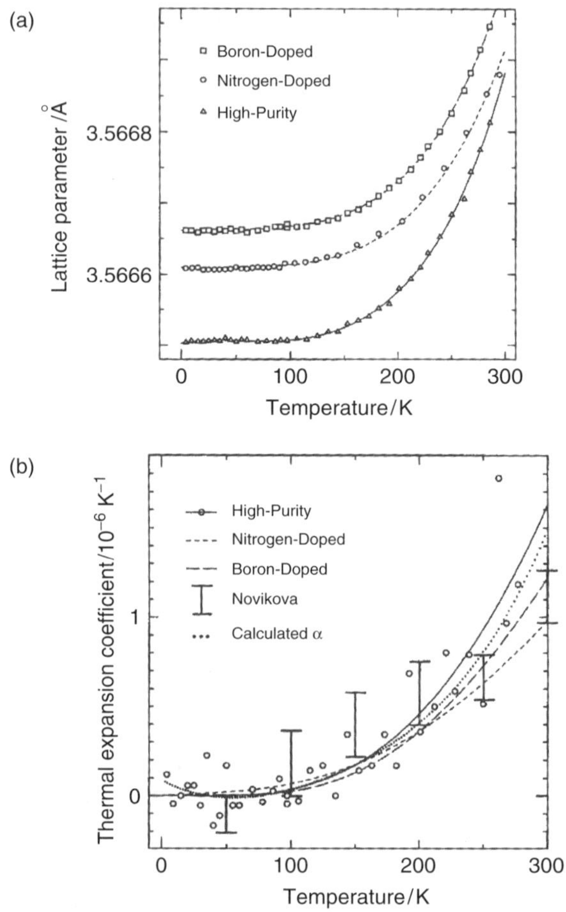

Figure 2.20 (a) Lattice parameters of diamond as a function of doping. (b) Corresponding linear CTE, α_a, as a function of temperature [111]. Reprinted with permission from [111] Copyright (2002) American Physical Society.

Commercial dilatometers are available, which determine the relative change in length of a specimen, $\Delta L/L_0$, where ΔL is the change from the initial length, L_0. Generally, the sample is in contact with a push-rod, which moves in response to the change in length of the sample. This change in length is detected electronically, typically using a linear variable differential transformer or a capacitor with a movable plate. For a temperature change, ΔT, the linear CTE is given by:

$$\alpha_\ell = \frac{1}{L_0}\left(\frac{\Delta L}{\Delta T}\right). \tag{2.51}$$

Because dilatometry is measured on bulk samples, it does not usually give the directional dependence of the CTE as diffraction techniques do. However, dilatometry does not require crystalline samples, so it can be used to determine CTEs for glasses and other non-crystalline materials as well as for composites.

The main difference in the CTE results obtained from dilatometry *versus* diffraction is that dilatometry provides the bulk thermal expansion information, which includes both intrinsic CTE (as determined by diffraction) and extrinsic effects. The latter can include changes in microstructure (*e.g.* microcracks, pores and vacancies) capable of modifying the dimensions of the specimen. In general, microcracks form on cooling and heal on heating, and there is therefore a contraction on heating and expansion on cooling. For ceramics, there is a threshold grain size below which microcrack formation plays a negligible role.[112] Therefore, CTE determination would be expected to be quantitatively similar for the diffraction and dilatometry of materials containing small, uniform grains. Microstructure can play such an important role in determining the thermal expansion of a material with very low thermal expansion that even the sign of the CTE can be different for results from diffraction compared with those from dilatometry; one example is $Al_2W_3O_{12}$.[113] The insights into microstructure provided by dilatometry can be used to determine the concentration of vacancies as a function of temperature, and hence the enthalpy of formation of vacancies.[114]

Dilatometry requires well-consolidated pellets, typically at least a few mm thick. It is essential that the pellets have flat ends that are parallel to one another and that they can be machined from a monolith (*e.g.* from a large piece of glass) or made by pressing (followed sometimes by sintering) powder. It is important to check, *e.g.* by XRD, that the structure of the material is not altered by the pressing and/or sintering of a pellet.

Commercial push-rod dilatometers can have $\Delta L/L_0$ accuracy of $\pm 1\%$ or better. A standard, such as α-alumina, should be tested for calibration in order to correct for any thermal expansion due to the sample holder and push rod. As with diffraction, accurate thermometry is required if accurate CTEs are to be obtained. A well-behaved standard of similar CTE to the material under study should be run in order to assess the accuracy of the CTE for a given set-up. The uncertainty for an unknown sample must be calculated by propagation of errors in both $\Delta L/L_0$ and T. The relative uncertainty can be quite small for large CTEs, but much greater for low CTEs. Dilatometric measurements should

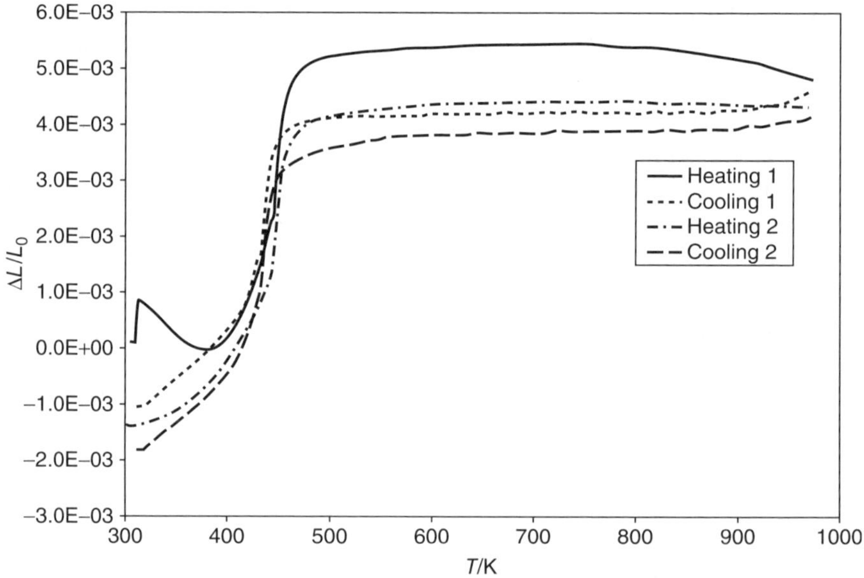

Figure 2.21 Dilatometric results for $In(HfMg)_{0.5}Mo_3O_{12}$.[115] Reprinted with permission from John Wiley and Sons.

be carried out both on heating and on cooling, possibly with a few thermal cycles in order to see the effect of thermal history. Generally, cooling results are more reliable as they have minimal contributions from changes in microstructure. As in diffraction, the temperature range over which the CTE has been determined should be reported.

The results of a dilatometric investigation of $In(HfMg)_{0.5}Mo_3O_{12}$ are shown in Figure 2.21.[115] At $T \approx 425\,K$, the dilatometric results clearly show the phase transition in $In(HfMg)_{0.5}Mo_3O_{12}$ from the low-temperature monoclinic phase, which shows positive thermal expansion, to a high-temperature orthorhombic structure with near-zero thermal expansion.

2.6 CONCLUSION

Thermal properties such as mass loss, endothermic and exothermic events (reactions and phase changes), heat capacity, thermal conductivity and thermal expansion can be determined for materials over a wide range of temperatures, from <1 to >2000 K. Some measurements can be performed on exceptionally small samples, down to 1 mg or even less, and with very high accuracy.

Thermal experiments can uniquely reveal information concerning matters as diverse as thermodynamic phase stability, structural phase transitions, chemical stability, the ability of a material to store and transport energy in the form of heat and the potential for thermal stress in a solid under a temperature gradient. In addition, the information gained from thermal property measurements can be used to assess theoretical models of structure and dynamics in solids.

REFERENCES

[1] M. Sorai, *Comprehensive Handbook of Calorimetry & Thermal Analysis*, John Wiley, Chichester, 2004.
[2] E. Mansfield, A. Kar, T. P. Quinn and S. A. Hooker, *Anal. Chem.*, **82**, 9977 (2010).
[3] American Society for Testing and Materials, Standard Practice for Calibration of Temperature Scale for Thermogravimetry, ASTM-E1582-04, 2004.
[4] P. S. Gill, S. R. Sauerbrunn and B. S. Crowe, *J. Therm. Anal.*, **38**, 255 (1992).
[5] S. C. Mojumdar, M. Sain, R. C. Prasad, L. Sun and J. E. S. Venart, *J. Therm. Anal. Calorim.*, **90**, 653 (2007).
[6] S. Yariv, I. Lapides and M. Borisover, *J. Therm. Anal. Calorim.*, **110**, 385 (2012).
[7] V. M. Mecea, *J. Therm. Anal. Calorim.*, **86**, 9 (2006).
[8] M. A. Cooper and V. T. Singleton, *J. Mol. Recognit.*, **20**, 154 (2007).
[9] M. Kaspar, H. Stadler, T. Weiss and Ch. Ziegler, *Fresen. J. Anal. Chem.*, **366**, 602 (2000).
[10] E. J. Severin and N. S. Lewis, *Anal. Chem.*, **72**, 2008 (2000).
[11] H. Ogi, Y. Fukunishi, T. Omori, K. Hatanaka, M. Hirao and M. Nishiyama., *Anal. Chem.*, **80**, 5494 (2008).
[12] S. Y. Kwon, J. C. Kim and B. I. Choi., *Meas. Sci. Technol.*, **19**, 115206 (2008).
[13] M. Jiang, J. Li, Y.-Q. Ho, Y. Xi, J.-F.Y and F-X. Zhang, *J. Chem. Eng. Data*, **56**, 1185 (2011).
[14] C. G. Armistead, A. J. Tyler, F. H. Hambleton, S. A. Mitchell and J. A. Hockey, *J. Phys. Chem.*, **73**, 3947 (1969).
[15] R. Mueller, H. K. Kammler, K. Wegner and S. E. Pratsinis, *Langmuir*, **19**, 160 (2003).
[16] M. L. Pope and M. D. Judd, *Differential Thermal Analysis – A Guide to the Technique and its Application*, Heyden, London, 1977.
[17] A. Janghorban, M. Lomello-Tafin, J. M. Moreau and Ph. Galez, *Intermetallics*, **18**, 2208 (2010).
[18] http://www.setaram.com/SETSYS-Evolution-DTA-DSC.htm (last accessed 28 October 2013).
[19] S. L. Boersma, *J. Am. Ceram. Soc.*, **38**, 281 (1955).
[20] E. S. Watson, M. J. O'Neill, J. Justin and N. Brenner, *Anal. Chem.*, **36**, 1233 (1964).
[21] R. Sabbah, A. Xu-wu, J. S. Chickos, M. L. Planas Leitão, M. V. Roux and L. A. Torres, *Thermochim. Acta*, **331**, 93 (1999).

[22] D. M. Price, *J. Therm. Anal.*, **45**, 1285 (1995).
[23] E. Gmelin and St. M. Sarge, *Pure Appl. Chem.*, **67**, 1789 (1995).
[24] G. D. Gatta, M. J. Richardson, S. M. Sarge and S. Stølen, *Pure Appl. Chem.*, **78**, 1455 (2006).
[25] H. K. Cammenga, W. Eysel, E. Gmelin, W. Hemminger, G. W. H. Hohne and S. M. Sarge, *Thermochim. Acta*, **219**, 333 (1993).
[26] S. I. Sineva and R. V. Starykh, *Inorg. Mater.* **46**, 1667 (2010).
[27] M. J. Richardson and N. G. Savill, *Thermochim. Acta*, **12**, 213 (1975).
[28] S. C. Mojumdar, J. Kozankova, J. Chocholusek, J. Majling and V. Nemecek, *J. Therm. Anal. Cal.*, **78**, 73 (2004).
[29] S. C. Mraw, *Specific Heats of Solids*, Hemisphere, New York, 1988.
[30] P. S. Gill, S. R. Sauerbrunn and M. Reading, *J. Therm. Anal.*, **40**, 931 (1993).
[31] M. A. White, *Thermochim. Acta*, **74**, 55 (1984).
[32] K. S. Shiral Fernando, M. J. Smith, B. A. Harruff, W. K. Lewis, E. A. Guliants and C. E. Bunker, *J. Phys. Chem. C. Letters*, **113**, 500 (2009).
[33] J. Meng, M. Levina, A. R. Rajabi-Siahboomi, A. N. Round, M. Reading and D. Q. M. Craig, *Pharm. Res.*, **29**, 2128 (2012).
[34] H. M. Pollock and A. Hammiche, *J. Phys. D: Appl. Phys.*, **34**, R23 (2001).
[35] R. K. Kummamuru, L. De La Rama, L. Hu, M. D. Vaudin, M. Y. Efremov, M. L. Green, D. A. LaVan and L. H. Allen, *Appl. Phys. Lett.*, **95**, 181911 (2009).
[36] M. A. White, *Physical Properties of Materials*, CRC Press, Boca Raton, FL, 2012.
[37] B. Fultz, *Progr. Mater. Sci.*, **55**, 247 (2010).
[38] D. W. Cooke, F. Hellman, J. R. Groves, B. M. Clemens, S. Moyerman and E. E. Fullerton, *Rev. Sci. Instrum.*, **82**, 023908 (2011).
[39] T. J. Quinn, *Physica Scripta*, **41**, 730 (1990).
[40] M. I. Bagatskii, V. V. Sumarokov and A. V. Dolbin, *Low Temp. Phys.*, **37**, 424 (2011).
[41] G. R. Stewart, *Rev. Sci. Instrum.*, **54**, 1 (1983).
[42] R. Bachmann, F. J. DiSalvo, Jr., T. H. Geballe, R. L. Greene, R. E. Howard, C. N. King, H. C. Kirsh, K. N. Lee, R. E. Schwall, H.-U. Thomas and R. B. Zubeck, *Rev. Sci. Instrum.*, **43**, 205 (1972).
[43] J. S. Hwang, K. J. Lin and C. Tien, *Rev. Sci. Instrum.*, **68**, 94 (1997).
[44] J. C. Lashley, M. F. Hundley, A. Migliori, J. L. Sarrao, P. G. Pagliuso, T. W. Darling, M. Jaime, J. C. Cooley, W. L. Hults, L. Morales, D. J. Thoma, J. L. Smith, J. Boerio-Goates, B. F. Woodfield, G. R. Stewart, R. A. Fisher and N. E. Phillips, *Cryogenics*, **43**, 369 (2003).
[45] C. A. Kennedy, M. Stancescu, R. A. Marriott and M. A. White, *Cryogenics*, **47**, 107 (2007).
[46] R. A. Marriott, M. Stancescu, C. A. Kennedy and M. A. White, *Rev. Scientific Instrum.*, **77**, 096108 (2006).
[47] E. Dachs and A. Benisek, *Cryogenics*, **51**, 460 (2011).
[48] P. Javorsky, F. Wastin, E. Colineau, J. Rebizant, P. Boulet and G. Stewart, *J. Nucl. Mater.*, **344**, 50 (2005).
[49] H. Suzuki, A. Inaba and C. Meingast, *Cryogenics*, **50**, 693 (2010).
[50] J. P. C. Ruff, J. P. Clancy, A. Bourque, M. A. White, M. Ramazanoglu, J. S. Gardner, Y. Qiu, J. R. D. Copley, M. B. Johnson, H. A. Dabkowska and B. D. Gaulin, *Phys. Rev. Lett.*, **101**, 147205 (2008).

[51] K. Morrison, K. G. Sandeman, L. F. Cohen, C. P. Sasso, V. Basso, A. Barcza, M. Katter, J. D. Moore, K. P. Skokov and O. Gutfleisch, *Int. J. Refrig.*, **35**, 1528 (2012).
[52] W. J. Parker, R. J. Jenkins, C. P. Butler and G. L. Abbott, *J. Appl. Phys.*, **32**, 1679 (1961).
[53] NIST Webbook: webbook.nist.gov (last accessed 28 October 2013).
[54] M. A. White, L. Qiu and G. Nolas, *Thermal Conductivity*, **27**, 72 (2004).
[55] L. Qiu and M. A. White, *J. Chem. Educ.*, **78**, 1076 (2001).
[56] R. G. Berman and T. H. Brown, *Contrib. Mineral. Petrol.*, **89**, 168 (1985).
[57] K. J. Miller, M. B. Johnson, M. A. White and B. Marinkovic, *Solid State Commun.*, **152**, 1748 (2012).
[58] V. Murashov and M. A. White, *Thermal Conductivity*, Academic/Plenum, New York, 2004.
[59] E. S. Toberer, L. L. Baranowski and C. Dames, *Annu. Rev. Mater. Res.*, **42**, 179 (2012).
[60] C. Wood, *Rep. Progress Phys*, **51**, 459 (1988).
[61] P. J. Price, *Philos. Mag.*, **46**, 1252 (1955).
[62] P. G. Klemens and M. Gell, *Mater. Sci. Eng. A*, **245**, 143 (1998).
[63] C. Kittel, *Introduction to Solid State Physics*, 7th edition, John Wiley, New York, 1996.
[64] A. L. Pope, B. L. Zawilski and T. M. Tritt, *Cryogenics*, **41**, 725 (2001).
[65] J. S. Tse and M. A. White, *J. Phys. Chem.*, **92**, 5006 (1988).
[66] C. A. Kennedy and M. A. White, *Solid State Commun.*, **134**, 271 (2005).
[67] R. R. Zarr, *ASHRAE Transactions 2001*, **107**, pt. 2 (2001).
[68] R. W. Powell, C. Y. Ho and P. E. Liley, *Thermal Conductivity of Selected Materials*, National Standard Reference Series – National Bureau of Standards: 8 (Category 5: Thermodynamic and Transport Properties), 1966.
[69] G. A. Slack, *Solid State Physics*, Academic Press, New York, 1979.
[70] American Society for Testing and Materials, ASTM C518–10, in *Annual Book of ASTM Standards, Vol. 4. 06*, West Conshohocken, PA.
[71] D. G. Cahill, *Rev. Sci. Instrum.*, **61**, 802 (1990).
[72] Thermal Transport Option for the Physical Property Measurement System, Quantum Design Corp., San Diego, CA (see also *Physical Property Measurement System: Hardware and Operation Manual*).
[73] J. Kopp and G. A. Slack, *Cryogenics*, **11**, 22 (1971).
[74] Purdue University, Thermophysical Properties Research Center, *Thermophysical Properties of Matter: The TPRC Data Series; A Comprehensive Compilation of Data, Vol. 1*, IFI/Plenum, New York, 1970.
[75] F. P. Incropera, D. P. DeWitt, T. L. Bergman and A. S. Lavine, *Introduction to Heat Transfer*, John Wiley, New Jersey, 2006.
[76] W. Hemminger and R. Jugel, *Int. J. Thermophys.*, **6**, 483 (1985).
[77] M. Li, H. Zhang and Y. Ju, *Rev. Sci. Instrum.*, **83**, 075106 (2012).
[78] ASTM C1044-12, '*Standard Practice for Using a Guarded-Hot-Plate Apparatus or Thin-Heater Apparatus in the Single-Sided Mode*', ASTM International; available at http://enterprise.astm.org/filtrexx40.cgi?+REDLINE_PAGES/C1044.htm (last accessed 28 October 2013).
[79] ASTM C 177–10: '*Standard Test Method for Steady-State Heat Flux Measurements and Thermal Transmission Properties by Means of the Guarded-Hot-Plate*

Apparatus,' in *Annual Book of ASTM Standards, Vol. 4. 06*, ASTM International, West Conshohocken, PA, 2012.
[80] B. M. Zawilski, R. T. Littleton and T. M. Tritt, *Rev. Sci. Instrum.*, **72**, 1770 (2001).
[81] L. Piraux, J.-P. Issi and P. Coopmans, *Measurement*, **5**, 2 (1987).
[82] M. B. Jakubinek, M. B. Johnson, M. A. White, C. Jayasinghe, G. Li, W. Cho, M. J. Schulz and V. Shanov, *Carbon*, **50**, 244 (2012).
[83] J. Pöhls, M. B. Johnson, M. A. White, R. Malik, B. Ruff, C. Jayasinghe, M. J. Schulz and V. Shanov, *Carbon*, **50**, 4175 (2012).
[84] O. Maldonado, *Cryogenics*, **32**, 908 (1992).
[85] N. R. Dilley, R. C. Black, L. Montes, A. Wilson and M. B. Simmonds, *Mat. Res. Soc. Symp. Proc.*, **691**, 85 (2001).
[86] P. G. Klemens, *High Temp.–High Press.*, **23**, 241 (1991) .
[87] F. I. Chu, R. E. Taylor and A. B. Donaldson., *J. Appl. Phys.*, **51**, 336 (1980).
[88] L. Vozár and W. Hohenauer, *High Temp.–High Press.*, **35/36**, 253 (2003/2004).
[89] F. Hemberger, H.-P. Ebert and J. Fricke, *Int. J. Thermophys.*, **28**, 1509 (2007).
[90] A. B. Donaldson and R. E. Taylor, *J. Appl. Phys.*, **46**, 4584 (1975).
[91] J. W. Vandersande and R. O. Pohl, *Rev. Sci. Instrum.*, **51**, 1964 (1980).
[92] J. Gembarovic and R. E. Taylor, *Rev. Sci. Instrum.*, **65**, 3535 (1994).
[93] T. Baba and A. Ono, *Meas. Sci. Technol.*, **12**, 2046 (2001).
[94] M. Li, Q. Xu and L. Wang, *Ceram. Int.*, **38**, 4357 (2012).
[95] S. E. Gustafsson, *Z. Naturf. A.*, **22**, 1005 (1967).
[96] E. E. Shpiel'rin, A. S. Umanskii and Y. A. Gorshkov, *High Temp.–High Pres.*, **7**, 361 (1975).
[97] S. E. Gustafsson, A. J. Hamdani and E. Karawacki, *J. Phys. E: Sci. Instrum.*, **12**, 387 (1979).
[98] S. E. Gustafsson, E. Karawacki and M. N. Khan, *J. Phys. D: Appl. Phys.*, **12**, 1411 (1979).
[99] U. Hammerschmidt and W. Sabuga, *Int. J. Thermophys.*, **21**, 217 (2000).
[100] O. Andersson and G. Johari, *J. Chem. Phys.*, **134**, 124903 (2011).
[101] B. E. Belkerk, M. A. Soussou, M. Carette, M. A. Djouadi and Y. Scudeller, *J. Phys. D: Appl. Phys.*, **45**, 295303 (2012).
[102] M. Okuda and S. Ohkubo, *Thin Solid Films*, **213**, 176 (1992).
[103] M. Okuda and S. Ohkubo. *Thin Solid Films*, **219**, 239 (1992).
[104] D. G. Cahill and R. O. Pohl, *Phys. Rev. B.*, **35**, 4067 (1987).
[105] S. N. Schiffres and J. A. Malen, *Rev. Sci. Instrum.*, **82**, 064903 (2011).
[106] C. P. Romao, K. J. Miller, C. A. Whitman, M. A. White and B. A. Marinkovic, in *Comprehensive Inorganic Chemistry II*, J. Reedijk and K. Poeppelmeier, Vol. 4, Elsevier, Oxford, p. 128–151 (2013).
[107] J. S. O. Evans, W. I. F. David and A. W. Sleight, *Acta Crystallogr., Sect. B: Struct. Sci.*, **55**, 333 (1999).
[108] H. M. Rietveld, *J. Appl. Crystallogr.*, **2**, 65 (1969).
[109] B. A. Hunter, *High Quality Mass Unit-Cell Refinement using Le Bail fitting and LHPM-Rietica*; available at http://www.ccp14.ac.uk/tutorial/lhpm-rietica/unitcell_refine_lebail.html (last accessed 28 October 2013).
[110] M. Dapiaggi and A. N. Fitch, *J. Appl. Crystallogr.*, **42**, 253 (2009).
[111] T. Sato, K. Ohasi, T. Sudoh, K. Haruna and H. Maeta, *Phys. Rev. B*, **65**, 092102 (2002).

[112] W. R. Buessem, in *Mechanical Properties of Engineering Ceramics*, W. W. Kreigel and H. Palmour, III, (Eds), Interscience, New York, 1961.
[113] J. S. O. Evans, T. A. Mary and A. W. Sleight, *J. Solid State Chem.*, **133**, 580 (1997).
[114] W. Sprengel, B. Oberdorfer, E.-M. Steyskal and R. Würschum, *J. Mater. Sci.*, **47**, 7921 (2012).
[115] K. J. Miller, C. P. Romao, M. Bieringer, B. A. Marinkovic, L. Prisco and M. A. White, *J. Amer. Ceram. Soc.*, **96**, 561 (2013).

3

Atomic Force Microscopy

Pablo Cubillas and Michael W. Anderson
Centre for Nanoporous Materials, School of Chemistry, The University of Manchester, Manchester, UK

3.1 INTRODUCTION

Atomic force microscopy (AFM) is a widely used technique in all fields of science and engineering, due to its ability to measure surfaces with sub-nanometre resolution and high accuracy. Atomic force microscopes have been used to observe individual atoms or even sub-atomic features in various types of material,[1−3] achieving greater resolution than the sister technique, scanning tunnelling microscopy (STM). Together, both techniques form the family known as scanning probe microscopy (SPM). The main advantage of AFM over STM is its ability to scan non-conductive samples in non-high-vacuum conditions, opening the door to the study of a great number of scientifically/technologically interesting samples, such as polymers, molecular and ionic crystals *etc.* In addition, AFM can be operated under all sorts of environmental conditions, including air, fluid, gas, vacuum and low (5 K) and high (1000 K) temperature.[4] The ability to scan in fluids has made AFM a frequently used technique in the biological and life sciences, being brought to bear in the study of a wide variety of samples, from living cells to DNA molecules.[5] Moreover, AFM is used widely to study the crystallisation and dissolution of all sorts of crystalline samples.[6]

Multi Length-Scale Characterisation, First Edition. Edited by Duncan W. Bruce, Dermot O'Hare and Richard I. Walton.

AFM operation can be divided into imaging (or topographic) and spectroscopy (force spectroscopy) modes. In imaging mode, a tip is scanned along the surface, producing three-dimensional (3D) images of surface structure or other dynamic properties, such as friction, elasticity, adhesion, chemical contrast, magnetic properties *etc.* In contrast, force spectroscopy mode measures interaction forces as a function of the tip–sample distance. This latter mode allows inter- and intramolecular interactions to be studied and has been used successfully in the study of polymers, living cells and interface phenomena.

3.2 HISTORY

The origin of AFM can be traced back to the development of the stylus profiler, first introduced in the 1920s.[7] Like AFM, this technique used a sharp tip attached to the end of a small bar to raster a surface, creating a height map. Development over the years included the introduction of an optical method to detect the motion of the tip as it scanned a surface,[7] which allowed for magnifications up to 1000×, and the use of a vibrating tip to minimise lateral forces and tip–sample contact.[8] This latter method improved the resolution of the device, but profilers nevertheless still suffer from unwanted tip–sample interactions as the probe touches the sample in an uncontrolled fashion.

In 1971, Young introduced the non-contact stylus profiler.[9] The operating principle of this instrument was based on the strong distance dependency of the electron field emission current generated between a sharp probe and a conductive sample. The probe was mounted on a piezoelectric material (commonly referred to simply as a 'piezo') that could move along x, y and z, with x and y being parallel to the sample surface and z perpendicular to it. Piezo movement along x and y was used to scan the sample and then a feedback control was used to keep the distance between probe and sample constant. Although representing an obvious improvement over the contact profiler, vibrations in the instrument limited its resolution.

In 1981, Binnig and Rohrer took these developments one step forward and created the scanning tunnelling microscope.[10] This instrument worked by monitoring the electron tunnelling current between a probe and a conducting sample at extremely small distances (0.3–1.0 nm). Tunnelling current is even more distance-dependant than field emissions as it varies exponentially, so only the very last atom of the tip will effectively interact with the sample. This fact gives the STM the ability to resolve individual atoms, as was demonstrated by Binnig *et al.* by resolving silicon atoms on a surface.[11,12] As in the stylus profiler, movement of the tip in x, y and z was carried out by means of piezoelectric elements and a feedback control was used to maintain a constant current by varying the tip–sample separation. Damping of vibrations to an acceptable level for atomic resolution was achieved initially by magnetic levitation of the device and later by suspending it from the ceiling with springs.

The main disadvantage of the STM is that it can only be used on conductive samples, which limits its application considerably. This fact led to the development of the AFM in 1986 by Binnig, Quate and Gerber.[13]

3.3 THE BASICS OF HOW AFM WORKS

AFM works by scanning a very sharp tip (tens of nanometres or less), which is attached to the end of a cantilever, over a surface with a very low force (on the order of nN). The deflection of the cantilever is then detected, recorded and used, through a feedback control mechanism, to drive a scanner and thus maintain the sample–tip force constant. A height image of the surface is then created from the movement of the scanner in the z-direction.[a] From these principles, it is easy to see that the most important parts in the construction and operation of an AFM are: the deflection detection method, the scanner, the feedback control mechanism and the tip or probe.

3.3.1 Instrument Architecture

The basic architecture of an AFM is shown in Figure 3.1. The main components are the scanner, the force sensor and the controller. In addition, commercial AFMs are normally supplied with sample stages and optical microscopes.

3.3.1.1 *Scanner*

This device provides movement on x and y (for sample scanning) and on z (for controlling the tip–sample distance). Normally, the scanner

[a] The z-direction is defined by convention as the vertical distance from the sample surface.

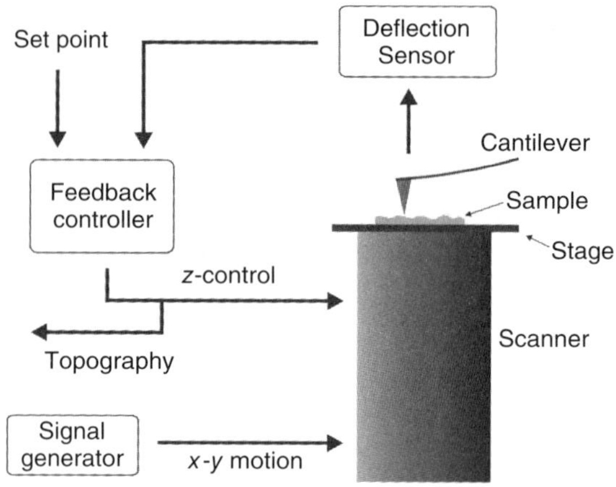

Figure 3.1 Schematic diagram showing the main parts of an atomic force microscope.

is composed of a ceramic piezoelectric material, which provides very accurate and precise control of movement when a particular voltage is applied. The scanner can have different configurations, with the most common being the tube scanner, the tripod scanner and the flexure scanner. Nowadays, the most common configuration is the tube scanner, which is used in more than 75% of commercial instruments. In spite of their widespread use, piezoelectric materials have a few disadvantages. The main one is that the displacement shows a non-linear behaviour with regard to the voltage applied, giving rise to hysteresis and creep, which translate into image artefacts (see Section 3.3.4.2). Non-linearities can be corrected by the so-called open- and closed-loop techniques. In the open-loop case, the scanner is calibrated using a standard, whereas in the closed-loop case, independent sensors are used to measure the piezo displacement and correct any deviation in real time. Another disadvantage of using piezoelectric tubes or tripods for scanners is that they have typical resonant frequencies of approximately 15 kHz[14] (although this is dependent on the size of the piezo element), which limits the maximum scan rate that can be achieved by a given scanner. Strategies for achieving higher scan rates vary from stacks of piezos to resonant scanners with tuning forks.[15,16] Depending on the type of microscope, the tip can be mounted on the scanner (tip-scanning AFM) or instead the sample can be mounted on the scanner (sample-scanning AFM). A signal generator that creates voltage ramps to define the position (scan area) of the scanner and its scan rate can be used to provide x–y motion of the tip or sample.

3.3.1.2 Deflection Detection Methods

In order for the AFM to achieve its high vertical resolution (<1 Å), lateral resolution cantilevers with very small spring constants (0.1–0.01 N m^{-1}) must be used so that very small tip–sample interactions can be measured (0.1 nN). This requires that sensors be able to detect very small displacements of the cantilever, making force sensors the most critical part of the AFM. In their original design, Binnig *et al.*[13] used a scanning tunnelling probe to measure the AFM probe displacement by measuring a tunnelling current between a tip and the cantilever. Since then, many other methods have been designed, including: capacitance detection,[17,18] piezoresistive detection,[19] crystal oscillation,[20] optical polarisation,[21] laser diode feedback[22] and optical beam detection.[23–26] The optical beam detection method, illustrated in Figure 3.2, is the most widely used in commercial systems. In this method, a laser beam is focused over the back of the cantilever (close to the tip position) and then reflected back into a mirror, which directs it onto a position-sensitive photodetector. The most common configuration of the photodetector is that of a four-segment photodiode. As the cantilever is deflected or torsioned during scanning, the position of the laser on the photodetector changes significantly. This can be monitored by measuring changes in the intensity signal reaching each quadrant of the detector. This geometry also allows decoupling changes in the laser position due to the vertical deflection or the lateral

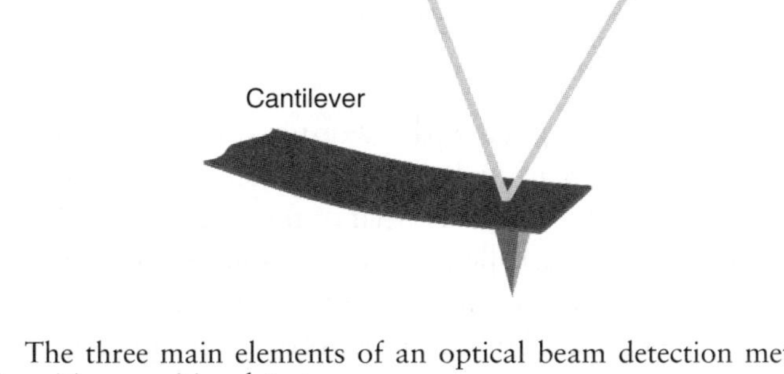

Figure 3.2 The three main elements of an optical beam detection method: laser, mirror and position-sensitive detector.

twisting of the cantilever. This is important as lateral twisting can be related directly to the friction between tip and sample (see Section 3.5.3).

3.3.1.3 Feedback Mechanism

As has been mentioned already, the success of the AFM as a high-resolution surface imaging method depends on its ability to keep a constant force between the tip and the sample. This is achieved by the use of a so-called proportional-integral-derivative (PID) feedback controller. This controller uses as input the signal coming out of the photodiode detector and compares it to a user-defined set point, equivalent to the tip–sample force, which must be kept constant. The difference between input and set point is the so-called error signal. The output signal of the feedback controller is then used to drive the piezo scanner in z in order to maintain a constant force (or deflection of the cantilever) and the value of this vertical displacement is used to create the height image. An output signal comprises three types of feedback: proportional, integral and derivative (not included in all commercial controllers). The proportional feedback, as its name implies, applies a correction that is directly proportional to the error signal. This correction factor works well in maintaining a constant value in linear systems but gives a residual error in non-linear systems. Therefore, one or two additional feedback mechanisms are used. The integral feedback monitors how the error signal changes with time (integrates the signal), *i.e.* to a 'cumulative' error, and applies a correction factor proportional to it. Finally, the derivative feedback looks at the rate of change of the error signal with time (the derivative). The proportional factors applied in each case are called gains. These are user-defined, as they will vary according to the scanning conditions, such as scan speed and set point.

3.3.1.4 Sample Stage

Almost all AFMs are equipped with a sample stage that allows for coarse movement along x and y in order to reposition the sample at regions of interest over the working scanning area of the microscope. In some cases, this movement can be controlled from a computer. Sample stages can be made simply to hold the sample in air and at room temperature, or can be equipped with heating/cooling elements or be designed to hold liquids. Fluid cells can be open to the atmosphere or closed and can normally be connected to fluid pumps (syringe or other types) so that liquids can be exchanged during AFM operation.

3.3.1.5 *Optical Microscope*

Most commercial AFMs can be equipped with an optical microscope. The type and geometry of optical microscope used depend on the type of AFM to which the microscope is coupled. Top-down microscopes are used for sample-scanning AFMs, but their total magnification is limited by the fact that there must be enough space between the microscope and the sample to allow the force sensor to be located, limiting their use to sample location and laser alignment. In the case of tip-scanning AFMs, an inverted optical microscope can be used. This allows working distances of a few mm to be achieved and therefore enables higher magnification. This kind of geometry is particularly useful for locating small crystal facets in crystal growth studies. It is also useful for biological applications, since very small cells can be located and additional optical microscopes, such as those working in fluorescence and confocal modes, can be integrated.

3.3.2 Basic Scanning Modes

As stated in the introduction, an AFM can be used in essentially two modes – imaging (or topographic) and non-topographic – each containing several sub-modes. More specialised modes will be described in more detail in Section 3.5, but the three basic modes of operation from which the others derive are introduced here.

3.3.2.1 *Forces in AFM*

When the AFM tip or probe approaches a surface, it is subjected to various forces, which produce a deflection of the cantilever. The most relevant forces when performing experiments in air are long-range, van der Waals interactions, short-range repulsive interactions, adhesion and capillary forces. In liquids, additional forces appear at the solid–liquid interface, such as solvation and electric double-layer interactions. The character of these forces can be repulsive or attractive, or can change according to the chemical nature of the liquid (ionic strength) or due to tip functionalisation. Figure 3.3 shows a so-called force–distance curve, where the force (obtained from the output voltage of the photodetector) is plotted as a function of the vertical distance between tip and sample. At long distances (hundreds of Å), attractive van der Waals interactions dominate and the tip is pulled towards the surface, bending the cantilever and therefore registering a change in the voltage (or force). As the distance is reduced, short-range repulsive forces take over (below 1 nm distance) and force the cantilever to bend in the opposite way. In this regime, sample compression will also occur, depending on the relative compliances between sample and tip. The more the cantilever is pushed against the sample, the higher the bending and therefore the higher the force applied. If the cantilever is then pulled away from the surface (dashed line), the force will decrease until the attractive regime is reached again. If capillary forces are present (due to water meniscus formation), the tip will be kept in contact with the surface across a longer distance until finally it 'snaps' back to its original 'relaxed' position.

According to the force–distance profile, three different scanning modes can be envisioned: contact, intermittent contact and non-contact (Figure 3.3).

3.3.2.2 Contact Mode

As the name indicates, in this mode the AFM operates in the repulsive regime and the tip is in direct contact with the surface. This was the

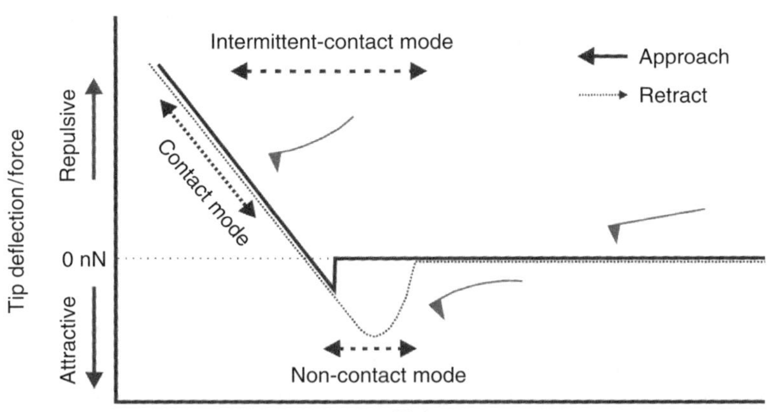

Figure 3.3 Ideal force–distance curve, showing the effect of repulsive and attractive forces on the tip deflection/force during approach and retraction. The operating regimes of the different scanning modes are also shown.

first mode developed for AFM, as it is the simplest both conceptually and from an operation point of view (feedback control, types of cantilever *etc.*). For these reasons, it is still one of the most heavily used scanning modes. Contact mode can be used in the constant deflection mode, where the feedback loop is active and is, therefore, trying to maintain a constant force by moving the sample or tip in the *z*-direction by means of the scanner. It can also be used in the constant-height mode, where the feedback loop is disengaged and the scanner therefore does not retract or extend. This mode has been used habitually for high-speed AFM applications. In contact mode, the total force exerted by the tip over the sample can be calculated easily using Hooke's law, *i.e.* by multiplying the vertical displacement (which can be calculated using the slope of the force–distance curve) by the vertical spring constant of the cantilever (see Section 3.3.3.4).

Although it is relatively easy to use and understand, contact mode has several drawbacks. Since the tip and sample are in contact and subject to a load, both can be damaged or otherwise modified during the scanning process; this issue can be minimised by imaging in liquid, where meniscus forces will not appear and a finer control on the load is therefore possible. Another problem is that lateral forces will develop as a result of the constant contact during scanning, which may produce more sample damage or generate artefacts. On the other hand, the study of these lateral forces can produce important information concerning physical properties and surface phenomena (see Section 3.5.3).

Regarding its application, contact mode has been used for submolecular resolution studies, although dynamic modes have recently taken over as the preferred tool for achieving real atomic resolution (see Section 3.4.5). It is also normally the mode of choice for applications in liquid and when dealing with hard samples (crystals, minerals, semiconductors *etc.*), although it has been shown to be useful when scanning softer samples such as polymers and biological samples.

3.3.2.3 Oscillating Modes

Oscillating modes were developed to overcome some of the main drawbacks of contact mode, namely the appearance of lateral forces and high loads over the sample. The benefits of oscillating the probe over the surface while scanning were first acknowledged by Binnig and Quate[13] in their original paper, but were not implemented fully in AFM designs until several years later,[27] mainly as a result of technical difficulties. Oscillating modes work by applying an external oscillation

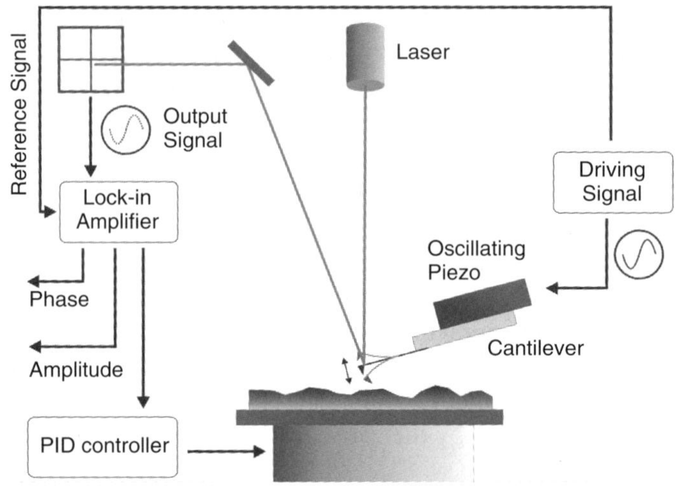

Figure 3.4 Schematic diagram showing the components utilised when AFM is operated in oscillating mode.

on the tip (close to its resonant frequency) by means of a small piezo element located at the cantilever holder (Figure 3.4) and then detecting the change in either the oscillation amplitude, the frequency or the phase produced as the tip approaches the sample (damping effect). These changes are recorded using the photodiode detector, analysed using a lock-in amplifier and then used in a feedback loop to adjust the *z*-position of either the tip or the sample to obtain a height image. Depending on the magnitude of the oscillation amplitude, two different methods can be differentiated: intermittent contact and non-contact.

3.3.2.3.1 Intermittent Contact Mode

In this mode, also known as tapping mode, the tip is oscillated with an amplitude large enough (1–100 nm) to move it from a non-force regime (far away from the sample), through the attractive zone and into the repulsive regime (see Figure 3.3). This pattern of movement implies that a brief contact with the sample will be produced, but effectively it eliminates the appearance of lateral forces while scanning. The tip will also be subjected to capillary forces when scanning in air (due to meniscus formation), which means in practice that cantilevers with higher stiffness or spring constants (tens of Nm^{-1}) than those used for contact mode must be used. Furthermore, the fact that the cantilever can 'pass through' the contamination layer makes it a more stable technique than its non-contact counterpart when scanning in air, and it is therefore more widely used under these conditions. In this mode, changes in the oscillation of the cantilever are normally detected by measuring the variation in the amplitude, and this is used as the feedback input signal. It is also common to detect the phase shift, but this is rarely used in the feedback loop.

The advantages of using intermittent contact mode include increased lateral resolution and reduced damage due to load, making the method easier to use with softer samples. As noted above, it is also more heavily used than the non-contact mode when scanning in air, as well as when scanning larger scan sizes and samples with high roughness, since the oscillating amplitude is larger and accommodates greater variations in height more easily. This technique is also used in operation in fluids, although this is more challenging because of modification of the cantilever resonant frequencies, dampening of the oscillation amplitude and the appearance of spurious 'resonant peaks' due to liquid vibration. Nonetheless, in recent years it has seen increased use in the biological application field.[5,28] Finally, since a brief tip–sample contact is established, this technique also offers the possibility of probing mechanical properties (see Section 3.5.1).

3.3.2.3.2 Non-Contact Mode

In this mode of operation, the tip usually oscillates with an amplitude of the order of 10 nm, meaning that it is affected only by long-range van der Waals and/or electrostatic forces (see Figure 3.3). As it is operated within the reach of the capillary forces (when scanning in air), cantilevers with very high stiffness and high resonance frequencies (300–400 kHz) are used in order to prevent the tip from snapping into the sample. In this operating mode, phase shift is normally the preferred method for detecting changes in the cantilever oscillation behaviour, although in some cases frequency changes are detected and used as a feedback signal, giving rise to the so-called frequency-modulation mode (see Section 3.4.3).

This method of operation can be used for any application, although operation in air is rather challenging due to the presence of capillary forces from the layer of adsorbed water. There are, however, several benefits. Since there is no sample–tip interaction and lateral forces are essentially non-existent, this method has the potential to achieve very high resolution, and in fact in a frequency-modulated version, it has been shown to produce higher resolution than STM[1] (see Section 3.4.5). It is also used widely when imaging soft samples in both air and fluid, and has the advantage of producing almost no tip wear, increasing the quality of the output data.

3.3.2.4 Force Spectroscopy

In this mode of operation, the probe does not move along the sample surface (therefore, no image of the topography is obtained) but moves instead along the *z*-axis (away from or towards the sample), and changes in deflection or the dynamics of oscillation (phase, frequency and amplitude) of the cantilever are measured, *i.e.* force–distance curves are obtained. Since the tip used in AFM is so small (tenths to a few nanometres in diameter), very small areas of the sample and, more importantly, very weak forces can be probed in this way. This technique allows information to be obtained concerning the mechanical properties of surfaces (such as adhesion and elasticity) and enables the probing of single molecule–sample or molecule–molecule interactions by means of tip functionalisation (see Section 3.4.1). In the latter case, the technique is referred as **chemical force microscopy**.

In reality, force–distance curves are more complex than that depicted in Figure 3.3. For example, in the absence of capillary forces (*e.g.* when the experiment is performed in a solution), the snap-on part of the curve may not easily be determined. In the compression part of the curve, the change in force as a function of distance will not be linear, except for very hard surfaces (Figure 3.5). In soft samples, this part of the curve will be very hard to quantify, as an increase in the force will lead to increases in the sample–tip contact area. Another feature of complex force–distance curves is the presence of two or more adhesion episodes

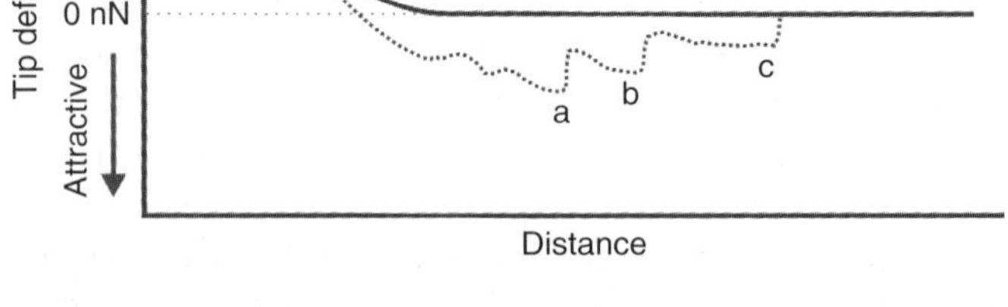

Figure 3.5 Real force–distance curve, showing the non-linear behaviour of the approach curve as the tip reaches the repulsive regime. The retract curve shows a complex behaviour, with multiple adhesion episodes.

when retracting the tip (Figure 3.5). This may correspond to different molecular contacts/interactions being overcome at different times and force levers.

An important issue with force spectroscopy is its reproducibility, as even measurements made in the same region or at the same spot on the surface can be different. So while data acquisition is fast, data processing can be slow and complex. The technique of conducting multiple force–distance curves over many different positions on a surface is normally called **force volume imaging** (see Section 3.5.4).

As noted above, force spectroscopy is used widely in nanomechanical studies, *e.g.* in determining stiffness (Young's modulus), viscoelastic response and adhesion.[29] Moreover, this technique has found its greatest application in the life sciences. A detailed account of studies in this field is beyond the scope of this chapter, but a few of the most common applications are: the study of molecular stretching (such as DNA) and protein unfolding;[5] studies of receptor–ligand-pair interactions (*e.g.* antibodies and antigens);[30] the study of the viscoelastic behaviour of cells under different physiological conditions or under the influence of specific drugs;[31] and the study of adhesion mapping over cell surfaces to determine receptor distribution.[32]

3.3.3 Cantilevers and Tips

The tip or probe is one of the more crucial parts in an AFM. This is because an AFM image is essentially the convolution of the tip–sample contact and so the sharpness of the tip will have a tremendous effect on the lateral resolution of the instrument (Figure 3.6). The specific design of the AFM probe will depend on the type of deflection detection method, but since the optical beam method is the most widely used, we will focus on cantilevers designed for this. In the optical beam method, the force sensor requires a sharp tip (with a typical radius between 10 and 30 nm) attached to the end of a long cantilever (tenths to hundredths of microns), itself attached to a chip ($3.5 \times 1.6\,mm^2$) or support that is attached to the cantilever holder (Figure 3.7). In addition, the cantilever is tilted 15° with respect to the horizontal to ensure that, when mounted on the AFM, the tip is the lowest part. Other requirements for AFM cantilevers are a high resonant frequency (so that the scan speed is not limited by it), low spring constant (to detect small forces) and high lateral spring constants (so lateral force effects are minimised). There are essentially two shapes

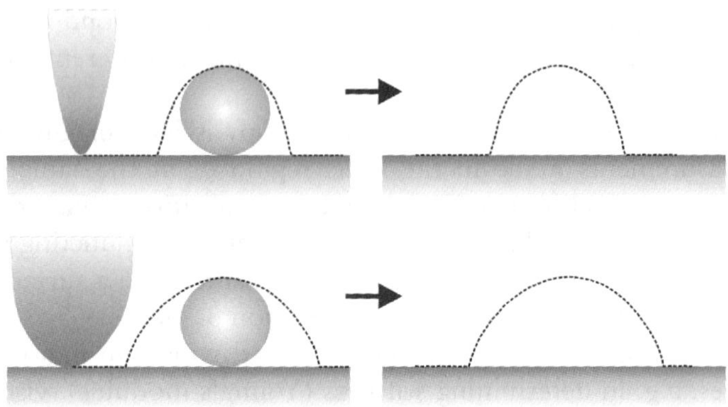

Figure 3.6 Two examples showing tip–sample convolution and the effect of tip sharpness on imaging. Small features will appear much wider than they are in reality when scanning with a blunt tip.

a cantilever may take: beam-like and V-shaped (Figure 3.7). The final shape, width and length of the cantilevers will vary according to a specific set of specifications, including spring constant, resonant frequency *etc.*

3.3.3.1 Materials

Initially cantilevers were fabricated manually from metal foils or small wires. For tips, either diamond crystals were attached to the ends of the foil cantilevers or the wires were etched to a sharp point electrochemically. These methods of probe fabrication were, however, unreliable and time consuming. Today, cantilevers are fabricated using microelectromechanical systems (MEMS) technologies, such as chemical vapour deposition (CVD) and micromachining, and can be made of different materials[4,33,34] depending on the type of application for which they are intended, including diamond, quartz forks, highly doped Si, tungsten, nickel and other metals. Normally, cantilevers are made of Si_3N_4 and Si. These two materials are chosen for their high yield strength and elastic modulus,[35–37] which translate into very small residual stress. Si_3N_4 offers the advantage of allowing very thin cantilevers to be made from it, so that it is the material of choice for low-spring-constant probes. On the other hand, Si_3N_4 tips are of lower quality than those made of Si. This fact has led to the appearance on the market of probes that combine an Si_3N_4 cantilever with a sharper Si tip. An issue with

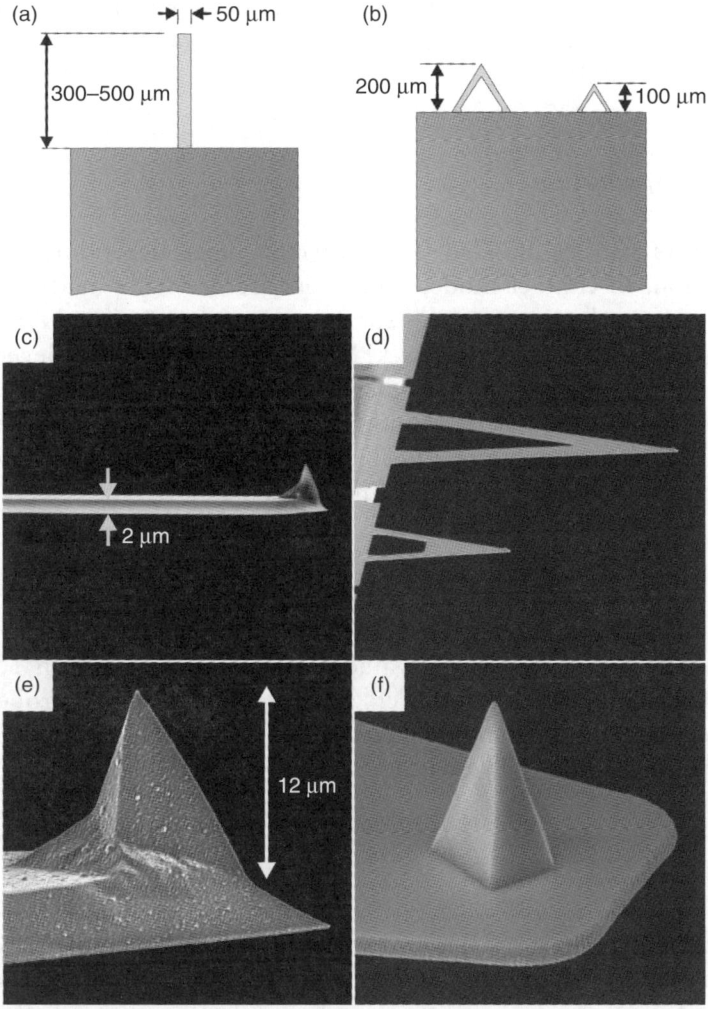

Figure 3.7 (a,b) Schematic diagrams and (c,d) SEM micrographs of rectangular (a,c) and triangular (b,d) cantilevers, showing their geometries and dimensions. (e,f) SEM micrographs showing the tip geometry in detail for rectangular (e) and triangular (f) cantilevers, reproduced with permission from Bruker corporation.

these two materials, and especially with Si, is that they are not very reflective to laser light, so manufacturers offer the option of coating the top of the cantilever (and sometimes the bottom as well) with metals, such as Au and Al. There can be other reasons for using coatings, for example to reduce the effect of fluids on tip morphology, to enhance the tip conductivity (for STM, electric force microscopy (EFM) and other applications) and to magnetise the probe (for magnetic force microscopy (MFM) and STM).

3.3.3.2 Application

Cantilevers can be divided into different types depending on their application. In contact mode, cantilevers are normally made of Si_3N_4 or Si and must have very small spring constants ($1.0–0.01\,N\,m^{-1}$) and low resonant frequencies (a few kHz). By comparison, cantilevers used in oscillating modes require higher spring constants ($>10\,N\,m^{-1}$) and higher resonant frequencies (hundreds of kHz). Another requirement is to have a high quality factor, Q, which is a measure of the hydrodynamic damping due to the environment and is calculated by:

$$Q = \frac{\omega_0}{(c/m)} \tag{3.1}$$

where ω_0 is the resonant frequency, c is the damping factor and m is the mass of the cantilever. The Q value normally varies between 100 and 1000. Oscillating-mode cantilevers are usually made of Si. Probes used in nanoscratching or nanoindentation (see Section 3.5.9) use diamond tips[38,39] or diamond-like coating[40] (Figure 3.8a).

3.3.3.3 Tip Shape

As noted above, tip shape is a critical factor in controlling the lateral resolution of an AFM and a great deal of effort has therefore been invested in manufacturing tips with very small radii (<10 nm) and high aspect

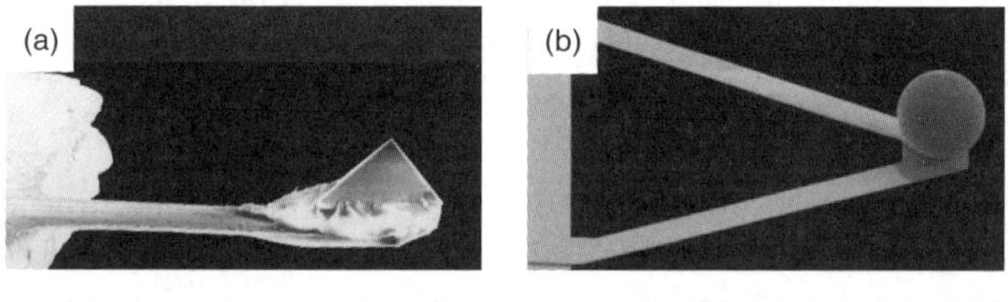

Figure 3.8 SEM micrographs showing detail of (a) a diamond tip, reproduced with permission from Givargizov *et al.* (1998) [40]. Copyright (1998) Elsevier, and (b) a colloid Probe, reproduced with permission from Hilal *et al.* (2009) [58]. Copyright (2009) Butterworth-Heinemann.

Figure 3.9 SEM micrographs of sharp tips produced by (a) oxide sharpening, reprinted with permission from [41] Copyright (1990) American Instititute of Physics. (b) focused ion-beam milling, reprinted with permission from [43] Copyright (1992) Elsevier Ltd., and (c) electron-beam deposition, reprinted with permission from [44] Copyright (1993) Institute of Physics.

ratios.[4,33,34] A common approach has been to use a post-fabrication process to enhance the sharpness of the tip (Figure 3.9). Such processes include:

- *Oxide sharpening*: This method can be applied to both Si_3N_4 and Si tips, although it yields better results with Si. In this latter case, Si is oxidised thermally to SiO_2 and the oxidation rate is decreased due to $Si{-}SiO_2$ stress in regions with higher curvature, leading to the formation of sharper tips. Subsequent etching of the SiO_2 with hydrogen fluoride results in the formation of high-aspect-ratio (HAR, 1 : 10) tips, with tip radii as low as 1 nm (Figure 3.9a).[41]
- *Focused ion-beam milling*: This technique is used to manufacture tips with high aspect ratios and is normally applied to silicon.[42,43] The resultant tips have a 3° angle at the walls of the pyramid (as opposed to 20–35° for conventional tips), which translates to a 1 : 10 aspect ratio (Figure 3.9b). Another advantage of this technique is that the tips can be milled in such a way that it they are perpendicular to the scanning surface, in contrast to normal tips, which are tilted due to the requirement that the cantilever be tilted by 15° (Figure 3.7e).
- *Electron-beam deposition*: In this technique, the cantilever is placed inside an SEM in such a way that the tip axis is parallel to the electron beam axis. In this arrangement, carbon derived from the dissociation of gases within the SEM vacuum chamber is deposited over the tip (Figure 3.9c).[44]

Another widely used method for producing sharp tips is to attach a carbon nanotube (CNT) to the end of the probe. This is treated in more detail in Section 3.4.2.

3.3.3.4 Spring Constants

In some AFM applications, such as force spectroscopy or friction force microscopy (FFM), knowing the exact force that is being applied to the sample is crucial. This can easily be calculated if the stiffness or vertical spring constant of the cantilever is known. Unfortunately, company-supplied values of batch microfabricated cantilevers can vary significantly as a result of small differences in cantilever thickness or the presence of defects. For these reasons, various methods have been proposed for the calculation or measurement of the vertical spring constant (as well as the lateral/torsional):

- *Geometric methods*: These use the measured dimensions of the cantilevers to calculate the spring constant. In the case of a rectangular cantilever, the spring constant can be calculated by means of a very simple equation[45]:

$$k_v = \frac{Et^3w}{4l^3} \tag{3.2}$$

  where k_v is the vertical spring constant, E is the Young's modulus, t is the thickness of the cantilever, w is the width and l is the length. For triangular-shaped cantilevers, the situation is more complex, and several authors have proposed formulae derived from different methods.[46] The main methods include the parallel-beam approximation, finite-element analysis and complex models that divide the cantilever into various shapes. Of these, one of the most widely used is the so-called Sader method,[47–49] which makes use of the measured physical geometry of the cantilever, the measured quality factor and the resonance frequency.
- *Added-mass method*: Also known as the Cleveland method, this relies on the change in resonant frequency experienced by a cantilever when a small mass is added to it.[50] Although reliable and non-destructive, it is still quite a challenging method as it requires very well characterised masses that can be added manually to, and removed from, the cantilever without breaking it.
- *Reference cantilever method*: This consists in pressing the cantilever to be investigated against a well-defined reference cantilever. Although easier to implement than the Cleveland method, it still requires the use of a well-defined cantilever and it can induce tip wear.[51]
- *Thermal noise method*: This relies on measuring the noise spectrum of the cantilever.[52] It uses the equipartition theorem to relate the

energy of the system due to thermal fluctuation ($\frac{1}{2}k_B T$, where k_B is the Boltzmann constant and T the temperature) to the measured energy resulting from the mean displacement of the cantilever ($\frac{1}{2}k_v <z>$, where $<z>$ is the mean cantilever displacement). Although reliable, this method requires application of correction values[53] and the use of a high-performance spectral analyser. Nevertheless, due to the ease of operation, it is one of the most widely used methods for determining spring constants and is included as standard in more recent AFM instruments.

3.3.3.5 Small Cantilevers

It has long been recognised that smaller cantilevers (a few tenths of a micron in length) can be very useful for certain applications. For example, a small cantilever will have a higher resonance frequency and so can be used at higher scan speeds.[54,55] Small cantilevers will also be better for force spectroscopy applications as they will have a smaller noise floor (since they will be less perturbed by the surrounding medium), leading to higher signal-to-noise ratios (SNRs).[56] Recently, it has also been shown that small cantilevers can be very useful in obtaining high-resolution images.[57] The widespread use of small cantilevers was very constrained until recent times, due to the lack of commercial manufacture. Now, new commercial AFM designs[57] that are optimised for the use of small cantilevers have appeared on the market, spurring probe manufacturers to commercialise them.

3.3.3.6 Colloid Probe

One way of obtaining probes with well-defined geometries is to substitute the protruding tip in the cantilever with a spherical colloid, which is normally referred to as a **colloid probe**. This kind of probe has many applications in the study of the mechanical properties of surfaces arising from their well-defined geometry, which allows for the application of continuum mechanical models by which to analyse the experimental data. Colloid probes are normally prepared by gluing a spherical colloid into a tipless cantilever by means of a micromanipulator,[58] although sintering can also be used (Figure 3.8b). The probe materials normally used are: silica,[59] polyethylene,[60] polystyrene[33] and gold.[33]

3.3.4 Image Artefacts

An AFM image is, in principle, a representation of the surface of a given material. Therefore, any feature of the image that does not correspond to the real surface can be considered an artefact. As will be shown later, there are many different types of artefact in AFM, some of them avoidable and some not. It is the task of a good microscopist to be able to interpret AFM images accurately, taking these artefacts into account. The origin of these artefacts can be categorised as follows:

3.3.4.1 Tip Artefacts

As has been noted before, an AFM image is a convolution of the probe geometry. Therefore, in order to obtain accurate images of a surface, the tip radius should be at least as sharp as the smallest surface feature. Nevertheless, tips can often become blunt or contaminated, giving rise to artefacts.

- *Tip shape issues*: The shape of an AFM probe can be defined by two parameters: its radius, which defines the end of the tip, and the angle cone, which characterises its overall sharpness or aspect ratio. For very small features (a few nanometres), the tip radius will play an important role. Blunt tips, therefore, will have a big effect on the lateral dimensions of small objects, amplifying their width significantly, which may create issues when they are close together (Figure 3.10).

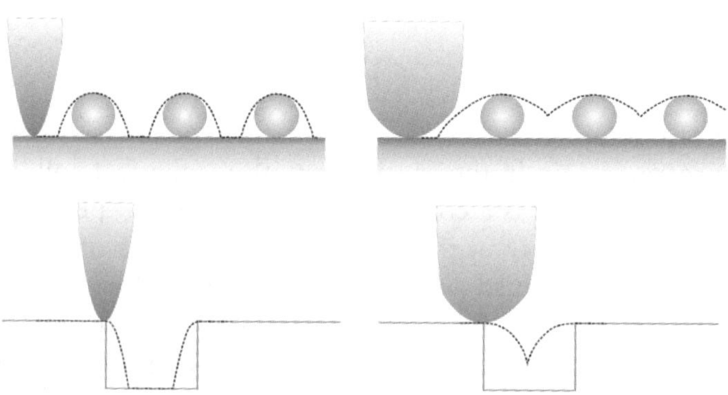

Figure 3.10 Schematic diagrams showing how a blunt tip will be unable to probe a surface decorated with small features or small holes/drops.

A blunt tip will also minimise the presence of small holes or drops in a surface (Figure 3.10). For larger features, it will be the aspect ratio of the tip that will have the most significant effect. In this case, the corners of the object will actually be imaging the side walls of the probe, producing artefact-dominated figures (Figure 3.11a). In some cases, the width of the cone tip will not allow the tip to probe depressions in a sample. HAR tips will help minimise this effect, although for some samples with extremely sharp features they may not be good enough. A good example of the increase in image quality with an HAR tip is given in Figure 3.11, which shows AFM images of an aluminium oxide membrane imaged with a conventional probe (Figure 3.11c) and with a HAR tip (Figure 3.11d).

- *Damaged/contaminated tips*: Sometimes tips can become contaminated or broken during (or prior to) scanning. Contamination is

Figure 3.11 (a) AFM deflection image, showing a zeolite A crystal. Only the centre square reflects the true topography of the crystal: the triangular, streaked features are the sides of the tip, 'scanned' by the edges of the tip. (b) High-resolution SEM image of a similar zeolite A crystal to that imaged in (a). The cubic shape of the crystal is apparent from this image, in contrast to the morphology show in image (a). (c) AFM height image of an aluminium oxide membrane taken with a conventional Si tip in intermittent contact mode. (d) AFM height image of the same sample, this time taken with an HAR tip in intermittent contact mode. The increase in resolution is evident.

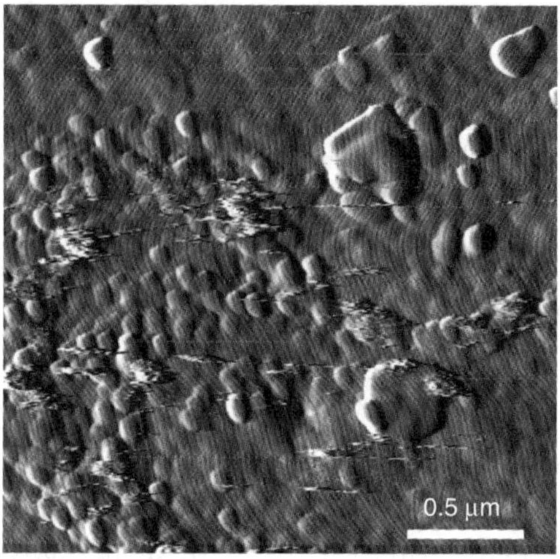

Figure 3.12 AFM error image of the surface of a ZSM-5 zeolite crystal. Triangular-shaped features over the surface were created by a blunt tip. Streaks across the image are due to contaminants being displaced by the tip during scanning.

especially common with biological samples or when scanning synthetic inorganic materials in air, where debris left from the synthesis can still be attached loosely to the sample. Broken and contaminated tips normally lead to the same types of artefact, which tend to be characterised by the presence of unexpected shapes that sometimes repeat themselves over the image (Figure 3.12). One way of determining whether certain features are real is to rotate the sample – if the features change orientation then they are real. On occasion, broken or faulty probes result in two tips, which will produce a repetition or doubling of most (if not all) of the small features in an image; they are, therefore, identified quite easily. Contamination issues can sometimes be solved by cleaning the tip using acid solutions[61,62] or by means of ultraviolet (UV) radiation.[63]

3.3.4.2 Scanner Artefacts

As noted above, scanners made of piezoelectric materials suffer from hysteresis and non-linearity when expanding/contracting. This behaviour will introduce artefacts into the images produced, although these can be reduced by using open-/closed-loop techniques (see Section 3.3.1.1). The geometry of the particular scanner will also play a part in determining the type of artefact observed. Some of the most typical artefacts found in tube scanners will be discussed here, as they are used mostly in commercial AFM instruments.

- *X–Y calibration/linearity*: Scanners in which the non-linear behaviour is not corrected in the x- and y-axes will produce images with features that appear contracted or expanded in different places (Figure 3.13). This can be corrected in real time by the use of a sensor, or off-line by the use of calibration standards. In addition, the calibration will allow the AFM to produce images with accurate measurements on the xy-plane.
- *Z calibration/linearity*: The displacement of the piezo scanner in the z-direction is also non-linear, which leads to incorrect height measurements in the images produced. This behaviour can be corrected by using sensors (linearised piezo) or calibration standards. The standards should have heights of the same order as those being investigated otherwise errors can still be produced in the measurements.
- *Scanner bow*: Due to their geometry, and with one side free and the other attached, tube scanners move the probe or sample in a curved motion (Figure 3.14). As a result, the height image displays a curvature or bow and z-measurements will vary as a function of image position. This artefact can be corrected only by post-scanning data analysis.
- *Scanner drift*: The phenomenon of creep in piezoelectric scanners is produced when the displacement does not follow the applied voltage but instead continues after the voltage has stopped (Figure 3.15).

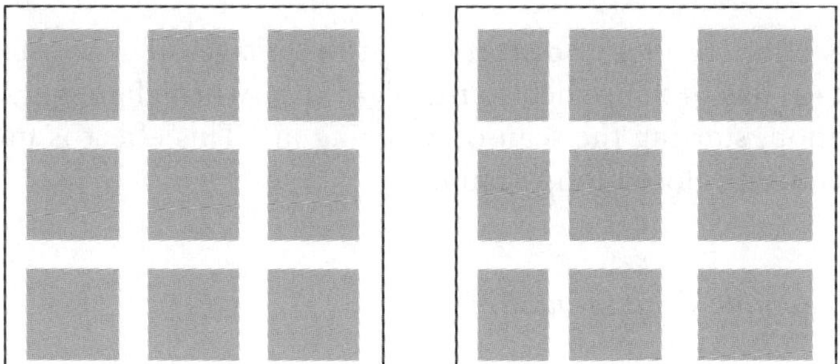

Figure 3.13 Scheme showing the effect of non-linear displacement in xy. The left-hand image shows the real sample, while the right-hand image shows how the square features become squashed or elongated when imaged with a non-calibrated scanner.

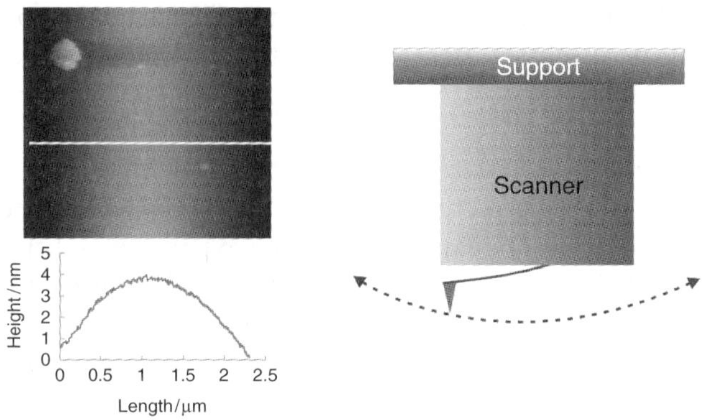

Figure 3.14 AFM height image of a crystal of MOF-5 showing the characteristic bow (shown in cross-section) caused by curved scanner motion (right).

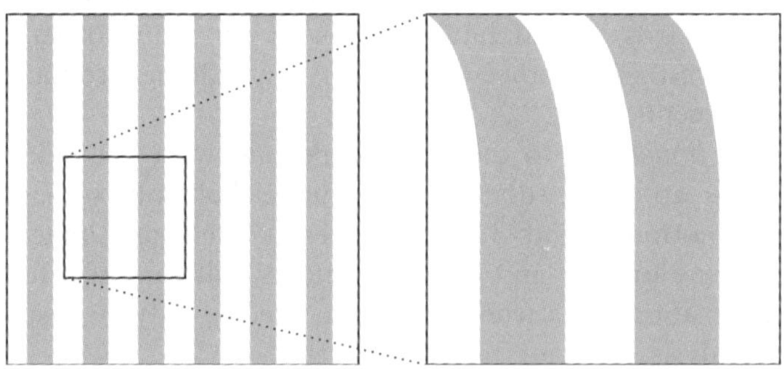

Figure 3.15 Scheme showing the effect of scanner drift when zooming in over a region of interest. This effects makes linear features (left-hand image) appear curved after zooming in. The effect tends to disappear over time, so features scanned later will show the correct shape (bottom of right-hand image).

This effect results in distortions in the image after a fast pulse of voltage has been applied to the piezo (*e.g.* when changing scanning position, starting the scan or zooming in). This effect is minimised when using closed-loop scanners.

3.3.4.3 Sample Contamination

Image artefacts can be created when contaminants/debris present on the scanned surface are moved around by the tip during scanning (see Figure 3.12). These artefacts are easily identified as they appear as white streaks over the surface that follow the tip displacement along the slow scan axis. Samples can be cleaned by means of ultrasonication, UV-light irradiation (especially good for organic contaminants) or the use of solvents.

3.3.4.4 Other Artefacts

Many other artefacts can appear in recorded images, originating from various sources, including: acoustic and mechanical noise, bad setting on the PID feedback control, sample drift due to thermal expansion, laser interferences, image processing *etc*. The reader is referred to reference [4] for a more detailed review of this topic.

3.3.5 Scanning Environment

One of the main advantages of AFM is its ability to perform observations in liquid, gas and vacuum environments. Scanning in liquids is very common and sometimes even beneficial, as adhesive forces resulting from meniscus formation are removed, allowing for a better control of the tip–sample forces and the possibility of scanning with very low forces. Operation must be carried out with great care as any spill of liquid into the AFM electronic or piezoscanner could damage the system severely. Currently, most commercial AFMs are equipped with either closed or open fluid cells, the former of which minimise the risk of spills and allow fluid to be replenished or circulated through the use of syringe and peristaltic pumps.

AFM experiments have been performed over a wide range of temperatures (5–1000 K), but when dealing with high temperatures, care must be taken to isolate the piezoelectric scanner, which can be done by using ceramics with very low thermal conductivity. Low-temperature experiments can induce condensation of water into electronic components, so they are normally performed under vacuum conditions.[64]

The combination of a fluid environment with temperature control was absent from commercial systems until very recently, which led to the development of custom-built systems that combined these capabilities, the most advanced example being the hydrothermal AFM (HAFM; see Section 3.4.6). In the last few years, several manufacturers have introduced various types of temperature-controlled, commercial fluid cells, but they are still limited to temperatures below 100 °C.

3.3.6 Sample Preparation

In most cases, sample preparation for AFM is very simple, as it does not require coatings or thin sections, unlike electron microscopy. In other cases, *e.g.* when imaging cells or nanoparticles or trying to achieve atomic resolution, sample preparation can be complex. In any case, a few common principles apply. Since it is a scanning technique in which there is usually some type of tip–sample contact, samples must be kept clean or attached to a surface (in the case of small particles). In addition, surface features must be smaller than the total available z-displacement for the piezoscanner. Finally, samples must be fixed to a sample holder in order to prevent movement and distorted images. Depending on the AFM geometry, sample size can also be restricted.

When dealing with very small particles, cells or DNA/proteins, very flat surfaces must be used, which can include mica, highly orientated pyrolytic graphite, glass or gold. These surfaces can be functionalised with self-assembled monolayers (SAMs) that hold the sample in place through the formation of chemical bonds. In other cases, glues or waxes can be used to fix the samples.

3.4 IMPORTANT DEVELOPMENTS IN AFM

3.4.1 Tip Functionalisation/Chemical Force Microscopy

Chemical force microscopy measurements are based on the interaction between two molecules, one attached to the tip and the other bound to a surface of interest. This allows specific interactions to be probed, *e.g.* between an antibody and an antigen. Depending on the type of study to be performed, the AFM tip can be functionalised in different ways. Normally this will require one or more steps, with each achieving some degree of functionalisation. Even before the functionalisation is begun, however, cantilevers must be selected that have a spring constant smaller than the interaction forces to be probed (typically $<0.1\,\mathrm{N\,m^{-1}}$), and care must be taken that the force attaching the ligand to the tip is stronger than the ligand–surface interaction.

The first step always consists in cleaning the tip, as oils and other residues from the manufacturing process may be present. This can be done by utilising the so-called 'piranha solution', a mixture of H_2SO_4

and H_2O_2 (70 : 30), although care must be exercised in applying such a mixture. The second step normally involves either direct amination by silanisation or esterification, or amination *via* the formation of thiol-based SAMs on gold-coated tips. The most common silanisation treatment is the use of the organosilane 3-aminopropytriethoxysilane (APTES).[65] This molecule reacts by forming strong Si–O–Si covalent bonds with the tip. Esterification reactions normally imply the reaction of the surface silanol groups with ethanolamine,[66] producing an amine-terminated tip, but a different choice of reagent may induce a different chemical termination. Finally, SAM formation by adsorption of alkanethiol molecules offers several advantages, including the formation of strong *S*–gold bonds. Nevertheless, this technique sometimes requires pre-deposition of gold over the cantilevers, which can be challenging technically.[67–69]

The third step in the functionalisation process involves the binding of the reactive chemical group and normally implies either the use of a SAM or the attachment of a polymer. Specific SAMs are normally attached concurrently to the previous step when using alkanethiol molecules. In this case, a small percentage of the SAM with the required active group will be mixed with a larger amount of SAMs with non-reactive groups, thereby providing control over the number of reactive species. Attachment of polymers, on the other hand, normally follows treatment by silanes or ethanolamine.[69] The most commonly used polymers are heterobifunctional polyethylene glycol and DNA.

Finally, after the addition of the SAM or polymer, the functionalisation is completed by the addition of a protein. Although this is not immediately obvious, the use of a SAM or polymer as a 'spacer' between tip and protein is necessary to minimise unwanted, non-specific interactions between tip and sample that could arise during the force experiment. Increasing the distance between the tip and the surface will, in principle, prevent this issue. Several techniques for the binding of different proteins have been devised.[69]

3.4.2 Nanotubes as Nanoprobes

Carbon nanotubes (CNTs) are graphene layers folded over into a tubular configuration. They are extremely thin (a few nanometres), have a high aspect ratio and possess excellent mechanical properties (low wear). For these reasons, CNTs have the potential to be an almost ideal AFM

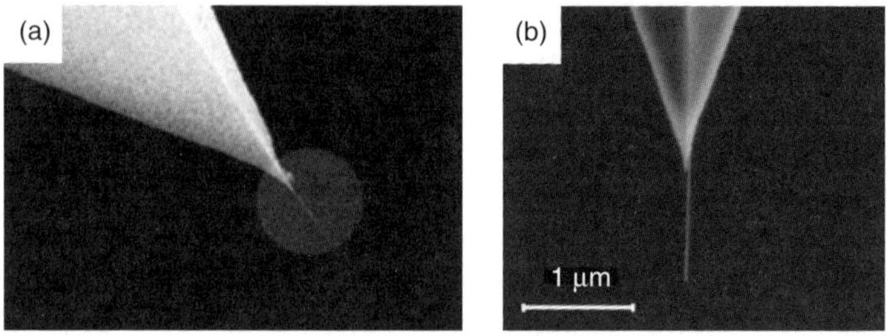

Figure 3.16 SEM micrograph showing (a) SWNT tips from nanosensors. Image provided from http://www.nanosensors.com/SpecialDevelopmentsList.pdf courtesy of Nanosensors. Copyright (2013) Nanoworld AG Date Last Accessed 26/05/2013. (b) MWNTs made in-house, reprinted with permission from [74] Copyright (2005) Wiley-VCH.

probe for many applications (Figure 3.16). In recent years, CNT probe technology has advanced considerably, to the point where they are produced commercially by some of the larger AFM probe manufactures (Figure 3.16a).[70] Fabrication methods include the manual attachment of the CNT to an AFM tip or growth by CVD. CNT probes are divided into single-walled (SWNT) and multi-walled nanotube (MWNT) types. MWNTs can have two different arrangements, depending on the fabrication method, being formed either of a series of concentric tubes (thermal CVD) or in a stacked-cone (or herringbone) structure (direct-current plasma-enhanced CVD).[34] The application of these types of CNT probe depends on their length and wall configuration. Long CNTs (a few hundredths of a micron) are very flexible and are therefore used only in biological applications. For high-resolution studies, SWNT offer a better performance, but they must be shorter than 50 nm.[71] SWNTs can achieve lateral resolutions below 2 nm.[72] In addition, CNT probes have been shown to have high wear resistance – much higher than conventional Si and Si_3N_4 probes.[73]

3.4.3 Frequency Modulation

Frequency modulation is a non-contact imaging mode in which the cantilever is operated like a self-driven oscillator, maintaining a constant amplitude and frequency. In this mode, the feedback parameter is a frequency shift between the resonant frequency of the cantilever away from the sample and the resonant frequency close to the surface; *i.e.* during the scan, the tip–sample distance is varied in order to maintain a given frequency shift.

The frequency shift can be described mathematically as the convolution between the tip–sample force gradient and a weight function. By using linear transformations, the convolution can be reversed and the tip–sample force gradient can be obtained from a plot of frequency shift *versus* distance.[75,76] This result indicates that very low amplitudes will result in higher sensitivity for tip–sample forces. Owing to the use of extremely stiff cantilevers (with spring constants of the order of thousands of Nm^{-1}), it is now possible to probe surfaces with very small amplitudes (a few nanometres) without suffering from snap-to-contact issues. This has allowed the realisation of true atomic resolution, initially under ultra-high-vacuum (UHV) conditions, but more recently at room temperature and even when operating in a liquid.[77,78]

3.4.4 Higher Harmonics

A new development in intermittent-contact mode involves the use of higher vibration modes to increase the spatial resolution and data acquisition time, as well material property imaging. Harmonics are components of the cantilever's oscillation that vibrate with a frequency equal to an integer multiple of the excitation frequency. The technique can be passive, where the higher harmonics are induced during the tip–sample interaction and then recorded and filtered, or active, where the cantilever is excited at multiple frequencies – or modes – at the same time.

Higher-harmonic AFM theory is based on the fact that the tip–sample interaction is not linear and that it is this non-linear behaviour that induces higher harmonic oscillation in the cantilever. The tip oscillation in non-contact mode can be described by a Fourier series.[79,80] Nevertheless, in conventional, intermittent-contact mode, only the first two coefficients (the fundamental and first-order harmonic) are recorded, but it is the higher-order coefficients (due to the higher harmonics) that contain the information related to the non-linear tip–sample interactions. Uncovering these interactions provides greater detail concerning the mechanical properties of specific materials and therefore allows for an improvement in the sensitivity to differences in properties.[81–83] This improved sensitivity has allowed the technique to achieve higher spatial resolution when imaging very soft materials at very low vibration amplitudes.[79,84]

3.4.5 Atomic Resolution

Ever since its invention, there has been a keen interest in using AFM to achieve atomic resolution. In the late 1980s and early 1990s, several authors published results claiming to have achieved this level of resolution.[85–87] However, the difficulty in reproducing these results and the absence of expected atomic defects in the published images led to many to believe that 'true' atomic resolution was beyond the capabilities of AFM. Nevertheless, in 1993 Ohnesorge and Binnig[88] demonstrated that atomic resolution was indeed possible, and since then many studies have expanded the types of condition under which it can be achieved.[89]

Although the nanoscale, lateral resolution of AFM is dictated by the tip shape for most applications, when dealing with atomic resolution only the atoms at the very end of the tip are considered. Therefore, the key to achieving high resolution is to be able to probe the chemical bonding forces between the outermost atom of the tip and those at the scanned surface, which in turn necessitates being able to control the tip–sample distance to an extraordinary degree. These requirements impose certain conditions that need to be attained,[90] *e.g.* the nature and identity of the tip apex cannot be altered during the scanning, long-range interaction forces must be minimised and a very robust feedback control has to be used in order to minimise issues associated with inherent drift. One way of achieving some of these requirements was to perform experiments under high vacuum and at very low temperatures. Nevertheless, the true breakthrough was achieved by using frequency-modulated AFM (FM-AFM).[91,92] As was explained in Section 3.4.3, this technique allows very precise control over tip–sample forces and enables very small forces to be probed; the technique has recently been expanded for operation under liquid.[77,78,93]

Although highly successful in providing atomic-resolution images, frequency modulation is still a complicated technique to operate and is not available commercially. This has led to other approaches in the quest for atomic resolution. One of these is the use of small cantilevers (see Section 3.3.3.5), which have a very low noise floor and, when used in low-drift/low-noise instruments, can achieve true atomic resolution.[57] Another recent development by a commercial AFM manufacturer is the

PeakForce™ technique, in which the cantilever is tapped at a frequency well below its resonance and every oscillation cycle is analysed. This allows the measurement and control, in a very precise manner, of the interaction forces between tip and sample. An additional benefit of this technique is that it allows all the conventional information from force spectroscopy to be collected at the same time, deriving mechanical information with an unprecedented degree of lateral resolution.[94] Further examples of atomic resolution in inorganic materials are discussed in Sections 3.6.2 and 3.6.3.

3.4.6 Hydrothermal AFM

As the use of AFM in the study of crystal growth and dissolution became more and more popular, the need to perform experiments away from ambient conditions developed. This is because many rates of growth and dissolution for oxides and silicates only fall within the bandwidth of conventional AFM measurements at temperatures higher than 100 °C and pressures above 1 atm. Although some commercial manufacturers offer heating devices, in most cases the range of temperatures attainable in fluid were only a few tens of degrees higher than room temperature. In addition, most AFM fluid cell designs are open to the environment or, if closed, are not strong enough to sustain pressures of more than 2–3 atm. This situation led to the development of HAFM by Higgins *et al.*[95] in 1998. This AFM design can achieve temperatures up to 150 °C and 12 atm pressure. The design was of the sample-scanning type and included the encapsulation of the piezoelectric actuator and stepper motor in a pressurised vessel. It also included a closed cell made of passivated titanium that was able to operate between pH 1 and 13. The fluid cell was designed so that the hydrodynamic conditions were fully understood,[96] an aspect that is routinely neglected in dissolution and crystal growth studies.

More recently, a new design of HAFM was developed to allow the study of growth and dissolution phenomena in the presence of supercritical CO_2 at pressures up to 100 atm and temperatures up to 80 °C.[97] The main concern in this design was the rapid change in fluid refractive index in response to small fluctuations in temperature and pressure, which could introduce significant noise in the optical detection mechanism. The technical solution was to enclose the whole AFM in an oven so that temperature fluctuations were kept to a minimum. Although noise levels are satisfactory, the design operating temperature is constrained to the operating temperature of some of the electronic devices, such as the photodiode detector.

3.4.7 Video-Rate AFM

One of the biggest drawbacks of the AFM is that the time it takes to scan a complete frame or image is, at best, of the order of a minute. This poses a severe limitation on the types of process that can be monitored adequately by the technique. In order to alleviate this situation, various approaches have been taken to increase the rate of scanning.

3.4.7.1 Small Cantilevers

As mentioned in Section 3.3.3.5, small cantilevers are ideally suited to fast scanning application, due to their intrinsic higher resonance frequencies.[55,56] In particular, these types of cantilever can achieve resonance frequencies of up to 2 MHz and still have a spring constant of around $2\,N\,m^{-1}$.[98] In addition, they give rise to very low noise.[56]

3.4.7.2 Non-Resonance Fast Scanners

Traditionally, the low resonant frequency of the scanner has been an impediment to the realisation of faster scanning speeds. One solution to this problem has been to use scanners made of stacks of small, piezoelectric elements,[55] allowing resonance frequencies above 240 Hz to be achieved.[14]

3.4.7.3 Resonant Fast Scanners

This approach uses resonant scanners in the shape of tuning forks, which can achieve scanning speeds up to 100 kHz,[16] although one major issue with these devices is that the scan speed cannot be adjusted. Recently, Picco *et al.*[15] demonstrated scan rates of up to 1000 frames per second by combining a flexure stage (made of a tandem of piezos) with a tuning fork arrangement. The flexural scanner performed the movement on the slow scan axis (at a rate of *ca* 1 kHz), whereas a tuning fork performed the fast scan up to speeds of 100 kHz.

In addition to these developments, new advances have been realised in the control system required to keep up with the increased data output. These include innovations in electronics,[14,99] control algorithms[100] and high-speed data acquisition.[101]

3.4.8 Active Cantilevers

As noted in Section 3.3.1.2, the optical beam detection method is the most common way of measuring cantilever deflection. It is not free of drawbacks, however, as it takes up a relatively large amount of space and the need to have a clear optical path influences the way the AFM is designed, which may not be optimal for a particular application. The laser light can also create artefacts in measurements (due to intensity fluctuations or unwanted reflections) and laser heating can cause an increase in the thermal noise of the cantilever. These limitations have led to the development of different types of so-called 'self-sensing' cantilevers.

3.4.8.1 Piezoresistive Cantilevers

In this type of cantilever, detection of the deflection is achieved by measuring the resistance change of the piezoresistive material of the cantilever. The first piezoresistive cantilevers were introduced by Tortonese *et al.*[102] in 1993, who fabricated cantilevers using p-type silicon, forming a resistor in a Wheatstone bridge. Since then, several authors have developed better and thinner cantilevers with pN sensitivity.[103–105] In addition, piezoresistive cantilevers have been developed for other applications, including capacitive, calorimetry and STM[106] and magnetometry.[107] Although useful in some applications, these types of cantilever still have the issue of needing to be driven externally.

3.4.8.2 Piezoelectric Cantilevers

This type of cantilever gives an output electric signal proportional to the applied strain or deflection. It was first introduced by Ioth and Suga in 1993.[108] In their design, the cantilever was made of two layers: one of zinc oxide, the other a lead zirconate titanate (PZT) film. A signal was passed through the PZT film to oscillate the cantilever, and when the oscillation amplitude changed due to being positioned near the surface, a change in the current flow was detected. Later developments have seen the use of silicon cantilevers 'sandwiched' between two thin films of PZT material.[109,110] One of the problems with this type of cantilever is the residual stress on the PZT or ZnO layers, which can result in bent cantilevers. However, this issue has recently been being solved by the use of low-stress ZnO films. Another concern has been applicability, since initially these cantilevers were only used in air, but application in liquids has been demonstrated recently.[111,112]

3.4.9 Dip-Pen Nanolithography

Dip-pen nanolithography (DPN) is a widely used surface modification technique, due to its ease of use and its applicability to a wide range of materials. The technique was introduced by Mirkin *et al.*[113] in 1999 and is based on the simple idea of transferring an ink into a surface by means of the capillary forces created during the contact of a very sharp tip with a surface, achieving resolution of up to 15 nm.[114] In this sense, DPN is a nanoscale analogue of a macroscopic pen. Some of the advantages of DPN over similar technologies are its ability to place selected molecules with sub-100 nm precision, its ability to deposit both organic and inorganic substances over all types of substrates and its ability to deposit molecules over existing layers. Another advantage is that molecules are deposited directly, so there is no need to expose them to harsh solvents, UV or electron-beam radiation. For these reasons, DPN is particularly attractive in biological studies.[114]

Among its many applications, DPN can be used to generate biomolecular nanoarrays for diagnostic applications and ligand–receptor studies.[114] When combined with chemical etching, DPN can be used to fabricate metal and semiconductor nanostructures,[115] and it is also used heavily in the template hierarchical assembly of materials.[116] Future developments are focused on incorporating DPN into high-throughput applications in biology and materials science.[116]

3.4.10 Scanning Near-Field Optical Microscopy

Scanning near-field optical microscopy (SNOM) works by passing light through a sub-wavelength aperture, which in turn illuminates a sample located at a distance much smaller than the wavelength of light (hence 'near-field'), allowing for much better resolution than conventional optical microscopes. The principle behind SNOM was first proposed by Synge in 1928,[117] but it could not be implemented technologically until 1984.[118]

An SNOM instrument works by using a point light source that is scanned very close to the surface under study, without touching it; the light reflected back from the surface (reflective samples) or transmitted (transparent samples) is then collected. There are two types of SNOM: those equipped with an aperture and those without one (known as 'apertureless'). Aperture-equipped SNOMs include the use of metal-coated optical fibre probes and hollow-catilevered probes (in which a laser is used as a source of light). In apertureless systems, a sharp tip acts to scatter light. The light source is maintained close to the surface using the AFM feedback control. Finally, light detection and collection can be achieved by several different methods, including an avalanche photodiode, photomultiplier tubes, charge-coupled devices and spectrometers.

An SNOM instrument can operate in four different modes:

(i) *Transmision*: In this mode, the sample is illuminated and the light passing through it is collected.
(ii) *Reflection*: Here only the light reflected back from the sample is collected.
(iii) *Collection*: The sample is illuminated with a macroscopic light source and the probe is used to collect the light reflected by it.
(iv) *Illumination/collection*: The probe is used to illuminate and collect the light from the sample.

Typical resolution varies between 50 and 100 nm, although sub-10 nm resolution has been demonstrated.[119] SNOM applications include studies on laser emission,[120] plasmonics,[121] photonic crystals and[122] quantum dots,[123] among many other areas.[124]

3.4.11 Raman/AFM

The combination of AFM with confocal Raman spectroscopy has been one of the major advancements in hybrid imaging/chemical analysis systems, as it opens the possibility of direct correlation between chemical information and physical, magnetic or electrical properties (as measured by AFM). Implementing true AFM/Raman compatibility in a single instrument is challenging for various reasons: for example, it can take a few minutes to create an AFM image, whereas Raman measurements can take several hours. There is therefore a need for very stable systems with low drift, in order to ensure that the Raman data correspond to a given surface feature. In addition, AFM and Raman have very different resolutions and, whereas AFM can achieve atomic resolution, Raman is diffraction limited, meaning it has a resolution of around 200 nm. This latter issue has been alleviated partly by the introduction of tip-enhanced Raman spectroscopy (TERS),[125] which offers a resolution on the nanometre scale.[126] TERS is a type of surface-enhanced Raman spectroscopy (SERS) that can be used on almost any type of sample,[127] unlike most forms of SERS, which require rough samples. In TERS, the sample and AFM (or STM) tip are both illuminated by a laser, and a near-field effect is produced in the vicinity of the tip that enhances Raman scattering. The curvature, shape and composition of the tip (Au- or Ag-coated tips must be used) determine the increase in scattering and the spatial resolution.

Both confocal Raman and TERS have been used to study a great variety of systems, including nanowires, CNTs, graphene, cells, polymers *etc*. They have also been used to study stress and corrosion in metals.

3.5 SPECIALISED SCANNING MODES

3.5.1 Phase Imaging

Phase imaging consists in measuring the change in phase produced when an oscillating probe is brought close to a surface. This phase change is related to the energy dissipation from the contact between tip and sample,[128,129] which in turn is controlled by a number of factors, including adhesion, viscoelasticiy and contact area.[27] This means that phase contrast will not only reflect variations in material properties but can also be linked to changes in topography (affecting the contact area).

3.5.2 Force-Modulation AFM

Force-modulation microscopy involves vibrating the cantilever or sample with a very small amplitude (a few nanometres) while scanning the sample with a constant (nominal) deflection, *i.e.* in contact mode or very close to it. Variations in amplitude and phase are measured at the same time as an image of the topography is created. For a soft sample, the cantilever will 'penetrate' more, increasing the sample contact area and thus requiring more energy to move the tip, resulting in a smaller amplitude of vibration. This technique can, therefore, monitor changes in the compliance of the tip–sample contact as compared to changes in the topography. By using contact mechanics theories such as the JKR (Johnson–Kendall–Roberts)[130] or DMT (Derjaguin, Müller and Toporov) models,[131] it is possible to calculate values of stiffness directly, provided that the cantilever stiffness (see Section 3.3.3.4) and the tip apex geometry are known.

3.5.3 Friction Force Microscopy

When the AFM tip is scanned over a surface, the tip will not only bend vertically (due to the repulsive or attractive forces) but will also be twisted laterally due to friction forces that oppose its displacement. This lateral twisting is recorded by the difference between the left and right halves of the photodiode detector. The output signal is normally referred to as the 'lateral signal' and its recording and use are normally referred as 'lateral force microscopy' (LFM). The total twisting of the cantilever is related directly to the friction between tip and sample, although it may also have a component resulting from the slope along which the tip is travelling.[132,133]

Friction force measurements often involve so-called 'friction loops', defined as the change in the lateral force signal as a function of the scanner lateral position. An example of a friction loop is shown in Figure 3.17a. Initially the tip will not be twisted, but after the scanning begins it will bend without moving until a certain value of twisting (recorded as voltage V_1 in the photodiode detector) is achieved. After this, the tip will slide over the sample. If the tip encounters an area of higher friction, it will bend a little more, recording a slightly higher voltage, V_2. After the trace scan is finished, the tip will stop and bend in the opposite direction as the scanner begins the retrace scan. Once a certain amount of bending has been reached, V_3, the tip will start moving again, and as it passes over the area of higher friction, it will deflect a little more, V_4. From Figure 3.17 it can be seen that the actual amount of bending that the tip is doing is defined as the difference between the two signals recorded while scanning in the opposite direction ($V_1 - V_3$

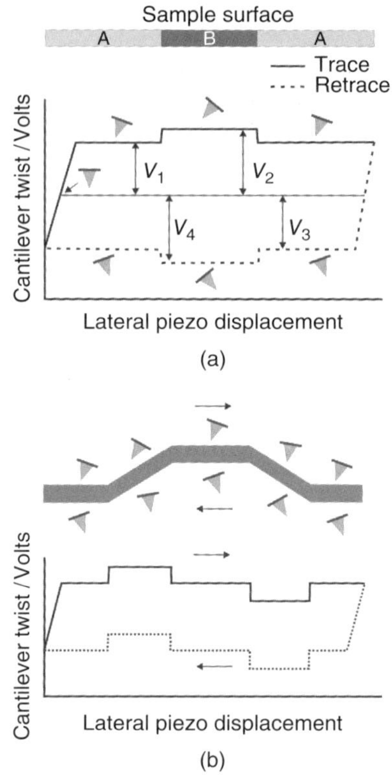

Figure 3.17 (a) Change in the lateral deflection signal as a function of lateral distance for a tip scanning a flat surface composed of two different materials, A and B. V_1 and V_2 show the respective increases in output lateral signal when the tip scans over A and B during the first (trace) pass. The tip deflection is represented schematically at each point. Voltages V_3 and V_4 represent the respective increases in lateral signal when the tip retraces the same materials, completing the so-called 'friction loop'. It can be seen graphically that the actual tip deflection will be equivalent to $\frac{1}{2}(V_1 - V_3)$ for A or $\frac{1}{2}(V_2 - V_4)$ for B. (b) Trace and retrace lateral signal output for a tip scanning a sample with changing slopes. It can be seen that calculation of true friction will eliminate the topographical effect on the lateral force signal.

and $V_2 - V_4$). From Figure 3.17, it is also obvious that the area of higher friction will result in a higher value of the voltage difference. Therefore, the real friction of a sample will be calculated by looking at the difference between the trace and retrace scans in the lateral images. Using this approach, differences in the lateral force signal due to sample slope will cancel out, as can be seen in Figure 3.17b.

FFM can, therefore, be used to observe variations in mechanical properties, and it has been heavily employed in the nanotribology field.[134–136] More recently, the technique has been used to infer information about chemical composition variations in mineral growth[137,138] and to calculate the energies of dissolution and growth processes.[139,140] In spite of this, FFM is still not applied widely as it requires a precise knowledge of the lateral or torsional spring constant of the cantilever, as well as calibration of the lateral sensitivity of the optical deflection sensor, which can both be rather complicated.[133,135]

3.5.4 Force Volume

Force volume combines conventional surface imaging with the acquisition of force–distance curves (see Section 3.3.2.4) in a 2D grid pattern. At every pixel of the image, the cantilever deflection and a force–distance curve will be recorded, from which information about adhesion and other mechanical properties can be obtained (see Section 3.3.2.4). This allows a user to directly overlay a height or topography image with an adhesion or elasticity map.[141,142] Height variations can affect the results from force–distance curves, so care must be taken when interpreting results from rough samples. Another disadvantage of the technique is that data acquisition is rather slow and uses a considerable amount of memory space, generally limiting the method to low-resolution images. Although still used widely, this technique could soon be superseded by more advanced modes, such a PeakForce™ (see Section 3.4.5), which is implemented at standard scanning rates and resolution.

3.5.5 Magnetic Force Microscopy

This mode of operation allows measurement of the presence and distribution of magnetic fields by use of a magnetic probe/tip. Magnetic tips consist of a standard Si or Si_3N_4 cantilever coated with a metal or alloy such as cobalt, iron, cobalt–nickel or cobalt–chromium. Because magnetic forces are weaker than normal tip–sample forces, this mode or operation uses the two-pass method. In this technique, the topography is first registered by either contact or intermittent contact mode and the tip is then lifted a few nanometres above the surface and a second scan is carried out. At this height difference, van der Waals interactions will be non-existent and magnetic forces will affect the bending (for a non-vibrating tip) or the phase and amplitude (for a vibrating probe).

The two-pass technique has several variations. The first to be implemented is the so-called **lifting mode**, in which the second pass is carried out at a constant height over the sample.[143] This method has the drawback that for samples with high roughness, there may still be an influence of long-range forces over the cantilever as it come closer to areas that are relatively high. To alleviate these problems, Lin *et al.*[144] proposed a variation of the method whereby the tip will 'follow' the previously recorded topography profile after moving above the surface by a few nanometres (Figure 3.18). This allows an almost constant tip–sample distance to be maintained and minimises greatly the effect of topography. Another way of avoiding the effects of long-range forces is to subtract the trace and retrace scans, which, in principle, will eliminate their influence on the final magnetic measurement. Other techniques allow the measurement of magnetic gradients by lifting the tip above the sample and measuring the magnetic field at various height positions.[145]

MFM has been used widely for industrial applications such as the study of data-storage materials, which require the formation of nanosized magnetic domains.[143,146] Other applications include the study of superconductors and, more recently, magnetic nanoparticles.[147]

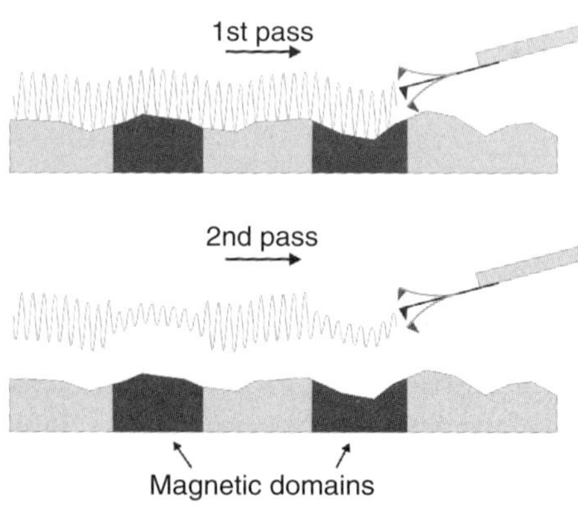

Figure 3.18 Schematic diagrams showing the two-pass technique described by Lin *et al.*[144] In the first pass, the tip records the surface topography in intermittent contact mode. Then, in the second pass, the tip is scanned a few nanometres above the sample, but 'following' the previously recorded topography. Magnetic domains will induce changes in the phase and amplitude of the recorded image.

3.5.6 Electric Force Microscopy

In this mode of operation, electrical fields can be measured in a way similar to that used for magnetic fields in MFM. In its most common inception, this mode operates with the two-pass technique and in a vibrating mode, and, as in MFM, the first pass consists of a measurement of the topography and the second pass uses this information to 'track' the topography at a distance of a few nanometres. The tip is biased with a direct current at the same time. Variations in the electrostatic forces between a tip and a biased sample as a function of distance (related, in turn, to variations in capacitance) will induce variations in the amplitude and phase of the vibrating cantilever. Recording these changes will allow inferences to be made concerning variations in the electrical field.

Like MFM, EFM is used widely in industry, especially within the consumer electronics field, as there is an ever-increasing need to be able to characterise the electronic characteristics of nanoelectrical assemblies adequately. Therefore, EFM has been used to detect trapped charges,[148] map electric polarisation and analyse electrical failures. Another interesting application has been the measurement of the electrical properties of CNTs.[149,150]

Over the years, a great number of EFM-derived techniques have been developed; some of the most important will now be described.

3.5.6.1 *Scanning Capacitance Microscopy*

This technique is used to measure changes in capacitance between a tip and a sample.[151] In this mode, the tip is biased and scanned using the two-pass method. Applications include the mapping of variations in electrical carrier concentrations across various types of sample, although it is more typically used in semiconductors.

3.5.6.2 *Conductive AFM*

This EFM mode is used to measure variations in conductivity across a surface. A conductive tip is scanned in contact mode across the sample surface while a voltage is applied between probe and sample. In this mode, the AFM is able to measure currents as low as a few pA and up to a few hundred nA. Conductive AFM is used to study a wide variety of systems, including conducting polymers,[152] semiconductors and nanotubes/nanofilaments.[153]

3.5.7 Kelvin-Probe Force Microscopy

This technique, first reported by Nonnenmacher *et al.* in 1991,[154] measures contact potential differences (CPDs) between a probe and a sample. CPDs depend on a number of parameters, including the work function, the presence of oxide or an adsorption layer and temperature changes. CPDs can be measured using the Kelvin method, whereby two conductors are positioned in a capacitor-type arrangement, with a small space between them. In Kelvin-probe force microscopy (KPFM), a conducting cantilever is used as a reference electrode, forming a capacitor with the surface. A DC potential bias, V_{DC}, is applied to the probe, along with a modulated AC signal, V_{AC}, operating at the resonant frequency (but not driving the cantilever). This means the potential of the cantilever changes at a rate equal to the resonant frequency, ω_0. When the surface potential is not the same as that of the probe, the cantilever will start to vibrate at a frequency, ω. The AFM electronics compare the two frequencies and adjust the voltage, V_{DC}, such that the electrostatic forces between tip and sample are zero and the cantilever vibration is thus minimised. This voltage will be equivalent to the surface potential. By calibrating the tip against a standard with a known work function, it is possible to obtain absolute values of the sample's work function.

KPFM can be used in lift or frequency-modulation mode. Its applications range from the study of semiconductors[155,156] and solar cells[157,158] to that of nanoparticles[159,160] and biofilms.[161]

3.5.8 Piezoresponse Force Microscopy

Piezoresponse force microscopy (PFM) is a contact-mode-derived technique based on the direct piezoelectric effect, *i.e.* the expansion/contraction of a material subjected to an electrical field. PFM allows measurement of the mechanical response of a sample when an electrical voltage is applied to its surface *via* the conductive tip of an AFM.

PFM works by applying a relatively high-voltage/high-frequency AC signal to the cantilever (sometimes with an additional DC component), causing the sample surface to expand and contract; this movement is then translated into a cantilever deflection. By means of a lock-in amplifier, the signal is demodulated and analysed, and the phase component of this signal yields information on the polarisation vector underneath the tip. In addition, by using a quadruple photodiode detector, the lateral and vertical components of the response can be decoupled, providing additional information on the electromechanical response vector.

The application of PFM has increased considerably over the years and now includes the study of ferroelectric materials[162] for non-volatile computer memories, piezoelectric materials for radiofrequency, electromechanically active materials for applications as actuators and sensors[163] and even electromechanical coupling in biological systems.[163]

3.5.9 Nanoindenting

In this technique, forces are measured when the tip approaches a surface, unlike in force spectroscopy (see Section 3.3.2.4), where they are measured as the tip moves away from the sample. Indentation is a common technique used to determine the mechanical properties of a material, such as modulus or hardness. Dedicated indentometers use a hard tip to probe a sample, but they are only sensitive enough to measure forces in the μN range, and they rely on optical methods to measure the holes or indents created. In contrast, an AFM offers the possibility of measuring forces down to the pN level in very small, user-selected areas and allows for very precise measurements of the indents. Moreover, the excellent control over the tip position allows indent patterns to be created and, therefore, the distribution of mechanical properties across a surface to be measured. However, a very good knowledge of the tip radius and spring constant is mandatory in order to obtain quantitative results, and data analysis also requires the use of continuum mechanical models such as the Hertz,[164] JKR[130] or DMT[131] models. Normally, very hard tips (such as diamond) are used for nanoindentation experiments, but some authors have used normal Si_3N_4 tips or even colloidal probes. Indentation has been used in the study of a multitude of materials, including polymer composites,[165] human hair[166] and even bacteria.[167]

3.6 APPLICATIONS

3.6.1 General Considerations

AFM has some distinct advantages over other types of microscopy in terms of applications, perhaps the most powerful of which is the ability to operate in gases or liquids as well as under high vacuum. As a consequence, dynamic processes can be probed that are completely inaccessible to electron microscopy techniques. The timescale of the dynamics is limited to some extent by the time taken to record a micrograph, but with fast scanning methodologies, this dynamic range is now quite large. Moreover, the ability to control the atmosphere above the sample also means that sensitive samples can be studied that might collapse or alter under conditions of high vacuum. The resolution of AFM is also important when considering suitable applications. A vertical resolution of 0.1 nm is readily achieved, but lateral resolution is subject to the mode of operation of the microscope and can be divided into three classes: (i) resolutions of *ca* 10 nm; (ii) lattice resolution; and (iii) true atomic resolution. Resolutions of *ca* 10 nm are readily achievable and can be described as the normal operation of an AFM. They are ideally suited to observing features with sizes that reflect the relative vertical and lateral resolution, such as terraces on a crystal surface, which are typically sub-nanometer in height but microns in width. Lattice resolution should, of course, not be confused with true atomic resolution; the former can be achieved on ordered, crystalline samples where the interaction footprint of the AFM tip is actually tens of lattice dimensions in width. Interference between the tip and the sample gives the impression of atomic resolution but in reality what is achieved is an impression of the repeating units in a perfect crystalline lattice. In order to achieve true atomic resolution capable of distinguishing point defects in a periodic lattice, it is necessary to overcome the long-range forces between tip and sample. This can be achieved in a variety of ways and results, in the limit, in resolution superior even to that of the STM, whereby individual atomic orbitals can be mapped. Thus, when looking at the primary literature, it is important to make the distinction between lattice resolution and true atomic resolution.

The other distinct advantage of AFM over electron microscopy is the ability to measure the sizes of surface features accurately, often to a degree that permits the features to be identified. However, the influence of the microscopy procedure upon the processes under observation should not be forgotten when deploying AFM. The force on the tip can be adjusted and it is therefore usually possible to determine the influence of the tip–sample interaction and thereby either to minimise it or else to take it into account in the interpretation of the data. However, this tip–sample interaction can also be put to good use, whether it be in the controlled rearrangement of a surface at the nanoscale, in promoting surface chemistry or in changing local surface temperatures. Finally, it is important to remember that the features observed by the AFM are a convolution of the AFM tip shape and the shapes of surface features and that the tip shape may change as a result of wear during the course of a lengthy experiment.

3.6.2 Atomic-Resolution, Non-Contact AFM of Metal Oxides

Achievement of true atomic resolution has two important pre-requisites. First, the AFM should be operated in a manner that allows the removal of attractive, long-range interactions, and second, attention must be given to ensuring careful sample preparation. Most work to date has been achieved by operating the AFM in a non-contact mode (nc-AFM), working at a constant frequency shift whereby the short-range, so-called chemical forces dominate the image. Sample preparation for an oxide surface must usually be in UHV in order to prevent unwanted surface contamination, which can involve sample cleaving, cleaning with ion bombardment and subsequent annealing. The quality of the images will be very dependent upon the care taken with the sample preparation.

Titania, TiO_2, is one of the most studied oxide surfaces, due to its importance as a photocatalyst. Figure 3.19 shows the (110) surface of TiO_2 (rutile form) recorded at room temperature a short time after sample preparation.[168] The image is dominated by a striped pattern with a separation of *ca* 0.65 nm, consistent with the crystal surface structure made from alternate rows of two-fold coordinated, bridging oxygen atoms [O(2c)] and five-fold coordinated titanium atoms [Ti(5c)]. In between the bright rows are isolated bright dots, resulting from a slightly reduced surface containing highly reactive oxygen vacancies, which react with residual water to form adjacent surface hydroxyl groups, known as d(OH) groups which in turn, over time, split into isolated single surface hydroxyl sites, s(OH). This is primarily what is observed in the AFM image. Different surface features can also be enhanced by changing the

Figure 3.19 nc-AFM image ($15 \times 20\,nm^2$) of the TiO_2 (110) surface, reprinted with permission from [78] Copyright (2008) American Institute of Physics.

polarity of the apex atom at the end of the tip, achieved by gently touching the tip onto the sample and depositing either $O^{\delta-}$ or $Ti^{\delta+}$. The former generates an image in which the Ti(5c) rows are imaged as bright and the O(2c) rows are imaged as dark, while the latter reverses this contrast (Figure 3.20). Due to the polar nature of the d(OH) and s(OH) surface groups, they can switch contrast upon switching polarity of the tip. A third mode, referred to as a neutral-tip mode (probably just a pure silicon tip) enhances the proton on top of the s(OH) groups lying directly above the O(2c) oxygen rows (Figure 3.20). By recording time-lapse images, it is also possible to observe the splitting of d(OH) groups into isolated s(OH)

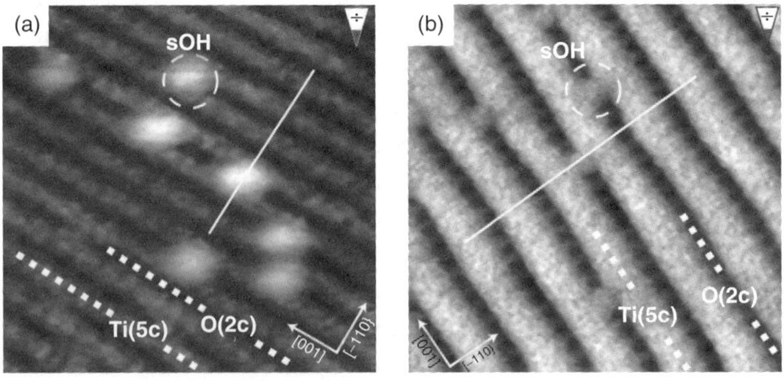

Figure 3.20 Constant-frequency-shift nc-AFM images of the TiO_2 (110) surface imaged with different tip polarities, reprinted with permission from [78] Copyright (2008) American Institute of Physics.

groups at the surface of TiO_2. Charged sub-surface defects can also be detected, as they have a long-range influence on the tip, providing that the tip is not charged, since a charged tip will interact predominantly with top-surface features.

Alumina, Al_2O_3, is a very important material for a wide variety of applications, from microelectronics, where it often acts as a substrate, to high-surface-area heterogeneous catalysis. Despite a variety of surface-sensitive tools, such as low-energy electron diffraction (LEED) and grazing-incidence electron diffraction (GIXD), having been used to characterise the important (0001) surface, a full understanding of the nature of the surface has eluded materials scientists. This is often because the preparation of a material with a well-defined structure and stoichiometry is not straightforward. In this regard, nc-AFM is ideal for elucidating fully the surface structure. nc-AFM reveals[169] that the surface structure consists of the known superstructure, which is substantially different to that proposed based on previous diffraction studies, and is able to pick out all the detailed subtleties of the atomic arrangement at the surface. Figure 3.21 shows the constant-height, atom-resolved nc-AFM image of the (0001) surface of Al_2O_3 prepared by reconstructing α-Al_2O_3, which is in turn prepared by heating a polished corundum single crystal to 1500–1600 K under UHV. The incredible detail shown in Figure 3.21 reveals not only the well-known $\sqrt{31} \times \sqrt{31}R9°$ superstructure but that within this superstructure there is a very pronounced internal pattern of substructures. Closer inspection, shown in Figure 3.22, reveals three distinct domain substructures. Owing to the picometer vertical resolution made possible by nc-AFM, the contrast in these images can be attributed to small variations in the atomic surface topography. The structure of the surface layer is both rotated and expanded to 3.01 Å compared to the underlying Al_2O_3, with an average sub-lattice of 2.75 Å. This implies that the stacking shifts between all possibilities of the Al–Al–O–R substrate and Al are not exclusively distributed on O_{sub} sites (= the Al hollow sites with a second layer of O directly underneath). Domain 1 Al is placed near the on-top (O_{top}) positions, while domain 2 has it placed near on-top Al sites and domain 3 has it placed near the hollow O_{sub} sites. Such inferences for this extremely important material are possible only with the advent of true atomic-resolution AFM and reveal the immense power of this technique.

nc-AFM has also revealed[170] that there is a correlation between the $\sqrt{31} \times \sqrt{31}R9°$ superstructure and a periodically modulated affinity for the adsorption of water molecules. This is probably in some way related to the different domain structures now known to exist, but it

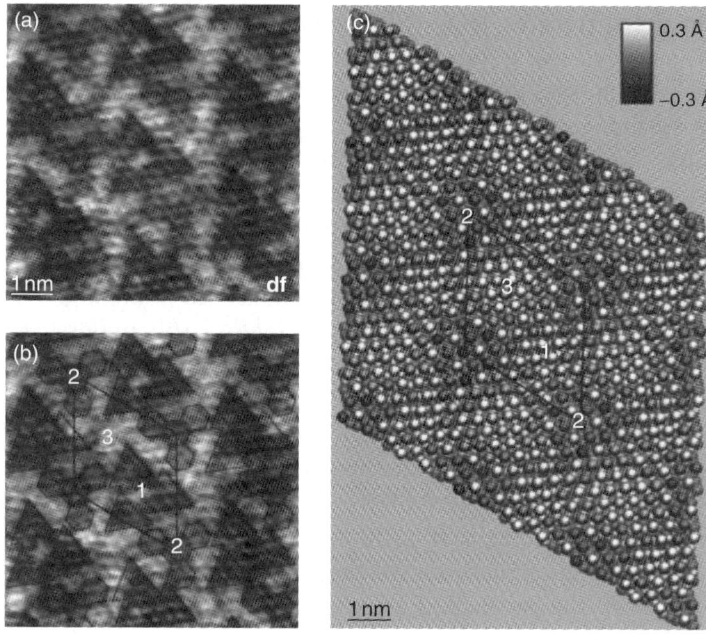

Figure 3.21 (a) Constant-height, atom-resolved nc-AFM image (40×40 Å^2) of the $\sqrt{31} \times \sqrt{31}R9°$ Al_2O_3 (0001) surface. (b) Superimposed black lines outline the three types of subdomain. (c) Calculated model of the structure, reproduced with permission from [169] Copyright (2009) APS.

demonstrates that there is still much work to be done on this eminently important material.

Zinc oxide, ZnO, is an important wide-bandgap semiconductor that is both tuneable and semi-transparent. The complex stabilisation mechanism for the polar-terminated ZnO(0001) surface has been investigated using nc-AFM.[171] For surfaces that contain a non-zero dipole moment perpendicular to the surface, there must be some rearrangement of the surface in order to compensate for this charge. This usually involves a significant change in both the surface geometry and electronic structure and the surface stoichiometry. ZnO crystallises in the hexagonal, wurtzite structure and the electrostatic potential at the surface increases with crystal thickness, resulting in substantial surface reconstruction, which can otherwise be achieved by a 25% deficiency of positive charge on the Zn-terminated faces or of negative charge on the O-terminated faces. This could be achieved by a simple transfer of charge from one face to the other, but in fact stabilisation involves substantial geometric reconfiguration and the creation of triangular surface features. It can be shown by nc-AFM that the surface reconstruction is influenced strongly

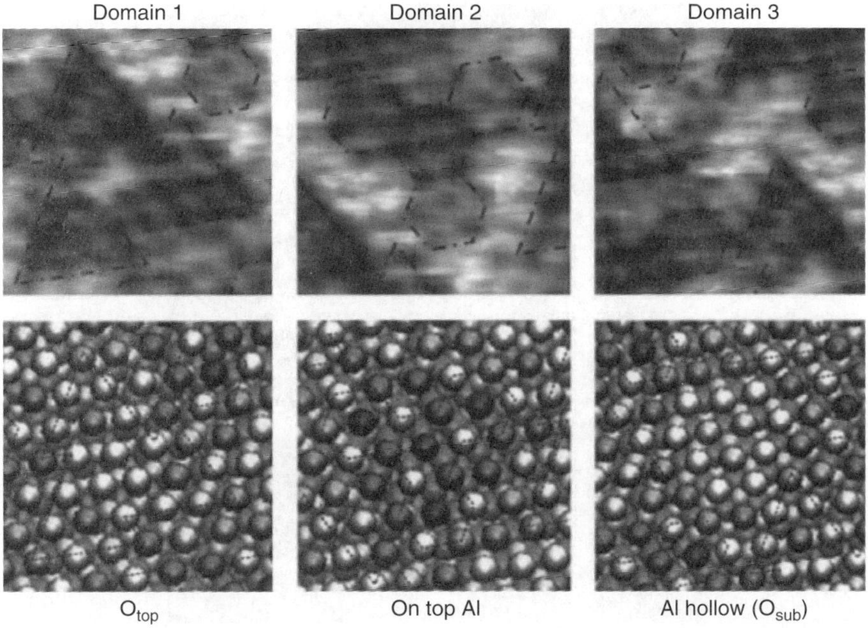

Figure 3.22 Three domains of the $\sqrt{31}$Å ~ $\sqrt{31}R9°$ Al_2O_3 (0001) surface sub-structure, reproduced with permission from [169] Copyright (2009) APS.

by preparation temperature, while at high temperatures there is an additional ordering among the triangular reconstructions, suggesting a repulsive electrostatic step–step interaction (Figure 3.23). The surface can be modelled according to alternating straight and sawtooth-like step edges (Figure 3.23), where structures with more highly coordinated

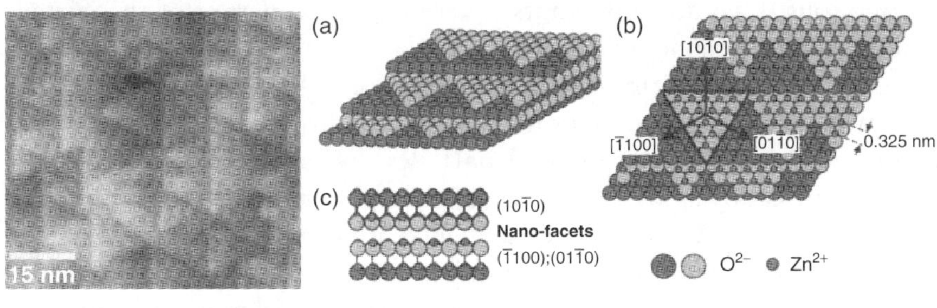

Figure 3.23 Left nc-AFM image of a Zn–ZnO(0001) surface prepared at an annealing temperature of 1390 K.[171] Model of the alternating straight and sawtooth step edges on Zn–ZnO(0001). (a) Perspective view. (b) Top view. (c) Side view of the three step edges marked by the triangle in (b), reproduced with permission from [171] Copyright (2008) APS.

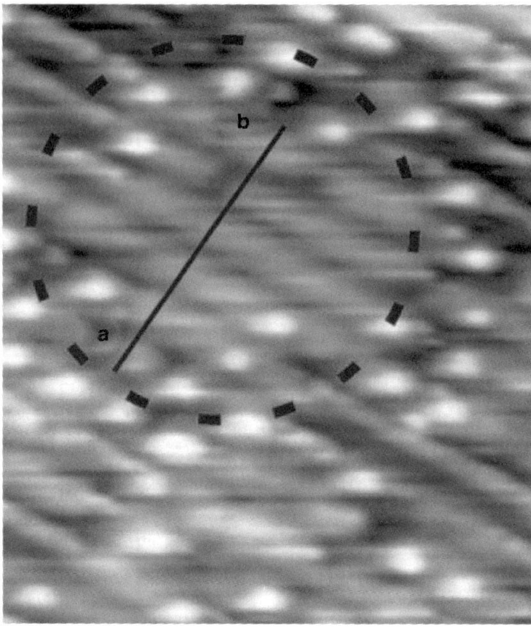

Figure 3.24 ZnO(0001), showing an area free of point defects, reprinted with permission from [172] Copyright (2009) American Chemical Society.

atoms are energetically more stable, resulting in oxygen-terminated step edges. This reconstruction requires that ZnO units at the surface are mobile, so that the step edges can arrange in equally spaced terraces. These triangular features result in the required 25% charge deficiency, due to the oxygen termination at the step edges. Further analysis of ZnO(0001) by nc-AFM[172] also examines the role of point defects at the surface (Figure 3.24), as well as the occurrence of missing rows of zinc ions. Overall, this paints a very complex picture of the surface reconstruction of zinc oxide.

Ceria (CeO_2) is an important material for solid-oxide fuel-cell technology and for the production of hydrogen in catalysis. It is used as a support for active noble-metal particles and as an oxygen transport medium in catalysis. It is an important component of the automotive catalytic converter, and as a consequence, its surface structure is crucial to performance. nc-AFM has been used to probe the structure[173] of the surface, which is shown to consist of hexagonal islands and pits with a triple-layer height consisting of O–Ce–O components. Kink sites and missing oxygen atoms can be identified, and these dominate the defective sites at the step edges. The AFM tip can also, in this case, interact with the surface, and oxygen atoms at the step edges can be removed by the tip of the microscope. Figure 3.25 shows a large-scale AFM image of CeO_2 (111), as well as a true atomic-resolution image of the stoichiometric oxygen-terminated surface. Figure 3.26 shows how the hexagonal surface structure results from adjacent stable steps having (001) and (110) faceting. This further results in a missing terminating oxygen atom at the

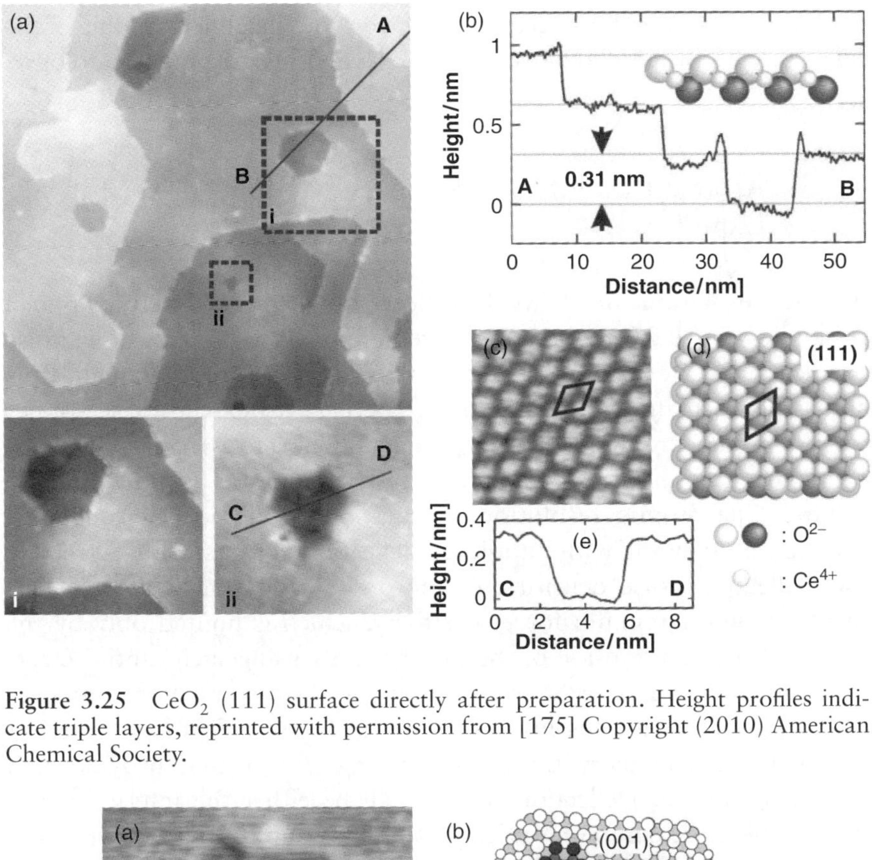

Figure 3.25 CeO_2 (111) surface directly after preparation. Height profiles indicate triple layers, reprinted with permission from [175] Copyright (2010) American Chemical Society.

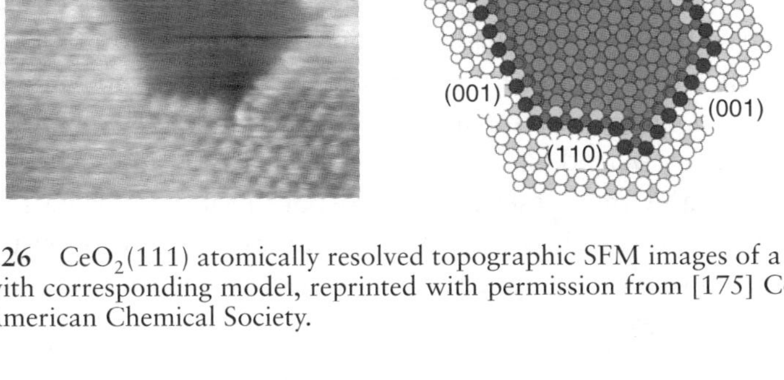

Figure 3.26 CeO_2(111) atomically resolved topographic SFM images of a hexagonal pit with corresponding model, reprinted with permission from [175] Copyright (2010) American Chemical Society.

vertex of the island. These missing oxygen atoms are highly likely to be implicated in the oxidation of CO to CO_2 *via* $CO + O_{lat} \rightarrow CO_2 + O_{vac}$ (where O_{lat} and O_{vac} represent a lattice oxygen and an oxygen vacancy, respectively), and understanding the atomic detail of such surfaces is crucial as a consequence. nc-AFM is the only method that can obtain this information for surfaces that are non-conducting. It is likely that in the near future, there will be a great deal of work on the understanding of the surface detail of metal oxides at atomic level using AFM methodology.

3.6.3 Atomic-Resolution, Frequency-Modulated AFM in Liquids

Atomic resolution in AFM has been almost exclusively confined to studies in UHV, owing to the requirements of a very low noise coefficient of the measuring device. However, in order to make this applicable to many real situations, it is crucial to extend these studies to systems under liquid. The experimental obstacles have now been overcome and modern commercial instrumentation is now able to achieve true atomic resolution in solution or liquid environments. The key is to work with cantilever deflection sensors with extremely low deflection noise densities, of the order of $17\,fm\,Hz^{-1/2}$.[174] This then achieves maximum force sensitivity, which is limited only by the thermal Brownian motion of the cantilever. By using such cantilevers, it is possible, using FM-AFM (see Section 3.4.3), to work with cantilever oscillation amplitudes as low as 0.2 nm in liquids. This has the effect of enhancing sensitivity for the short-range forces that provide high spatial resolution. Deflection sensors can be either the interferometry type or the more common optical beam deflection (OBD). Fukuma and Jarvis in particular have developed methods using the OBD strategy that employ very small cantilevers with a focus lens of sufficient strength to obtain a very small laser spot and to collect the reflected laser beam.[174] Figure 3.27 shows a typical set-up for work under liquids capable of achieving atomic resolution, while Figure 3.28 shows the result of using such a set-up on muscovite mica taken in water.[77] An atomic-scale structure with a periodicity of 0.52 nm is clearly visible, as are surface corrugations and the presence of adsorbate molecules. The magnified image also reveals bright spots on the hexagons of mica, consistent with Al^{3+} ions caused by the protrusion of electron density at these sites.

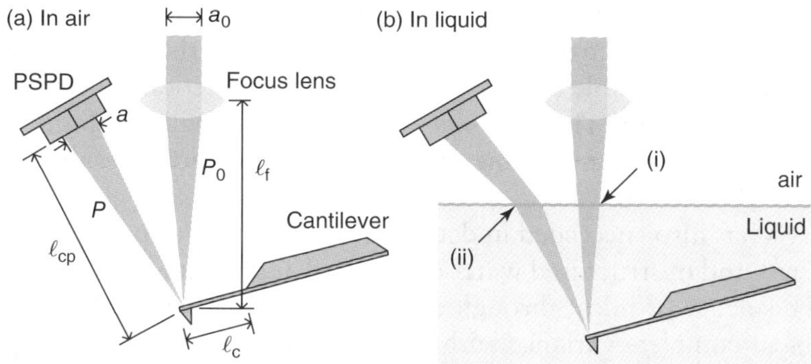

Figure 3.27 Typical experimental set-up for a cantilever deflection sensor using the OBD method (a) in air and (b) in liquid, reprinted with permission from [174] Copyright (2006) American Institute of Physics.

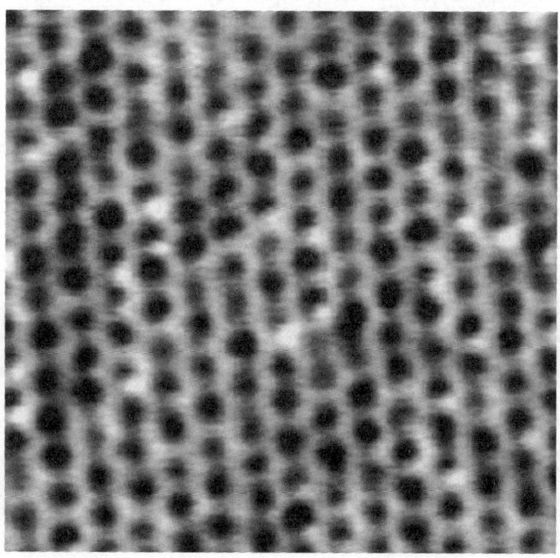

Figure 3.28 FM-AFM images of the cleaved (001) surface of muscovite mica, taken in water, reprinted with permission from [77] Copyright (2005) American Institute of Physics.

These non-periodic features are clear evidence that true atomic resolution has been achieved by FM-AFM, even when the *Q* factor in water is as low as 30. These images were recorded with a relatively large frequency shift, greater than +200 Hz at an amplitude of 0.20 nm, indicating a relatively large tip–sample interaction force. With such conditions, there is some asymmetry in the contrast in the fast scan direction, which indicates the influence of frictional interactions between tip and sample. Indeed, by using precise control in the vertical position of the tip at the point of contact it is possible to measure a range of interactions, including the friction interaction. These results show that the small-amplitude operation is sufficient to overcome the large-frequency noise resulting from the low *Q* factor in water.

Jarvis has also succeeded in determining the distribution of Mg^{2+} *versus* Ca^{2+} and of structured water distributed at the liquid–solid interface on the surface of mica, through direct imaging.[175] These distributions can be at complete variance with the distribution within the bulk phase and determine the function for biological applications and for colloidal suspensions. As the tip approaches the surface, it experiences a discrete layering of structured water; it is consequently possible to obtain an image at a number of quasi-stable distances from the crystal surface, all of which have an equivalent frequency shift or force. The quantised distance of these discrete locations is 2.8 Å, corresponding to the size of a water molecule. The images are, therefore, either direct tip–mica images or images mediated through a layer of water. Figure 3.29 shows four images of mica recorded in different electrolyte solutions, all displaying the honeycomb arrangement expected on a mica surface. Two arrangements are observed: Figure 3.29a shows oxygen atoms with a ditrigonal array, while Figure 3.29b shows a hexagonal array of Si or Al atoms. As

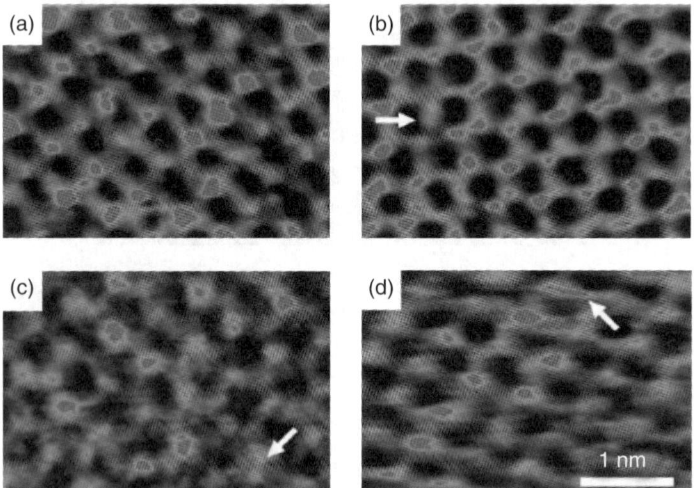

Figure 3.29 FM-AFM images of the cleaved surface of muscovite mica, taken in electrolyte solutions:[175] (a) LiCl, (b) NaCl, (c) $MgCl_2$; (d) $CaCl_2$. All share the scale bar shown in image. Reprinted with permission from [175] Copyright (2010) American Chemical Society.

the tip cannot interact directly with the Si or Al atoms located below the oxygen atoms, the hexagonal arrangement in fact comprises the cations, which are located above the oxygen atoms with the same symmetry as the Si and Al. The protruding features are distributed randomly, 50 pm above the oxygen surface of the mica, and the variation in the size of the protrusions is probably affected in large part by whether there is an underlying Si or Al. This information is not accessible by any other surface characterisation tool and indicates that in the future it should be possible to locate non-periodic ions and small molecules at the surfaces of underlying crystalline layers. This could be particularly useful for the study of small template molecules adsorbed at the surfaces of nanoporous crystals such as zeolites, for example, allowing the precise role and location of these important structure-directing agents to be determined. Such atomic resolution is also likely to be crucial in understanding the role of point defects at surfaces that are not readily visible to any other tool, whether microscopic or spectroscopic. AFM has the unique ability to record, with precision, single defects with vanishingly low concentrations. Diffraction techniques are not appropriate for such studies unless the structure of interest is periodic, but the functionality of many materials is the result of aperiodic defects housed in an underlying crystalline background. Therefore, unless the concentrations are sufficient to render spectroscopic techniques suitable, AFM becomes the technique of choice for surfaces of non-conducting samples, especially under liquid.

3.6.4 Crystal Growth

The study of crystal growth has been transformed by the development of AFM techniques. Studies of the growth of crystals have been carried out for over one hundred years, producing a wealth of information. However, in all these studies it has never been possible to observe crystals growing unit cell-by-unit cell, molecule-by-molecule. Crystal growth models at the nanoscale have been developed based upon circumstantial evidence gleaned from the shapes and sizes of the final crystals formed. Knowing this nanoscopic detail is vital to understanding not only how crystals grow perfectly but also, more importantly for today's applications, (i) how crystal growth goes wrong and defects are incorporated and (ii) how crystals switch growth modes at the unit-cell scale, forming crystal intergrowths that might ultimately be controlled at this scale, thereby fashioning crystals with multiple functionality. In the last 10 years, crystal growth studies have been transformed with the development of scanning probe microscopies that, for the first time, permit *in situ* observation of molecular detail in solution under growth conditions. STM readily achieves atomic resolution, but this is usually under high-vacuum conditions and requires conducting samples. AFM is the technique of choice for observing processes under solution conditions on any crystal type. This technique readily achieves Å-level vertical resolution and nanometre lateral resolution, ideal for observing terrace growth at crystal surfaces, as the terraces reflect a similar aspect ratio. However, it is also possible to achieve true atomic resolution, and indeed sub-atomic resolution, by careful choice of cantilever and operation of the microscope. Ultimately, AFM has superior lateral resolution to STM, although achieving this under solution conditions is at the forefront of current technology and requires considerable development. As a consequence of these and other breakthroughs in technologies, it is now possible to map precisely how crystals develop, with the prospect of creating designer crystals honed precisely to end-user requirements. There are a number of international groups who are pursuing molecular-scale crystal growth studies, examining nanoporous materials,[176–181] proteins and molecular crystals,[182] calcite and other biominerals[183,184] and minerals in general.[185–187] There are also groups working on materials as diverse as macromolecules, viruses,[188] lysozyme,[189] polymers,[190] silicon carbide[191] and insulin.[192]

Work on nanoporous materials reported the first use of AFM to study the growth mechanism in a zeolite.[198] Since that first publication, the methodology has been developed to use a variety of techniques, in order to probe the rather complex train of events that occurs during the crystallisation of a nanoporous material. The approach combines both experimental and theoretical methods. Experimentally, it is crucial to follow both solution speciation and the development of the solid-state phase. The former has been targeted by a combination of mass spectrometry (MS) and NMR spectroscopy.[199–201] These techniques are complementary, with MS delivering fast, temporal identification of solution species with limited speciation resolution and NMR spectroscopy providing accurate speciation identification at the expense of a longer spectral collection times. Together, they can disentangle the rather complex solution chemistry of oxide precursors. Development of the solid, crystalline phase can then be studied by a combination of high-resolution transmission electron microscopy (HRTEM), high-resolution scanning electron microscopy (HRSEM) and AFM. HRTEM is used to identify the termination structure at the crystal surface and the type and number of internal defect structures.[202] Understanding the incorporation of defects is critical in understanding crystal growth, as any model developed must be able to account for not only the regular crystal structure but also the inclusion of well-defined defects. In framework structures, the defect signatures are particularly varied and are unique to different crystal systems, and, consequently, act as very important guides to the growth process. HRSEM has also been developed such that it has been shown to be possible to observe sub-nanometer growth features at the surfaces of non-conducting framework materials without having to coat the surfaces with a conducting film.[203,204] It has been demonstrated using a combination of HRSEM and AFM on the same crystals that any electron-beam damage is confined to sub-surface layers and, therefore, does not interfere with delicate surface structures.[203] Reference [205] gives a link to a video clip that shows both HRSEM and AFM measurements taken at the same place on the same crystal of STA-7 – a nanoporous silicoaluminophosphate (quite a technical achievement); the grey image is the HRSEM and the orange image the

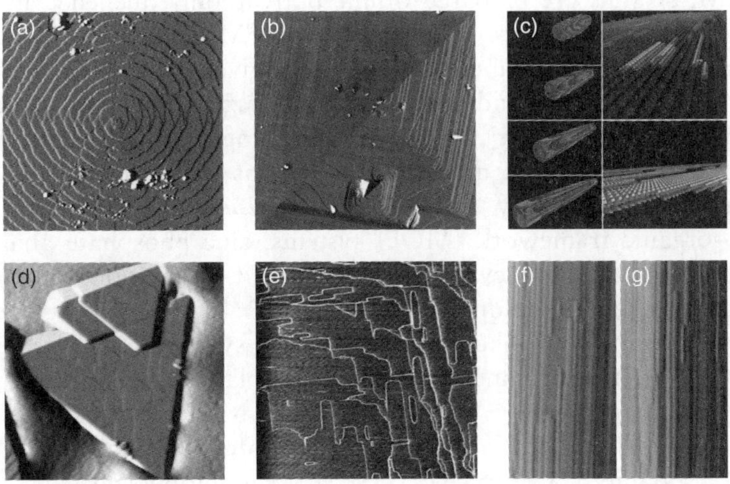

Figure 3.30 AFM images of: (a) interlaced spiral on aluminophosphate STA-7, reprinted with permission from [193] Copyright (2009) American Chemical Society; (b) zeolite A, recorded by both AFM and HRSEM, reprinted with permission from [194] Copyright (2008) Royal Society of Chemistry; (c) Monte Carlo simulation of LTL, reprinted with permission from [139] Copyright (2010) American Chemical Society; (d) *in situ* $NaZnPO_4$ growth FAU structure, reprinted with permission from [195] Copyright (2011) American Chemical Society; (e) exothermic nanochemistry highlighted by LFM on $NaZnPO_4$ with SOD structure, reprinted with permission from [196] Copyright (2012) American Chemical Society; (f,g) *in situ* dissolution of zeolite L, reprinted with permission from [197] Copyright (2008) Wiley-VCH.

AFM. This allows a rapid screen of complex crystal surfaces that may be twinned or highly intergrown, which would be rather difficult to tackle using AFM (see Figure 3.30b).

Another advantage of this technique over AFM is the enhanced lateral resolution achieved at the terrace edges, which is superior when using HRSEM. It has also been demonstrated, for the first time, that by slicing micron-sized crystals with an Ar ion beam it is possible, using HRSEM, to observe internal faulting structures that hold clues to the crystal growth history.[206] However, AFM is the most versatile and powerful technique for exploring crystal growth from a solution environment.[6,195,196,207–216] Nanoporous materials grow in environments ranging from those that are quite mild and benign to those that are highly aggressive. In the extreme, this ranges from near room temperature in an aqueous solution at pH 7 or a simple organic solvent environment at 1 atm pressure to solutions in excess of pH 13 at 220 °C and at tens of atmospheres pressure. In order to cover this complete range of conditions for AFM measurement, a dual strategy has been developed. For conditions of temperature and pressure inaccessible to AFM, crystals are prepared offline but carefully quenched in order to preserve their surface structure. The AFM is then performed on crystals in an *ex situ* manner. This strategy gives the greatest flexibility in terms of structure and conditions, although some very rapid and crucial steps can be lost during the quenching step. Therefore, where possible, *in situ* studies are preferred. It has been shown that some systems are particularly amenable to *in situ* study, including most metal–organic framework (MOF) systems, zinc phosphate analogues of aluminosilicate framework structures and dissolution rather than growth studies of aluminosilicates.[139,195,211,212,214–217] Extensive studies to date have looked at: untemplated syntheses such as zeolites A[208,218–221], Y[198,222] and L;[139,197,223] templated preparations such as MFI/MEL intergrowths[224] and SSZ-42;[225] MOFs such as HKUST-1, ZIF-8 and MOF-5;[212–214,216] and nanoporous phosphates such as $ZnPO_4$ sodalite,[196,215] STA-7,[193] Mn aluminophosphates[226] and SAPO-34.[227] The following references, which provide links to short videos from the authors' laboratory, illustrate some of the power of AFM to tackle crystal growth problems, and all illustrate firsts within the field: the Frank–Reed dislocation loop on ZIF-8;[228] spiral and birth-and-spread growth on ZIF-8;[229] interlaced spiral growth on zincophosphate (SOD) (100);[229] multiple, interlaced spiral growth on zincophosphate (SOD) (100);[230] spiral birth-and-spread growth on zincophosphate (SOD) (111);[231] zeolite A dissolution;[232] zeolite L dissolution, with lateral deflection highlighting where the chemistry is taking place through changes in friction;[233] birth-and-spread growth on zincophosphate (FAU) (111);[234] and anisotropic friction and growth, highlighted in lateral force (ZnPO–SOD).[235]

A particularly important recent publication[139] has further demonstrated that the nanoscopic region, where chemical activity takes place, can be directly targeted and highlighted by monitoring the energy imparted from the crystal surface to the AFM tip during a local exothermic or endothermic process (see Figure 3.30e and reference [233]). This is a tremendous breakthrough and is a general phenomenon that has now been observed across many crystal systems, offering tremendous potential for delineating the different energetics at different growth sites discerned at the nanoscale. The same paper also shows, for the first time, extraction of accurate activation energies for specific nanoscopic processes during the chemical transformation of a zeolite surface. Until now it was only possible to determine activation energies for bulk (average) crystal growth phenomena experimentally. By having access to activation energies for specific processes, the possibility of turning some processes off or on selectively has been revealed. Such control is exactly what is required in order to tune defect structures and defect concentration, intergrowth structures and ultimately gross morphological change, such as in the production of single-crystal thin films.

This same paper[139] also made an exciting discovery with enormous potential ramifications for crystal growth studies in general. AFM works by tracking a sharp needle (tip) across a surface with atomic precision, thereby tracking the surface topography – essentially creating a topographic map on which the contours can be measured in nanometres. The force between this tip and the surface can be controlled to the level of nN and, depending on the force, the tip can play a role in the crystal growth process. In usual operation, this interference is eliminated, so that the underlying fundamental crystal growth process is revealed. However, it is also possible to use the tip in a proactive manner to unstitch the crystal, nanoscopic unit by nanoscopic unit. As each unit is decoupled from the crystal, an associated energy is imparted to the AFM tip, which can be measured. Further, the persistence of each unit before it is decoupled gives a measurement of the stability of the unit. This is somewhat akin to the use of AFM to unfold proteins *via* so-called 'nanofishing' and extract the associated energies[236] that has revolutionised the study of protein folding. Thus, a technique is now available that can probe all the individual sub-steps in a crystal growth process and should be applicable across all crystal classes. A detailed, coarse-grained Monte Carlo approach to the modelling of zeolite L growth that predicts accurately, for the first time, crystal habit, aspect ratio and surface structure in terms of energies of attachment and supersaturation, has also proved possible (see references [237] and [238] for 2D and 3D videos, respectively). This permits the calculation of both metastable, local-equilibrium crystal shapes and surface structures and the effect of raised or lowered supersaturation. Consequently, it is now possible to follow the course of a crystallisation from high supersaturation, monitoring habit and surface topography *via* HRSEM and AFM, to near equilibrium, and to model the process in order to extract the relevant crystal attachment energies.

3.6.5 Atom Manipulation with AFM

AFM is capable of manipulating atoms at surfaces, even at room temperature, in a manner similar to STM. The first example of such manipulation was achieved in 2005, when it was shown to be possible to manipulate Sn atoms on a Ge(111) surface (Figure 3.31).[239] Clean Si tips and a clean Ge(111) surface were prepared in UHV and reconstructed Ge(111)-c(2×8) surfaces were produced by argon sputtering techniques; Ge(111)-c(2×8) was chosen because it represents a surface with rather weak bonding of adatoms to the substrate. Sn adatoms were added to the surface and it was found to be possible to induce their directional diffusion by scanning at stronger interaction forces in specific tip–scan directions. The manipulation was carried out as follows: the tip–scan direction was aligned between an Sn and a Ge adatom; the tip was lifted about 1 Å above the surface; the Sn atom was moved by the tip with successive reductions in tip–surface distance until it replaced the Ge

Figure 3.31 Topographic nc-AFM image of Sn on Ge (111), reprinted with permission from [239] Copyright (2005) Nature Publishing Group.

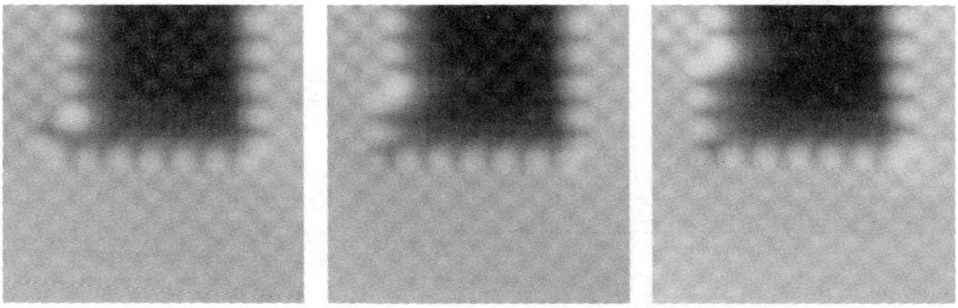

Figure 3.32 Movement of the kink site on KBr along a step edge, induced by nc-AFM, reprinted with permission from [240] Copyright (2009) Nature Publishing Group.

atom. The process can be reversed. It is possible to control the manipulation well, and artificial nanostructures, such as that shown in Figure 3.31, can be created. In a different type of manipulation, shown in Figure 3.32, a kink site on KBr(100) was dragged along the surface at a tip–surface temperature of 80 K. By performing successive line scans over the kink and increasing the tip–surface force progressively, the kink was forced to migrate along the step edge.[240]

3.6.6 Data Storage

An extension of the concept of atom manipulation at surfaces is the ability to store data in a manner that allows read/write/erase functions. Using an AFM device, it is possible to achieve 1 Tb in^{-2} or more while performing every read/write/erase function. This is to be compared with magnetic storage devices, which currently work on an areal density of *ca* 10 Gb in^{-2}; however, magnetic storage devices can easily attain data rates of 1 Gb s^{-1}, whereas with AFM cantilevers the data rate is limited by the mechanical resonant frequency of a few Mb s^{-1}. The solution to this is to use arrays of cantilevers operating in parallel in what is known as a micro-electro-mechanical system (MEMS). Different types of MEMS device have been developed, and one, called the Millipede,[241] utilises vast 32×32 AFM cantilever arrays with a thermomechanical sensor that has read/write/erase functionality. Figure 3.33 shows an example of the Millipede set-up that can achieve areal densities approaching 1 Tb in^{-2}. The information is stored as either an indentation or the lack of an indentation by writing on a nanometre-thick polymer layer. The thermochemical writing is achieved by a simultaneous application of force and local heating (to 400 °C) on the polymer, and each cantilever

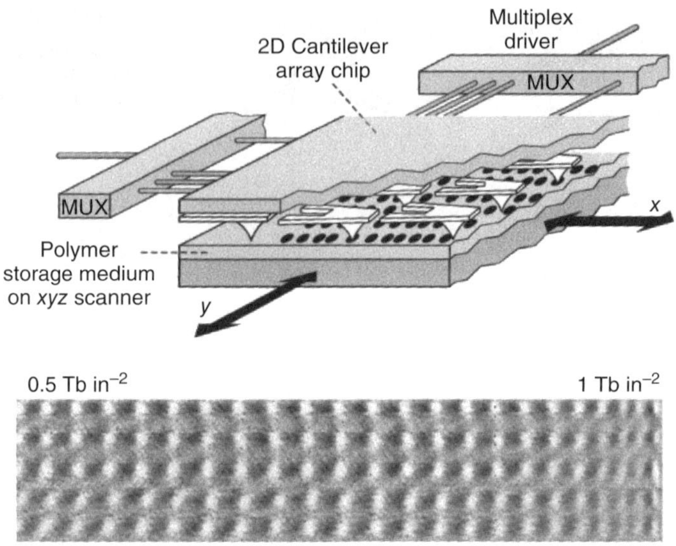

Figure 3.33 MEMS storage device, achieving areal densities approaching 1 Tb in^{-2}, reprinted with permission from [241] Copyright (2003) IEEE.

can write over a total area of $100 \times 100\,\mu m^2$. Figure 3.33 also shows the result of a single-lever experiment, where indentations are spaced 25 nm apart, although there is some degradation in the read/write quality. In order to read the data, the thermochemical sensor is used in reverse. Operating at a temperature below that which will soften the polymer (350 °C), thermal sensing is achieved as the thermal conductance between the heater platform and the polymer changes according to the distance between them.

Erasure of the data is then achieved by creating a ring around an indentation such that the depth of the previous indentation decreases markedly. By adjusting the distances properly, it is therefore possible to achieve an erase function either at the line or even at the bit level.

However, the AFM tip can respond not only to topographic information but also to changes in the surface, such as electrical and optical properties.[242] One possible scheme is based upon charge storage in a nitrideoxide semiconductor, whereby the tip is used as a moving electrode.[243] Another method utilises a conducting tip and injects current into a chalcogenide film, thereby changing the material from an amorphous to a crystalline phase,[244] as a consequence of which the resistivity of the film changes, which can be read by the tip.

3.6.7 Oxide Epitaxial Overgrowths

The epitaxial overgrowth of calcite in particular has been studied in detail by AFM. This system is particularly amenable to study by *in situ* AFM as most of the growth occurs at room temperature from clear solutions on large substrates. A variety of solid solutions consisting of Ba^{2+}, Sr^{2+}, Mn^{2+}, Cd^{2+} or Mg^{2+} have been grown from multicomponent aqueous solutions.[245] The differences in the ionic radii of these divalent cations lead to a lattice mismatch between the growing layer and the underlying calcite substrate, and this in turn causes a deviation from the expected equilibrium partitioning and consequent compositional zoning. The AFM images reveal a very active role for the underlying lattice in the growth process, in addition to the role of the solution. Figure 3.34 shows an AFM growth sequence for the {10–14} surface of calcite in contact with a strontium concentration of 4.0 mmol dm^{-3}, demonstrating a thickening of the advancing steps followed by eventual dissolution and step retreat. In a separate study,[246] growth on calcite in the presence of cobalt showed that growth of the first layer proceeds by step growth

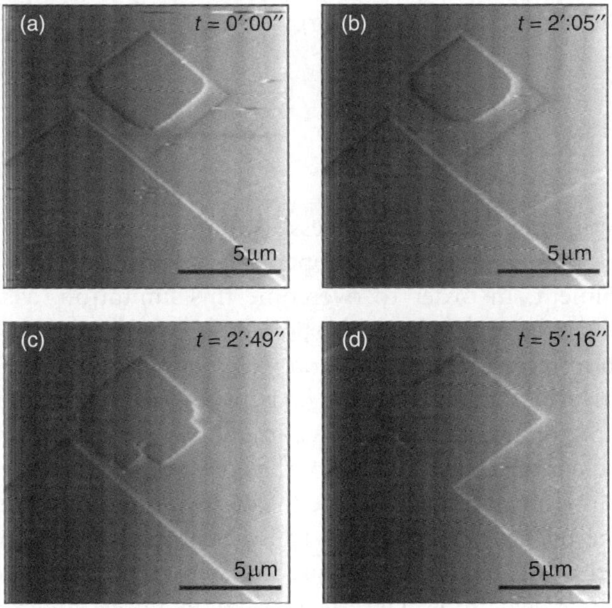

Figure 3.34 Calcite overgrowth from 4.00 mmol dm^{-3} $SrCO_3$ solution, reprinted with permission from [245] Copyright (2006) Elsevier Ltd.

faster than growth in pure $CaCO_3$ solution. Subsequent growth is then substantially slower, even at a constant supersaturation. The surface acts like a 'template' and the original surface topography is mimicked. By carrying out dissolution experiments, it was possible to show that the Co^{2+} was sorbed preferentially at the negative (acute) kink sites, which led to monomolecular, semi-triangular etch pits. Heteroepitaxial growth on both $CaCO_3$ and $MgCO_3$ by exposure to aqueous O_2 in order to form the oxide showed that the former does not promote growth whereas the latter does.[247] Rhombohedral oxide islands form on terraces with a 90° rotation relative to the crystallographic axis of the substrate. These islands grow to several microns in size but are self-limited to only 2–3 nm in height. They are unable to grow across terrace steps or down into etch pits. The self-limiting nature of these islands can be explained in terms of the free energy of the epitaxial layers. In a similar fashion, anglesite ($PbSO_4$) has been grown on the (001) facet of celestite ($SrSO_4$) and monitored *in situ* by AFM.[248] At the initial concentrations used for the heteroepitaxial growth, only two anglesite monolayers were formed and subsequent multilayer growth was strongly inhibited. AFM revealed that this inhibition was caused by an in-plane strain generated by the very small lattice mismatch (1.1%) between the two minerals. By increasing the supersaturation in order to achieve multilayer growth, it was possible to determine an in-plane strain energy of $11.4 \pm 0.6\,mJ\,m^{-2}$.

3.6.8 Hydrothermal AFM

Many growth and dissolution processes of interest occur at temperatures and pressures outside the normal operating conditions of a commercial AFM instrument. In order to overcome this limitation, a small number of scientists have developed bespoke HAFM instruments capable of extending both pressure and temperature ranges (see Section 3.4.6). One example is the thermal dehydration on gypsum (010) surfaces[249] at temperatures ranging from 25 to 130 °C. It is shown that dehydration takes place by nucleation and spreading of the etch-pit pattern, where the etch pits spread laterally in an unrestricted fashion. Conversely, in the vertical direction the pit growth is limited to a few microns. Another application in which HAFM is becoming used increasingly is the study of reactions in supercritical fluids. The effects of, for example, supercritical CO_2 ($scCO_2$) on the calcite mineral surface have been investigated by HAFM at pressures up to 100 atm and temperatures up to 350 K.[97] Images of the freshly cleaved surfaces and of surfaces aged under $scCO_2$ reveal both atomic resolution and real-time, *in situ* processes and demonstrate the versatility of this approach in the investigation of mineral surfaces.

REFERENCES

[1] F. J. Giessibl, S. Hembacher, H. Bielefeldt and J. Mannhart, *Science*, **289**, 422 (2000).

[2] F. J. Giessibl and C. F. Quate, *Physics Today*, **59**, 44 (2006).

[3] L. Gross, F. Mohn, P. Liljeroth, J. Repp, F. J. Giessibl and G. Meyer, *Science*, **324**, 1428 (2009).

[4] P. Eaton and P. West, *Atomic Force Microscopy*, Oxford University Press, Oxford, 2010.

[5] V. J. Morris, A. P. Gunning and A. R. Kirby, *Atomic Force Microscopy for Biologists*, Imperial College Press, London, 2010.

[6] C. M. Pina and G. Jordan, in *Nanoscopic Approaches in Earth and Planetary Sciences, Vol.* **8**, F. E. Brenker and G. Jordan (Eds), European Union of Mineralogy and the Mineralogical Society of Great Britain and Ireland, 2010.

[7] G. Shmalz, *Verein Deutscher Ingenieure*, 1461 (1929).

[8] H. Becker, O. Bender, L. Bergmann, K. Rost and A. Zobel, US Patent 2728222 (1955).

[9] Y. Russell, W. John and S. Fredric, *Rev. Sci. Instrum.*, **43**, 999 (1972).

[10] G. Binnig, H. Rohrer, C. Gerber and E. Weibel, *Phys. Rev. Lett.*, **49**, 57 (1982).

[11] G. Binnig and H. Rohrer, *Helv. Phys. Acta*, **55**, 726 (1982).

[12] G. Binnig, H. Rohrer, C. Gerber and E. Weibel, *Phys. Rev. Lett.*, **50**, 120 (1983).

[13] G. Binnig, C. F. Quate and C. Gerber, *Phys. Rev. Lett.*, **56**, 930 (1986).

[14] T. Ando, N. Kodera, E. Takai, D. Maruyama, K. Saito and A. Toda, *Proc. Natl. Acad. Sci.*, **98**, 12468 (2001).

[15] L. M. Picco, L. Bozec, A. Ulcinas, D. J. Engledew, M. Antognozzi, M. A. Horton and M. J. Miles, *Nanotechnology*, **18**, 044030/1 (2007).

[16] A. D. L. Humphris, M. J. Miles and J. K. Hobbs, *Appl. Phys. Lett.*, **86**, 034106 (2005).

[17] N. Gabi, R. C. Sidney, M. M. Gary, H. Don and C. M. Mate, *Rev. Sci. Instrum.*, **61**, 2296 (1990).

[18] T. Goddenhenrich, H. Lemke, U. Hartmann and C. Heiden, *J. Vac. Sci. Technol. B*, **8**, 383 (1990).

[19] U. Stahl, C. W. Yuan, A. L. de Lozanne and M. Tortonese, *Appl. Phys. Lett.*, **65**, 2878 (1994).

[20] P. Zhiqiang and W. Paul, *Appl. Phys. Lett.*, **86**, 014107 (2005).

[21] C. Schonenberger and S. F. Alvarado, *Rev. Sci. Instrum.*, **60**, 3131 (1989).

[22] D. Sarid, D. Iams, V. Weissenberger and L. S. Bell, *Opt. Lett.*, **13**, 1057 (1988).

[23] M. Gerhard and M. A. Nabil, *Appl. Phys. Lett.*, **57**, 2089 (1990).

[24] M. Gerhard and M. A. Nabil, *Appl. Phys. Lett.*, **53**, 1045 (1988).

[25] S. Alexander, L. Hellemans, O. Marti, J. Schneir, V. Elings, P. K. Hansma, L. Matt and G. John, *J. Appl. Phys.*, **65**, 164 (1989).

[26] M. Gerhard and M. A. Nabil, *Appl. Phys. Lett.*, **56**, 2100 (1990).

[27] R. Garcia, *Amplitude Modulation Atomic Force Microscopy*, Wiley-VCH, Weinheim, 2010.
[28] P. Parot, Y. F. Dufrêne, P. Hinterdorfer, C. Le Grimellec, D. Navajas, J.-L. Pellequer and S. Scheuring, *J. Mol. Recognit.*, **20**, 418 (2007).
[29] J. Drelich and K. L. Mittal (Eds), *Atomic Force Microscopy in Adhesion Studies*, VSP, Leiden–Boston, 2005.
[30] M. Ludwig, M. Rief, L. Schmidt, H. Li, F. Oesterhelt, M. Gautel and H. E. Gaub, *Appl. Phys. A*, **68**, 173 (1999).
[31] H. W. Wu, T. Kuhn and V. T. Moy, *Scanning*, **20**, 389 (1998).
[32] K. Hyonchol, H. Arakawa, T. Osada and A. Ikai, *Colloids Surf. B: Biointerfaces*, **25**, 33 (2002).
[33] A. Yacoot and L. Koenders, *J. Phys. D: Appl. Phys.*, **41**, 103001 (2008).
[34] J. Hafner, E. Chen, R. Lal and S. Jin, in *Springer Handbook of Nanotechnology*, B. Bhushan (Ed), Springer-Verlag, Berlin, 2010.
[35] S. Habermehl, *J. Appl. Phys.*, **83**, 4672 (1998).
[36] A. Khan, J. Philip and P. Hess, *J. Appl. Phys.*, **95**, 1667 (2004).
[37] J. J. Wortman and R. A. Evans, *J. Appl. Phys.*, **36**, 153 (1965).
[38] K. Unno, T. Shibata and E. Makino, *Sens. Actuators A*, **88**, 247 (2001).
[39] P. Niedermann, W. Hänni, D. Morel, A. Perret, N. Skinner, P. F. Indermühle, N. F. de Rooij and P. A. Buffat, *Appl. Phys. A: Mater. Sci. Process.*, **66**, S31 (1998).
[40] E. I. Givargizov, A. N. Stepanova, E. S. Mashkova, V. A. Molchanov, F. Shi, P. Hudek and I. W. Rangelow, *Microelectron. Eng.*, **41–42**, 499 (1998).
[41] R. B. Marcus, T. S. Ravi, T. Gmitter, K. Chin, D. Liu, W. J. Orvis, D. R. Ciarlo, C. E. Hunt and J. Trujillo, *Appl. Phys. Lett.*, **56**, 236 (1990).
[42] M. J. Vasile, D. Grigg, J. E. Griffith, E. Fitzgerald and P. E. Russell, *J. Vac. Sci. Technol. B*, **9**, 3569 (1991).
[43] H. Ximen and P. E. Russell, *Ultramicroscopy*, **42–44**, 1526 (1992).
[44] K. I. Schiffmann, *Nanotechnology*, **4**, 163 (1993).
[45] C. M. Harris and A. G. Piersol, *Shock and Vibration Handbook*, McGraw-Hill, New York, 2002.
[46] C. A. Clifford and M. P. Seah, *Nanotechnology*, **16**, 1666 (2005).
[47] J. E. Sader, J. W. M. Chon and P. Mulvaney, *Rev. Sci. Instrum.*, **70**, 3967 (1999).
[48] J. E. Sader, I. Larson, P. Mulvaney and L. R. White, *Rev. Sci. Instrum.*, **66**, 3789 (1995).
[49] J. E. Sader, *J. Appl. Phys.*, **84**, 64 (1998).
[50] J. P. Cleveland, S. Manne, D. Bocek and P. K. Hansma, *Rev. Sci. Instrum.*, **64**, 403 (1993).
[51] R. S. Gates and M. G. Reitsma, *Rev. Sci. Instrum.*, **78**, 086101 (2007).
[52] J. L. Hutter and J. Bechhoefer, *Rev. Sci. Instrum.*, **64**, 1868 (1993).
[53] H. J. Butt and M. Jaschke, *Nanotechnology*, **6**, 1 (1995).
[54] P. K. Hansma, G. Schitter, G. E. Fantner and C. Prater, *Science*, **314**, 601 (2006).
[55] G. E. Fantner, G. Schitter, J. H. Kindt, T. Ivanov, K. Ivanova, R. Patel, N. Holten-Andersen, J. Adams, P. J. Thurner, I. W. Rangelow and P. K. Hansma, *Ultramicroscopy*, **106**, 881 (2006).
[56] M. B. Viani, T. E. Schaffer, A. Chand, M. Rief, H. E. Gaub and P. K. Hansma, *J. Appl. Phys.*, **86**, 2258 (1999).
[57] Asylum Research, http://www.asylumresearch.com (last accessed 28 October 2013).

[58] N. Hilal, D. Johnson, W. R. Bowen and P. M. Williams, in *Atomic Force Microscopy in Process Engineering*, Butterworth-Heinemann, Oxford, 2009.
[59] M. D. Louey, P. Mulvaney and P. J. Stewart, *J. Pharm. Biomed. Anal.*, **25**, 559 (2001).
[60] J. Nalaskowski, J. A. Drelich, J. Hupka and J. D. Miller, *Langmuir*, **19**, 5311 (2003).
[61] L. Sirghi, O. Kylin, D. Gilliland, G. Ceccone and F. Rossi, *J. Phys. Chem. B*, **110**, 25975 (2006).
[62] Y.-S. Lo, N. D. Huefner, W. S. Chan, P. Dryden, B. Hagenhoff and T. P. Beebe, *Langmuir*, **15**, 6522 (1999).
[63] T. Arai and M. Tomitori, *Appl. Phys. A: Mater. Sci. Process.*, **66**, S319 (1998).
[64] M. Morgenstern, A. Schwarz and U. D. Schwarz, in *Springer Handbook of Nanotechnology*, B. Bhushan (Ed), Springer-Verlag, Berlin, 2010, p. 663.
[65] C. K. Riener, C. M. Stroh, A. Ebner, C. Klampfl, A. A. Gall, C. Romanin, Y. L. Lyubchenko, P. Hinterdorfer and H. J. Gruber, *Anal. Chim. Acta*, **479**, 59 (2003).
[66] P. Hinterdorfer, K. Schilcher, W. Baumgartner, H. J. Gruber and H. Schindler, *Nanobiology*, **4**, 177 (1998).
[67] I. Touzov and C. B. Gorman, *J. Phys. Chem. B*, **101**, 5263 (1997).
[68] H. F. Knapp and A. Stemmer, *Surf. Interface Anal.*, **27**, 324 (1999).
[69] C. D. Blanchette, A. Loui and T. V. Ratto, in *Handbook of Molecular Force Spectroscopy*, A. Noy (Ed), Springer-Verlag US, New York, 2008.
[70] Nanosensors, http://www.nanosensors.com (last accessed 7 October 2013).
[71] S. Akita, H. Nishijima, Y. Nakayama, F. Tokumasu and K. Takeyasu, *J. Phys. D: Appl. Phys.*, **32**, 1044 (1999).
[72] C. V. Nguyen, K.-J. Chao, R. M. D. Stevens, L. Delzeit, A. Cassell, J. Han and M. Meyyappan, *Nanotechnology*, **12**, 363 (2001).
[73] N. R. Wilson and J. V. Macpherson, *Nat. Nano.*, **4**, 483 (2009).
[74] B. C. Park, K. Y. Jung, W. Y. Song, B. H. O and S. J. Ahn, *Adv. Mater.*, **18**, 95 (2006).
[75] F. J. Giessibl, *Appl. Phys. Lett.*, **78**, 123 (2001).
[76] U. Durig, *New J. Phys.*, **2**, 5 (2000).
[77] T. Fukuma, K. Kobayashi, K. Matsushige and H. Yamada, *Appl. Phys. Lett.*, **86**, 193108 (2005).
[78] F. Takeshi, K. Kei, M. Kazumi and Y. Hirofumi, *Appl. Phys. Lett.*, **86**, 193108 (2005).
[79] F. J. Giessibl, *Surf. Interface Anal.*, **38**, 1696 (2006).
[80] O. Sahin, C. F. Quate, O. Solgaard and F. J. Giessibl, in *Springer Handbook of Nanotechnology*, B. Bhushan (Ed), Springer-Verlag, Berlin, 2010.
[81] R. W. Stark and W. M. Heckl, *Rev. Sci. Instrum.*, **74**, 5111 (2003).
[82] J. R. Lozano and R. Garcia, *Phys. Rev. Lett.*, **100**, 076102 (2008).
[83] R. Proksch, *Appl. Phys. Lett.*, **89**, 113121 (2006).
[84] S. Patil, N. F. Martinez, J. R. Lozano and R. Garcia, *J. Mol. Recognit.*, **20**, 516 (2007).
[85] H. Hartman, G. Sposito, A. Yang, S. Manne, S. A. C. Gould and P. K. Hansma, *Clays Clay Miner.*, **38**, 337 (1990).
[86] P. E. Hillner, A. J. Gratz, S. Manne and P. K. Hansma, *Geology*, **20**, 359 (1992).
[87] L. Scandella, N. Kruse and R. Prins, *Surf. Sci. Lett.*, **281**, 331 (1993).
[88] F. Ohnesorge and G. Binnig, *Science*, **260**, 1451 (1993).

[89] Y. Gan, *Surf. Sci. Rep.*, **64**, 99 (2009).
[90] U. D. Schwarz, H. Holscher and R. Wiesendanger, *Phys. Rev. B*, **62**, 13089 (2000).
[91] T. R. Albrecht, P. Grutter, D. Horne and D. Rugar, *J. Appl. Phys.*, **69**, 668 (1991).
[92] F. J. Giessibl, *Science*, **267**, 68 (1995).
[93] M. J. Higgins, J. E. Sader and S. P. Jarvis, *Biophys. J.*, **90**, 640 (2006).
[94] Bruker, http://www.bruker.com (last accessed 7 October 2013).
[95] S. R. Higgins, C. M. Eggleston, K. G. Knauss and C. O. Boro, *Rev. Sci. Instrum.*, **69**, 2994 (1998).
[96] S. R. Higgins, L. H. Boram, C. M. Eggleston, B. A. Coles, R. G. Compton and K. G. Knauss, *J. Phys. Chem. B*, **106**, 6696 (2002).
[97] A. S. Lea, S. R. Higgins, K. G. Knauss and K. M. Rosso, *Rev. Sci. Instrum.*, **82**, 043709 (2011).
[98] T. Gutsmann, G. E. Fantner, J. H. Kindt, M. Venturoni, S. Danielsen and P. K. Hansma, *Biophys. J.*, **86**, 3186 (2004).
[99] N. Kodera, H. Yamashita and T. Ando, *Rev. Sci. Instrum.*, **76**, 053708 (2005).
[100] G. Schitter, F. Allgower and A. Stemmer, *Nanotechnology*, **15**, 108 (2004).
[101] G. E. Fantner, P. Hegarty, J. H. Kindt, G. Schitter, G. A. G. Cidade and P. K. Hansma, *Rev. Sci. Instrum.*, **76**, 026118 (2005).
[102] M. Tortonese, R. C. Barrett and C. F. Quate, *Appl. Phys. Lett.*, **62**, 834 (1993).
[103] I. Shiraki, Y. Miyatake, T. Nagamura and K. Miki, *Rev. Sci. Instrum.*, 77 (2006).
[104] J. A. Harley and T. W. Kenny, *Appl. Phys. Lett.*, **75**, 289 (1999).
[105] G. Villanueva, J. A. Plaza, J. Montserrat, F. Perez-Murano and J. Bausells, *Microelectron. Eng.*, **85**, 1120 (2008).
[106] T. Gotszalk, P. Grabiec and I. W. Rangelow, *Ultramicroscopy*, **82**, 39 (2000).
[107] H. Takahashi, K. Ando and Y. Shirakawabe, *Ultramicroscopy*, **91**, 63 (2002).
[108] T. Itoh and T. Suga, *Nanotechnology*, **4**, 218 (1993).
[109] T. Itoh, C. Lee and T. Suga, *Appl. Phys. Lett.*, **69**, 2036 (1996).
[110] S. Watanabe and T. Fujii, *Rev. Sci. Instrum.*, **67**, 3898 (1996).
[111] B. Rogers, L. Manning, T. Sulchek and J. D. Adams, *Ultramicroscopy*, **100**, 267 (2004).
[112] C. Lee, T. Itoh, T. Ohashi, R. Maeda and T. Suga, *J. Vac. Sci. Technol. B*, **15**, 1559 (1997).
[113] R. D. Piner, J. Zhu, F. Xu, S. Hong and C. A. Mirkin, *Science*, **283**, 661 (1999).
[114] D. S. Ginger, H. Zhang and C. A. Mirkin, *Angew. Chem., Int. Ed.*, **43**, 30 (2004).
[115] Y. Xia, X.-M. Zhao, E. Kim and G. M. Whitesides, *Chem. Mater.*, **7**, 2332 (1995).
[116] K. Salaita, Y. Wang and C. A. Mirkin, *Nat Nano*, **2**, 145 (2007).
[117] E. H. Synge, *Philos. Mag.*, **6**, 356 (1928).
[118] A. Lewis, M. Isaacson, A. Harootunian and A. Muray, *Ultramicroscopy*, **13**, 227 (1984).
[119] F. Zenhausern, Y. Martin and H. K. Wickramasinghe, *Science*, **269**, 1083 (1995).
[120] H. Gai, J. Wang, Q. Tian, W. Xia and X. Xu, *Appl. Opt.*, **46**, 7746 (2007).
[121] Q. Wang, J. Bu and X. C. Yuan, *Opt. Express*, **18**, 2662 (2010).
[122] R. Geiss, S. Diziain, R. Iliew, C. Etrich, H. Hartung, N. Janunts, F. Schrempel, F. Lederer, T. Pertsch and E. B. Kley, *Appl. Phys. Lett.*, **97**, 131109.
[123] Y. Toda, M. Kourogi, M. Ohtsu, Y. Nagamune and Y. Arakawa, *Appl. Phys. Lett.*, **69**, 827 (1996).
[124] G. Kaupp, in Atomic Force Microscopy, Scanning Nearfield Optical Microscopy and Nanoscratching, Springer, Berlin Heidelberg, New York, 2006.

[125] R. M. Stockle, Y. D. Suh, V. Deckert and R. Zenobi, *Chem. Phys. Lett.*, **318**, 131 (2000).

[126] J. Steidtner and B. Pettinger, *Phys. Rev. Lett.*, **100**, 236101 (2008).

[127] B. Pettinger, P. Schambach, C. J. Villagomez and N. Scott, *Annu. Rev. Phys. Chem.*, **63**, 379 (2012).

[128] N. F. Martinez and R. Garcia, *Nanotechnology*, **17**, S167 (2006).

[129] R. García, J. Tamayo and A. S. Paulo, *Surf. Interface Anal.*, **27**, 312 (1999).

[130] K. L. Johnson, K. Kendall and A. D. Roberts, *Proc. R. Soc. London A*, **324**, 301 (1971).

[131] B. V. Derjaguin, V. M. Muller and Y. P. Toporov, *J. Colloid Interface Sci.*, **53**, 314 (1975).

[132] R. G. Cain, S. Biggs and N. W. Page, *J. Colloid Interface Sci.*, **227**, 55 (2000).

[133] M. Munz, *J. Phys. D: Appl. Phys.*, **43**, 063001 (2010).

[134] I. Szlufarska, M. Chandross and R. W. Carpick, *J. Phys. D: Appl. Phys.*, **41**, 123001 (2008).

[135] R. W. Carpick and M. Salmeron, *Chem. Rev.*, **97**, 1163 (1997).

[136] B. Bhushan, *Springer Handbook of Nanotechnology*, Springer-Verlag, Berlin, 2010.

[137] P. Cubillas and S. Higgins, *Geochem. Trans.*, **10**, 7 (2009).

[138] X. Hu, P. Cubillas and S. R. Higgins, *Langmuir*, **26**, 4769 (2010).

[139] R. Brent, P. Cubillas, S. M. Stevens, K. E. Jelfs, A. Umemura, J. T. Gebbie, B. Slater, O. Terasaki, M. A. Holden and M. W. Anderson, *J. Am. Chem. Soc.*, **132**, 13858 (2010).

[140] H. Holscher, D. Ebeling and U. D. Schwarz, *Phys. Rev. Lett.*, **101**, 246105 (2008).

[141] C. Rotsch and M. Radmacher, *Langmuir*, **13**, 2825 (1997).

[142] N. Chaiyut, T. Amornsakchai and S. Thanawan, *Polym. Test.*, **26**, 396 (2007).

[143] S. Porthun, L. Abelmann and C. Lodder, *J. Magn. Magn. Mater.*, **182**, 238 (1998).

[144] C. W. Lin, F. R. F. Fan and A. J. Bard, *J. Electrochem. Soc.*, **134**, 1038 (1987).

[145] S. Hosaka, A. Kikukawa, Y. Honda and T. Hasegawa, *Appl. Phys. Lett.*, **65**, 3407 (1994).

[146] H. J. Mamin, D. Rugar, J. E. Stern, B. D. Terris and S. E. Lambert, *Appl. Phys. Lett.*, **53**, 1563 (1988).

[147] S. Schreiber, M. Savla, D. V. Pelekhov, D. F. Iscru, C. Selcu, P. C. Hammel and G. Agarwal, *Small*, **4**, 270 (2008).

[148] J. E. Stern, B. D. Terris, H. J. Mamin and D. Rugar, *Appl. Phys. Lett.*, **53**, 2717 (1988).

[149] M. Zdrojek, T. Melin, H. Diesinger, D. Stievenard, W. Gebicki and L. Adamowicz, *J. Appl. Phys.*, **100**, 114326 (2006).

[150] Y. Yaish, J. Y. Park, S. Rosenblatt, V. Sazonova, M. Brink and P. L. McEuen, *Phys. Rev. Lett.*, **92**, 046401 (2004).

[151] Y. Martin, D. W. Abraham and H. K. Wickramasinghe, *Appl. Phys. Lett.*, **52**, 1103 (1988).

[152] J. Waddell, R. Ou, C. J. Capozzi, S. Gupta, C. A. Parker, R. A. Gerhardt, K. Seal, S. V. Kalinin and A. P. Baddorf, *Appl. Phys. Lett.*, **95**, 233122/1 (2009).

[153] F. Nardi, D. Deleruyelle, S. Spiga, C. Muller, B. Bouteille and D. Ielmini, *J. Appl. Phys.*, **112**, 064310/1 (2012).

[154] M. Nonnenmacher, M. P. O'Boyle and H. K. Wickramasinghe, *Appl. Phys. Lett.*, **58**, 2921 (1991).

[155] A. C. Narvaez, T. Chiaramonte, K. O. Vicaro, J. H. Clerici and M. A. Cotta, *Nanotechnology*, **20**, 465705/1 (2009).
[156] M. Tanimoto and O. Vatel, *J. Vac. Sci. Technol., B*, **14**, 1547 (1996).
[157] C. S. Jiang, Z. G. Li, H. R. Moutinho, L. Liang, A. Ionkin and M. M. Al-Jassim, *J. Appl. Phys.*, **111**, 083704/1 (2012).
[158] J. R. O'Dea, L. M. Brown, N. Hoepker, J. A. Marohn and S. Sadewasser, *MRS Bull.*, **37**, 642 (2012).
[159] C. R. Barry, J. Gu and H. O. Jacobs, *Nano Lett.*, **5**, 2078 (2005).
[160] J. Park, D. Bang, K. Jang, S. Haam, J. Yang and S. Na, *Nanotechnology*, **23**, 365705/1 (2012).
[161] B. Moores, F. Hane, L. Eng and Z. Leonenko, *Ultramicroscopy*, **110**, 708 (2010).
[162] J. F. Scott, *Ferroelectric Memories*. Springer, Berlin, 2000.
[163] S. V. Kalinin, B. J. Rodriguez, S. Jesse, E. Karapetian, B. Mirman, E. A. Eliseev and A. N. Morozovska, *Annu. Rev. Mater. Res.*, **37**, 189 (2007).
[164] H. J. Hertz, *J. Reine Angew. Math.*, **92**, 156 (1882).
[165] P. Eaton, F. F. n. Estarlich, R. J. Ewen, T. G. Nevell, J. R. Smith and J. Tsibouklis, *Langmuir*, **18**, 10011 (2002).
[166] B. Bhushan and C. LaTorre, in *Springer Handbook of Nanotechnology*, B. Bhushan (Ed), Springer-Verlag, Berlin, 2010.
[167] N. P. Mortensen, J. D. Fowlkes, C. J. Sullivan, D. P. Allison, N. B. Larsen, S. R. Molin and M. J. Doktycz, *Langmuir*, **25**, 3728 (2009).
[168] G. H. Enevoldsen, A. S. Foster, M. C. Christensen, J. V. Lauritsen and F. Besenbacher, *Phys. Rev. B*, **76**, 205415 (2007).
[169] J. V. Lauritsen, M. C. R. Jensen, K. Venkataramani, B. Hinnemann, S. Helveg, B. S. Clausen and F. Besenbacher, *Phys. Rev. Lett.*, **103**, 076103 (2009).
[170] C. Barth and M. Reichling, *Nature*, **414**, 54 (2001).
[171] F. Ostendorf, S. Torbruegge and M. Reichling, *Phys. Rev. B*, **77** (2008).
[172] S. Torbruegge, F. Ostendorf and M. Reichling, *J. Phys. Chem. C*, **113**, 4909 (2009).
[173] S. Torbruegge, M. Cranney and M. Reichling, *Appl. Phys. Lett.*, **93** (2008).
[174] T. Fukuma and S. P. Jarvis, *Rev. Sci. Instrum.*, **77**, 043701 (2006).
[175] S.-H. Loh and S. P. Jarvis, *Langmuir*, **26**, 9176 (2010).
[176] M. W. Anderson, J. R. Agger, N. Hanif, O. Terasaki and T. Ohsuna, *Solid State Sci.*, **3**, 809 (2001).
[177] M. W. Anderson, *Curr. Opin. Solid State Mater. Sci.*, **5**, 407 (2001).
[178] M. W. Anderson, T. Ohsuna, Y. Sakamoto, Z. Liu, A. Carlsson and O. Terasaki, *Chem. Commun.*, **907** (2004).
[179] M. W. Anderson, J. R. Agger, L. I. Meza, C. B. Chong and C. S. Cundy, *Faraday Discuss.*, **136**, 143 (2007).
[180] M. W. Anderson, L. I. Meza, J. R. Agger, M. P. Attfield, M. Shoaee, C. B. Chong, A. Umemura and C. S. Cundy, in *Turning Points Solid-State, Materials and Surface Science*, K. D. M. Harris and P. P. Edwards (Eds), Royal Society of Chemistry, Cambridge, 2008.
[181] P. Cubillas and M. W. Anderson, in *Zeolites and Catalysis: Synthesis, Reactions and Applications*, J. Cejka, A. Corma and S. I. Zones (Eds), Wiley-VCH, 2010.
[182] T. A. Land, J. J. De Yoreo and J. D. Lee, *Surf. Sci.*, **384**, 136 (1997).
[183] C. A. Orme, A. Noy, A. Wierzbicki, M. T. McBride, M. Grantham, H. H. Teng, P. M. Dove and J. J. DeYoreo, *Nature*, **411**, 775 (2001).

[184] L. E. Wasylenki, P. M. Dove, D. S. Wilson and J. J. DeYoreo, *Geochim. Cosmochim. Acta*, **69**, 3017 (2005).
[185] M. Kowacz, M. Prieto and A. Putnis, *Geochim. Cosmochim. Acta*, **74**, 469 (2010).
[186] C. M. Pina, A. Putnis and J. M. Astilleros, *Chem. Geol.*, **204**, 145 (2004).
[187] E. Ruiz-Agudo, C. V. Putnis, L. Wang and A. Putnis, *Geochim. Cosmochim. Acta*, **75**, 3803 (2011).
[188] A. J. Malkin, Y. G. Kuznetsov, R. W. Lucas and A. McPherson, *J. Struct. Biol.*, **127**, 35 (1999).
[189] J. H. Konnert, P. D'Antonio and K. B. Ward, *Acta Crystallogr., Sect. D: Biol. Crystallogr.*, **D50**, 603 (1994).
[190] M. Schultz and M. J. Miles, *J. Polym. Sci., Part B: Polym. Phys.*, **36**, 2311 (1998).
[191] J. A. Powell and D. J. Larkin, *Phys. Status Solidi B*, **202**, 529 (1997).
[192] C. M. Yip and M. D. Ward, *Biophys. J.*, **71**, 1071 (1996).
[193] P. Cubillas, M. Castro, K. E. Jelfs, A. J. W. Lobo, B. Slater, D. W. Lewis, P. A. Wright, S. M. Stevens and M. W. Anderson, *Cryst. Growth Des.*, **9**, 4041 (2009).
[194] S. M. Stevens, P. Cubillas, K. Jansson, O. Terasaki, M. W. Anderson, P. A. Wright and M. Castro, *Chem. Commun.*, **3894** (2008).
[195] P. Cubillas, M. A. Holden and M. Anderson, *Cryst. Growth Des.*, **11**, 3163 (2011).
[196] M. A. Holden, P. Cubillas, M. P. Attfield, J. T. Gebbie and M. W. Anderson, *J. Am. Chem. Soc.*, **134**, 13066 (2012).
[197] R. Brent and M. W. Anderson, *Angew. Chem., Int. Ed.*, **47**, 5327 (2008).
[198] M. W. Anderson, J. R. Agger, J. T. Thornton and N. Forsyth, *Angew. Chem., Int. Ed.*, **35**, 1210 (1996).
[199] M. Haouas, D. P. Petry, M. W. Anderson and F. Taulelle, *J. Phys. Chem. C*, **113**, 10838 (2009).
[200] D. P. Petry, M. Haouas, S. C. C. Wong, A. Aerts, C. E. A. Kirschhock, J. A. Martens, S. J. Gaskell, M. W. Anderson and F. Taulelle, *J. Phys. Chem. C*, **113**, 20827 (2009).
[201] D. P. Petry, S. C. C. Wong, M. Haouas, F. Taulelle, S. J. Gaskell and M. W. Anderson, *Stud. Surf. Sci. Catal.*, **174**, 941 (2008).
[202] B. Slater, C. R. A. Catlow, Z. Liu, T. Ohsuna, O. Terasaki and M. A. Camblor, *Angew. Chem. Int. Ed.*, **41**, 1235 (2002).
[203] S. M. Stevens, P. Cubillas, K. Jansson, O. Terasaki and M. W. Anderson, *J. Phys. Chem. C*, **113**, 18441 (2009).
[204] S. M. Stevens, K. Jansson, N. S. John, O. Terasaki, M. W. Anderson, M. Castro, P. A. Wright and P. Cubillas, *Stud. Surf. Sci. Catal.*, **174B**, 775 (2008).
[205] http://www.chemistry.manchester.ac.uk/groups/cnm/images/afm/movie2.mov (last accessed 7 October 2013).
[206] R. Brent, S. M. Stevens, O. Terasaki and M. W. Anderson, *Cryst. Growth Des.*, **10**, 5182 (2010).
[207] J. J. De Yoreo, L. A. Zepeda-Ruiz, R. W. Friddle, S. R. Qiu, L. E. Wasylenki, A. A. Chernov, G. H. Gilmer and P. M. Dove, *Cryst. Growth Des.*, **9**, 5135 (2009).
[208] L. I. Meza, M. W. Anderson and J. R. Agger, *Chem. Commun.*, **2473** (2007).
[209] L. I. Meza, M. W. Anderson, J. R. Agger and C. S. Cundy, *Stud. Surf. Sci. Catal.*, **170A**, 177 (2007).
[210] L. I. Meza, M. W. Anderson, J. R. Agger, C. S. Cundy, C. B. Chong and R. J. Plaisted, *J. Am. Chem. Soc.*, **129**, 15192 (2007).
[211] L. I. Meza, M. W. Anderson, B. Slater and J. R. Agger, *Phys. Chem. Chem. Phys.*, **10**, 5066 (2008).

[212] N. S. John, C. Scherb, M. Shoaee, M. W. Anderson, M. P. Attfield and T. Bein, *Chem. Commun.*, **6294** (2009).
[213] P. Y. Moh, P. Cubillas, M. W. Anderson and M. P. Attfield, *J. Am. Chem. Soc.*, **133**, 13304 (2011).
[214] M. Shoaee, M. W. Anderson and M. P. Attfield, *Angew. Chem., Int. Ed.*, **47**, 8525 (2008).
[215] M. A. Holden, P. Cubillas and M. W. Anderson, *Chem. Commun.*, **46**, 1047 (2010).
[216] P. Cubillas, M. W. Anderson and M. P. Attfield, *Chem. Eur. J.*, **18**, 15406 (2012).
[217] M. Shoaee, J. R. Agger, M. W. Anderson and M. P. Attfield, *CrystEngComm*, **10**, 646 (2008).
[218] J. R. Agger, N. Pervaiz, A. K. Cheetham and M. W. Anderson, *J. Am. Chem. Soc.*, **120**, 10754 (1998).
[219] J. R. Agger, C. B. Chong and M. W. Anderson, *Stud. Surf. Sci. Catal.*, **154B**, 1282 (2004).
[220] P. Cubillas, S. M. Stevens, N. Blake, A. Umemura, C. B. Chong, O. Terasaki and M. W. Anderson, *J. Phys. Chem. C*, **115**, 12567 (2011).
[221] A. Umemura, P. Cubillas, M. W. Anderson and J. R. Agger, *Stud. Surf. Sci. Catal.*, **174A**, 705 (2008).
[222] J. R. Agger and M. W. Anderson, *Stud. Surf. Sci. Catal.*, **142A**, 93 (2002).
[223] R. Brent and M. W. Anderson, *Stud. Surf. Sci. Catal.*, **174B**, 909 (2008).
[224] N. S. John, S. M. Stevens, O. Terasaki and M. W. Anderson, *Chem. Eur. J.*, **16**, 2220 (2010).
[225] M. W. Anderson, N. Hanif, J. R. Agger, C. Y. Chen and S. I. Zones, *Stud. Surf. Sci. Catal.*, **135**, 167 (2001).
[226] L. I. Meza, J. R. Agger, N. Z. Logar, V. Kaucic and M. W. Anderson, *Chem. Commun.*, **2300** (2003).
[227] B. Holme, P. Cubillas, J. H. Cavka, B. Slater, M. W. Anderson and D. Akporiaye, *Cryst. Growth Des.*, **10**, 2824 (2010).
[228] http://www.chemistry.manchester.ac.uk/groups/cnm/images/afm/movie8.mov (last accessed 28 October 2013).
[229] http://www.chemistry.manchester.ac.uk/groups/cnm/images/afm/movie9.mov (last accessed 28 October 2013).
[230] http://www.chemistry.manchester.ac.uk/groups/cnm/images/afm/movie11.mov (last accessed 28 October 2013).
[231] http://www.chemistry.manchester.ac.uk/groups/cnm/images/afm/movie12.mov (last accessed 28 October 2013).
[232] http://www.chemistry.manchester.ac.uk/groups/cnm/images/afm/movie3.mov (last accessed 28 October 2013).
[233] http://www.chemistry.manchester.ac.uk/groups/cnm/images/afm/movie13.mov (last accessed 28 October 2013).
[234] http://www.chemistry.manchester.ac.uk/groups/cnm/images/afm/movie14.mov (last accessed 28 October 2013).
[235] http://www.chemistry.manchester.ac.uk/groups/cnm/images/afm/movie15.mov (last accessed 28 October 2013).
[236] R. B. Best, D. J. Brockwell, J. L. Toca-Herrera, A. W. Blake, D. A. Smith, S. E. Radford and J. Clarke, *Anal. Chim. Acta*, **479**, 87 (2003).
[237] http://www.chemistry.manchester.ac.uk/groups/cnm/images/afm/movie7.mov (last accessed 28 October 2013).

[238] http://stream.manchester.ac.uk/Play.aspx?VideoId=9942 (last accessed 28 October 2013).

[239] Y. Sugimoto, M. Abe, S. Hirayama, N. Oyabu, O. Custance and S. Morita, *Nat. Mater.*, **4**, 156 (2005).

[240] O. Custance, R. Perez and S. Morita, *Nat. Nanotechnol.*, **4**, 803 (2009).

[241] E. Eleftheriou, T. Antonakopoulos, G. K. Binnig, G. Cherubini, M. Despont, A. Dholakia, U. Durig, M. A. Lantz, H. Pozidis, H. E. Rothuizen and P. Vettiger, *IEEE Trans. Magn.*, **39**, 938 (2003).

[242] H. J. Mamin, R. P. Ried, B. D. Terris and D. Rugar, *Proc. IEEE*, **87**, 1014 (1999).

[243] R. C. Barrett and C. F. Quate, *J. Appl. Phys.*, **70**, 2725 (1991).

[244] H. Kado and T. Tohda, *Appl. Phys. Lett.*, **66**, 2961 (1995).

[245] J. M. Astilleros, C. M. Pina, L. Fernandez-Diaz, M. Prieto and A. Putnis, *Chem. Geol.*, **225**, 322 (2006).

[246] S. J. Freij, A. Putnis and J. M. Astilleros, *J. Cryst. Growth*, **267**, 288 (2004).

[247] Y. S. Jun, T. A. Kendall, S. T. Martin, C. M. Friend and J. J. Vlassak, *Enviro Sci Tech*, **39**, 1239 (2005).

[248] C. M. Pina and A. Rico-Garcia, *Surf. Sci.*, **603**, 2708 (2009).

[249] G. Jordan and J. M. Astilleros, *Am. Mineral.*, **91**, 619 (2006).

4

Gas Sorption in the Analysis of Nanoporous Solids

Philip L. Llewellyn
Aix-Marseille University & CNRS, Laboratoire MADIREL, Marseille, France

4.1 INTRODUCTION

When molecules in the gaseous or liquid phase approach a solid surface, their concentration at the surface will increase. This retention of the fluid by the superficial atoms of the solid is known as adsorption.[1] From this, it can be appreciated that the more solid surface that is available, the more fluid molecules can be retained. This phenomenon has applications in such spheres as heterogeneous catalysis, where the surface is used as a reaction area.

In order to increase the reactivity of these solids it is necessary to increase their surface area. One way of achieving this is grinding, but a better option is to make use of any available porosity. A nonporous solid might have some tens of square metres of surface per gram but porous materials such as zeolites and activated carbons will have hundreds or even thousands of square metres per gram.

The porosity can be used as an extra reaction surface, but can also be used in its own right as a storage medium for controlled release (*e.g.* drugs, fertilisers, hydrogen) or as a nanofilter for separation (CO_2 recovery, drug or water purification). It can, therefore, be appreciated that adsorption is found in applications ranging from the petroleum industry

Multi Length-Scale Characterisation, First Edition. Edited by Duncan W. Bruce, Dermot O'Hare and Richard I. Walton.

to agriculture, pharmacy and medicine. Closer to home, adsorbents are used as water filters, stain removers, alcohol tests, stoppers in medicine tubes, in extractor hoods and even in certain medical pills and capsules.

This interest explains why much research effort is being applied both experimentally and theoretically to the development of novel, porous materials. In the case of adsorption-based processes, it is important to gain a deeper understanding of the adsorption phenomena under conditions close to those in applications. In the separation application known as pressure swing adsorption (PSA),[2] for example, process conditions can occur up to pressures of around 30 bar at 50 °C, while some hydrogen storage applications can reach 200 bar at 77 K.

It can thus be appreciated that the surface area and porosity (pore size, pore volume) of an adsorbent are critical to any given application. Therefore, prior to any evaluation of a porous material with respect to application, it is essential to characterise it in terms of these parameters. Such a characterisation permits an initial screening of products or can be used to show trends if a series of materials is under investigation. These experiments are usually carried out with gases such as nitrogen and argon at the liquid nitrogen temperature (77.4 K), using readily commercially available instruments (see reference [3] for links to some websites). Such instruments are quite simple to use and can usually provide data of high quality. Other gases (such as Kr and CO_2) and conditions (87 K, 273 K *etc.*) have also been proposed,[4] but they are less common and as such the next few pages will be devoted to the use of nitrogen at 77 K for the characterisation of porous solids. Given the popularity of commercial adsorption instruments, the theoretical exploitation of these raw data is more or less automatic and as such is something of a 'black box', which can lead to poor exploitation and misinterpreted results. This chapter aims therefore to give the reader a 'practical guide' to the interpretation of adsorption data.

Experimental data have been obtained in-house for four types of ordered porous material:

- A silico-alumino zeolite known as NaX (or 13X) (Figure 4.1a).[5] This material is constructed from corner-shared SiO_4 and AlO_4 tetrahedra to give what is known as the faujasite structure.[6] There is a negative charge on the framework due to the incorporation of Al^{3+}, which in this case is compensated for by a Na^+ cation.
- Two ordered mesoporous oxides: MCM-41[7] and SBA-15.[8] These materials contain amorphous silica walls and their porosity is ordered in a honeycomb fashion (Figure 4.1b). The difference

between these two materials lies in their different synthesis conditions, which lead to different pore wall thicknesses and pore sizes.

- A metal-organic framework (MOF) known as MIL-100(Fe).[9] This material is constructed from iron-oxo clusters linked together *via* trimesic acid ligands (Figure 4.1c). MOF materials have long-range order much like that of zeolites. In particular, the MIL-100(Fe) MOF has an expanded MTN zeolite type structure.[6]

An extensive study of the calculations related to adsorption data cannot be conducted here; instead, the author has chosen those to be used and has selected those that can generally be applied in any materials laboratory. For the reader looking for more detailed descriptions or for complementary information, the following references are recommended: [2, 10–14].

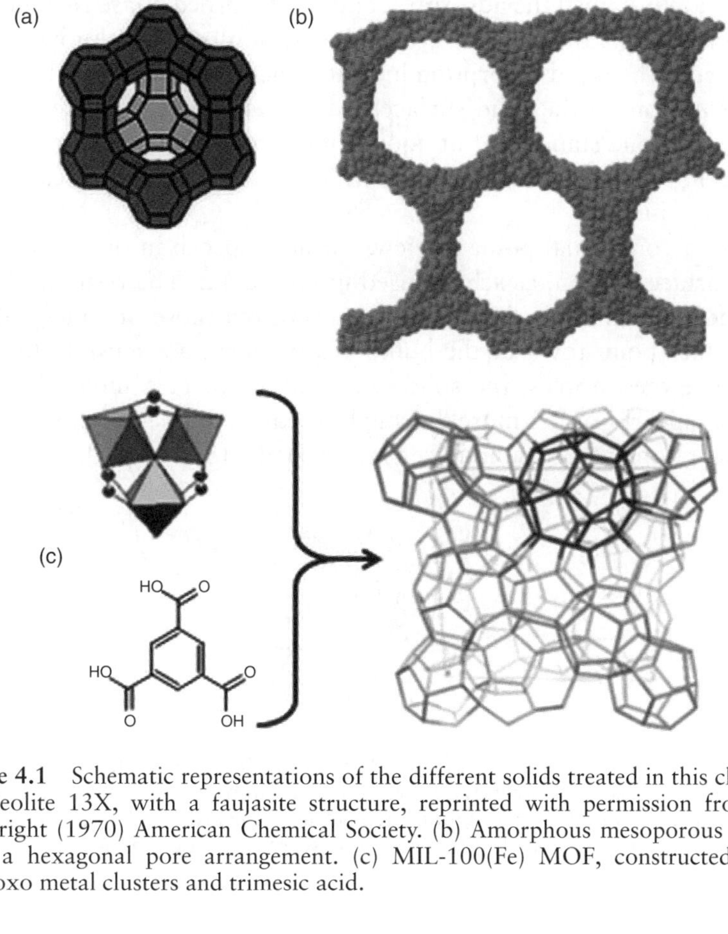

Figure 4.1 Schematic representations of the different solids treated in this chapter. (a) Zeolite 13X, with a faujasite structure, reprinted with permission from [5] Copyright (1970) American Chemical Society. (b) Amorphous mesoporous oxide, with a hexagonal pore arrangement. (c) MIL-100(Fe) MOF, constructed from iron-oxo metal clusters and trimesic acid.

4.2 WHAT IS ADSORPTION, WHY DO FLUIDS ADSORB AND HOW CAN ADSORPTION PHENOMENA BE VISUALISED?

Adsorption is a general phenomenon that occurs every time a gas or liquid (fluid) comes into the vicinity of a solid surface. The fluid is retained by the superficial atoms of the solid and is concentrated at the surface. The word adsorption indicates that this is a surface phenomenon. It should not be confused with the word absorption, which means that the fluid penetrates the solid mass. In some cases, it is difficult to distinguish between the two phenomena. Where this is the case, the general term **sorption** is used.

Adsorption can thus be defined as 'the enrichment of a solid surface by an interface of liquid or gas'.[10] The fluid (gas or liquid) available for adsorption is named the **adsorptive** and the adsorbed phase is named the **adsorbate**, while the solid on which adsorption occurs is known as the **adsorbent**. The word **desorption** indicates the liberation of fluid retained by adsorption on the solid surface. Adsorption is often treated from a thermodynamic standpoint at equilibrium and it can be forgotten that adsorption is a dynamic process in which diffusional aspects can sometimes be critical.

From a conceptual point of view, surfaces adsorb in order to reduce their surface tension, as schematised in Figure 4.2. The resultant forces of a molecule in the bulk of a solid are zero but those of a molecule on the surface point towards the bulk, creating a surface tension. In order to reduce these forces, the solid will readily accept a molecule on its surface (adsorption). A more thorough treatment of the thermodynamics of adsorption can be found in references [10], [13] and [15].

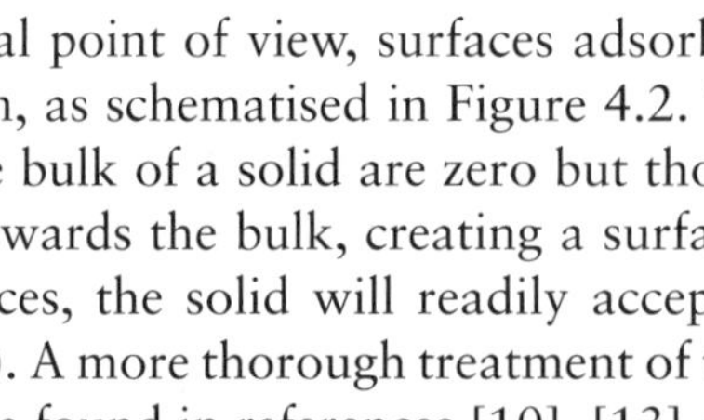

Figure 4.2 Scheme explaining why solids adsorb. The resultant forces of an atom in the bulk of a solid (light grey) are zero, but an atom on the solid surface (dark grey) is only stabilised by a resultant force pointing towards the bulk. This leads to a solid surface tension, which is reduced on adsorption of a gas molecule.

The interaction of a fluid with the solid can involve van der Waals-type forces, in which case one talks of physical adsorption or **physisorption**. Such forces are analogous to those in vapour condensation and do not change the properties of the solid or fluid. Where electron transfer occurs and the properties of the solid are changed, one talks of chemical adsorption or **chemisorption**.

The forces involved in physisorption are such that, at the equilibrium of charge distribution, there is no transfer or sharing of electrons among the participating atoms. What is more, the integrity of each species involved in a physisorption process is maintained. In principle, it is possible to calculate these forces of interaction by determining the quantum mechanical state of the system, but many approximations are usually made. For ease of conception, and also of computation, the interactions involved in adsorption are split up, even if to do so is artificial. The most commonly held view of the total adsorption interaction, ϕ, was first proposed by Barrer:[16,17]

$$\phi = \phi_D + \phi_R + \phi_P + \phi_{f\mu} + \phi_{\overline{F}Q} + \phi_{SP} \tag{4.1}$$

The first three terms of this expression – the dispersion, ϕ_D, the repulsion, ϕ_R, and the polarisation, ϕ_P, interactions – occur for every adsorbate–adsorbent system and can consequently be considered 'non-specific' interactions. The final three terms – the dipole, $\phi_{F\mu}$, quadrupole, $\phi_{\overline{F}Q}$, and self-potential, ϕ_{SP}, interactions – are unique to the nature of the adsorbate–adsorbent system in question and as such are often termed 'specific' interactions.

Molecules such as carbon monoxide and water possess a permanent dipole, while carbon dioxide and nitrogen possess a quadrupole; the 'polar' molecules will induce specific interactions with a surface. Any surface heterogeneities such as hydroxyl groups, cations or other functionalisation may also induce 'specific' interactions with an adsorbing molecule.

In a general sense, the Lennard–Jones potential function can be used to obtain a global picture of physisorption[18]:

$$\phi(\mathrm{r}) = \mathrm{B}r^{-12} - \mathrm{A}_{(6-12)}r^{-6} \tag{4.2}$$

where A and B are constants and B(**r**) is the interatomic distance at distance r. This is more often written in the form:

$$\phi(r) = 4\varepsilon \left[\left(\frac{\sigma}{r}\right)^{12} - \left(\frac{\sigma}{r}\right)^{6} \right] \tag{4.3}$$

where ε is the depth of the potential well and σ is the distance at which the potential is zero (exponents are not zero). Although this is an approximate potential, it has the features needed to describe the interactions between two closed-shell atoms: a strong repulsive short-range interaction, a long-range van der Waals attraction and a potential well. ϕ is the depth of the potential well, σ is the finite distance at which the interparticle potential is zero and r is the distance between the particles. The first term, r^{-12}, describes Pauli repulsion at short ranges, which results from overlapping electron orbitals. The second term, r^{-6}, describes van der Waals or dispersion force attraction at long ranges. Tables are available for these values[19] for different species. This function can be drawn as a plot of ϕ_i against r, as shown in Figure 4.3 (left).

The depth of the potential well describes the maximum interaction and it can be seen that as the two molecules approach each other, the attraction will increase. The 12-6 exponents used for the interaction between two atoms can be modified to 10-4 exponents for the interaction of a molecule with a flat surface or to 9-3 exponents for the interaction of a molecule within a slit-shaped pore. This latter expression has been used to construct Figure 4.3 (right). The different curves are plotted as the distance between two plates forming the pore wall decreases. It can be seen that at some point, the potentials overlap, leading to enhanced interactions within the pores. This illustrates the relatively strong interactions that arise in such narrow pores.

As will be shown below, a specific type of adsorption mechanism occurs in these 'narrow' pores, and it is possible to classify solids with

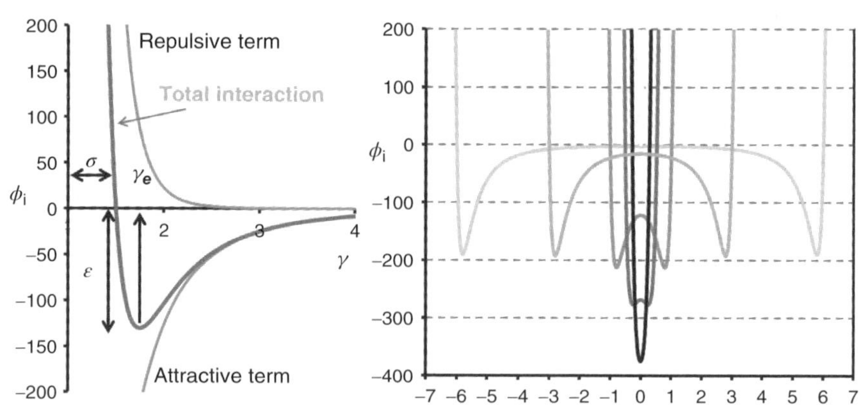

Figure 4.3 Plot of the Lennard–Jones potential function: left, between two molecules; right, for a molecule in a slit-shaped pore of different sizes.

Table 4.1 IUPAC classification of pore sizes.

Denomination	Dimensions/nm	N_2/pore size ratio
Micropore	0.4–2.0	1–5
Mesopore	2.0–50	5–125
Macropore	>50	>125
Nanopore	<100	

respect to the adsorption mechanisms of a probe of given size in pores of different dimensions. With this in mind, the IUPAC defines pore sizes[20], that are related to the adsorption phenomena of small molecules such as nitrogen or argon at cryogenic temperatures (77 K or 87 K), as shown in Table 4.1.

Independent of this IUPAC classification, the general term **nanopore** is increasingly used to describe materials with pores smaller than 100 nm.[21]

From a schematic standpoint, how can different adsorption mechanisms be pictured? A simple way of viewing what occurs is to consider that adsorption obeys the rules of thermodynamics and that while entropic effects can play a role, enthalpies of adsorption mostly dictate the outcome.

From an energetic standpoint:

(i) The first fluid molecules that come into contact with any adsorbent surface will adsorb at the most energetic sites, if these exist at the surface. These specific adsorption sites can be chemical, such as:
- hydroxyl groups, often found on silica surfaces;
- cations (Na^+, Ca^{2+}, K^+ *etc.*), which compensate for the negative charge of the zeolite framework;
- coordinatively unsaturated sites (also known as free metal sites in MOFs);
- functional groups ($-NH_2$, $-NO_2$, $-COOH$, Cl, Br *etc.*) in organic ligands;
- metal nanoparticles, which can be stabilised in some pores.

Spectroscopic methods such as Fourier transform–infrared (FTIR) are often used to follow the adsorption and interactions of small concentrations of probe species at these sites. This adsorption will mostly occur at pressures well below atmospheric.

(ii) Probe molecules fill the narrowest pores, known as micropores, at low partial pressures before covering the rest of the adsorbent surface. Two types of pore filling can be distinguished. The smaller

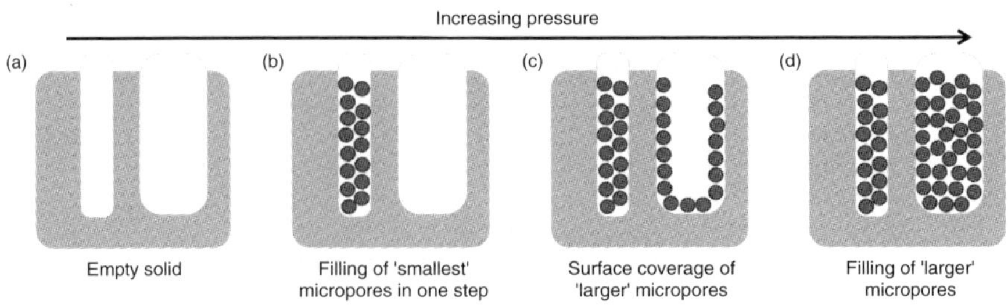

Figure 4.4 Schematic illustration of micropore filling mechanisms.

micropores (sometimes known as ultra-micropores) will fill at the lowest pressures in a single step for a given class (Figure 4.4b). Slightly larger micropores (super-micropores) will fill in two steps: initially the surface will be covered (Figure 4.4c) and then the rest of the available pore volume will be filled (Figure 4.4d).

(iii) The next region of note is where the probe molecules cover the surface with a single, statistical, molecular layer. When measuring the adsorption energies, this region corresponds to a clear decrease in the energies, which tend towards the enthalpy of liquefaction.

(iv) The adsorbate molecules will then continue with the gradual edification of a multi-molecular layer, which can occur up to the final experimental pressure. This multilayer is assumed to be in a liquid-like state.

(v) In samples with mesopores, these will fill – depending on their size – during the edification of the multilayer (Figure 4.5). It is suggested that this capillary condensation process occurs at some point *via* a bridging of two liquid films from opposing pore walls (Figure 4.5e), followed by the formation of a meniscus and filling (Figures 4.5f,g). Capillary condensation is considered not to be at equilibrium during adsorption. Indeed, this vapour–liquid transition is considered to be delayed as the adsorption film is in a metastable state. After the pore filling, multilayer formation of a liquid adsorbate will continue on any remaining external surface, which can be followed by condensation between particles when saturation is approached. A good overview of this process is given in reference [22] and on the associated website.[23]

(vi) On desorption, the mesopores will empty and the mechanism will depend on the size of the pore opening. Above a critical pore diameter (*ca* 3.6 nm for N_2 at 77 K) it is understood that mesopores will empty *via* a receding meniscus (Figure 4.6b), which is associated

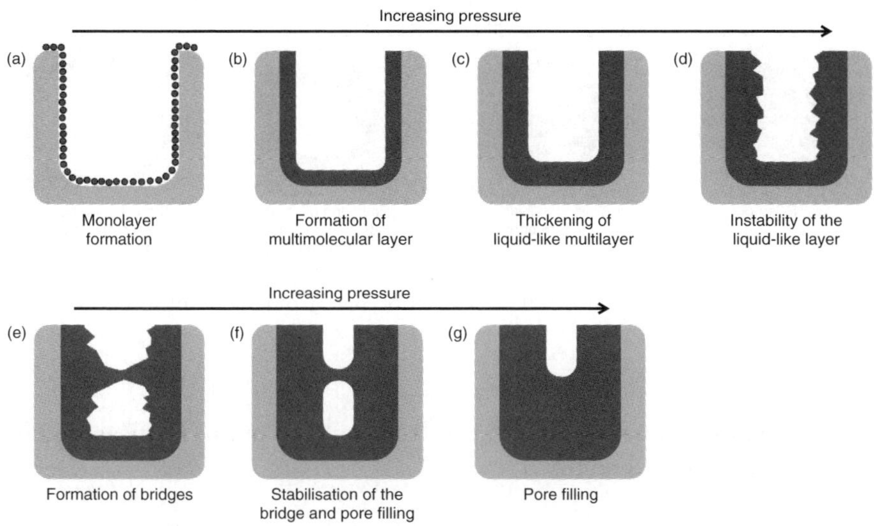

Figure 4.5 Schematic illustration of mesopore filling.

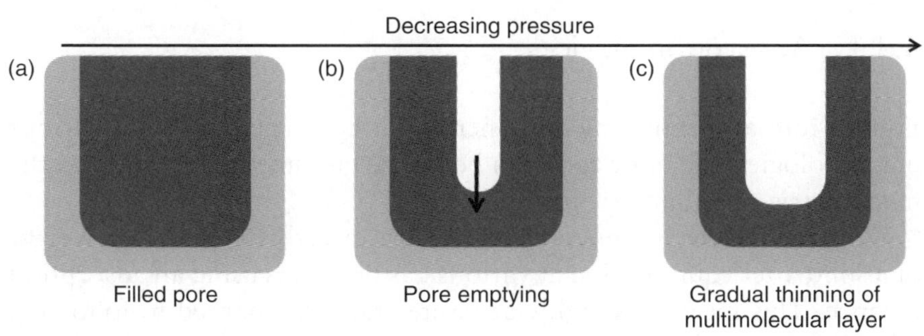

Figure 4.6 Schematic illustration of emptying from an 'open' mesopore.

with the equilibrium vapour–liquid transition. However, it may be that the pore entrances (necks) are smaller than the main pore body (Figure 4.7), in which case spontaneous cavitation (or bubble formation, Figure 4.7b) is expected in the pore bodies and pore emptying will occur through the necks without emptying of the entrances (Figure 4.7c).

At this point, from a conceptual viewpoint we know that adsorption occurs, why it occurs and how it occurs in different types of pore. We know that different forces arise with different probe molecules and that the pore size is important, since enhanced interactions occur in smaller

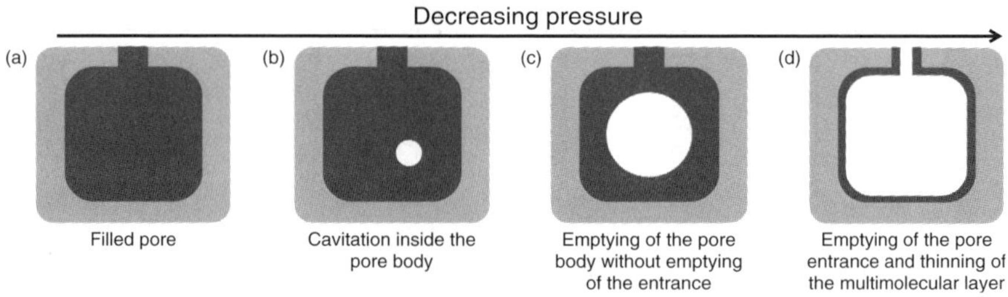

Figure 4.7 Schematic illustration of emptying from an 'ink-bottle-shaped' mesopore.

pores. Section 4.3 will deal with how the gas uptake can be measured in powders and porous solids.

4.3 ADSORPTION EXPERIMENTS

4.3.1 Adsorption Devices

Equilibrium adsorption measurements can be carried out with various methodologies, the most common being the gravimetric method and the manometric or volumetric method.

With **gravimetry**, it is possible to measure directly the change in weight of a solid after equilibration at various gas pressures using an adsorption balance. This method is highly accurate and well adapted to following adsorption at ambient temperature and above, so long as the buoyancy effect is well estimated or measured. However, for experiments at 77 K, gravimetry has some drawbacks in terms of accurate sample temperature control and measurement. This can be understood, as while the sample should be at 77 K, the balance mechanism itself is at room temperature, which leads to significant temperature gradients. Furthermore, the sample holder is inside a tube and not in direct contact with the cryogenic bath, which again leads to variations in temperature between the sample and the bath, especially for the first adsorption points, which are essentially under various states of vacuum.

The **volumetric** or **manometric** method, while being slightly less accurate than the gravimetric method, involves a simpler (and cheaper) experimental set-up, which, among other things, allows for direct contact between the sample and the cryostat, permitting better control of the sample temperature. The fact that adsorption is favoured at lower temperatures often means that the extra accuracy obtained with gravimetry is superficial for experiments at 77 K.

The term **volumetry** is still widely used instead of **manometry** to describe modern experiments even though there is no measurement of a change in volume. This misuse dates back to former experiments in which mercury was used to measure differences in the volume of the gas phase. All modern equipment uses sensitive pressure gauges; a schematic representation of a manometric device is described in Section 4.3.2. Commercial devices are almost all variations on this scheme with different methods of gas introduction/removal and several pressure gauges to allow for accurate measurement of low pressures.

A number of other devices that can be used to measure adsorption isotherms. A quartz microbalance is an interesting method: the sample is attached to an oscillating quartz crystal, with the frequency of the oscillation being a function of the weight, thus giving a very sensitive measurement of the amount adsorbed. However, this method is usually used in the liquid or vapour phase. It is also possible to follow the uptake in flat, thin films using ellipsometry, in which a polarised, elliptical light beam is shone onto the surface and the reflected beam is measured. This method allows the direct measurement of film thickness on a flat surface and has recently also been used to follow uptake in porous solids. Again, this method is currently limited to liquid- and vapour-phase adsorption, although current trials are adapting it to gas adsorption at 77 K.[24] Further techniques, such as chromatography, can also give details of adsorption phenomena, although these are outside the scope of this chapter.

4.3.2 Experimental Protocol

The general experimental protocol is relatively simple. After weighing the sample into the measuring cell, there are essentially four steps:

(i) activation;
(ii) dead-space volume calibration;
(iii) adsorption;
(iv) desorption.

The ideal amount of sample depends on the approach used. For a manometry experiment, around 40 m^2 of surface is ideal and so for a sample with around 500 $m^2\,g^{-1}$ of surface area, this translates to 80 mg of sample in the cell. However, for low-surface-area samples, this can translate to a large amount of powder, which can lead to diffusion problems in the sample bed, and for very high-surface-area samples, a minimum of material should be used to avoid sampling problems.

The activation, or outgassing, should ensure a defined and reproducible sample state prior to the experiment – the aim is to remove any physisorbed species (*e.g.* contaminants, water) without structural degradation. In some cases, however, a study of the effect of different sample activation temperatures may be of interest, for example to observe the effect of sample dehydroxylation. It is recommended that a thermogravimetric analysis experiment be carried out on a small amount of sample prior to the adsorption analysis so that the sample is subjected to outgassing up to a temperature region in which the mass curve is horizontal. The example in Figure 4.8, for a NaX zeolite, shows a quasi-horizontal region between 623 K (400 °C) and 1073 K (800 °C), and the fact that the sample is often outgassed under vacuum and that the thermogravimetric curve is obtained in an inert atmosphere should be taken into account. As a general rule of thumb, therefore, a temperature should be chosen towards the start of the plateau, which corresponds to 473 K (200 °C) for the NaY sample in Figure 4.8. Other samples may be less stable, as is the case for the MIL-100Fe sample (Figure 4.8, right). Here, the weight loss above 473 K (200 °C) corresponds to the degradation of the structure, and as such, outgassing in the region between 373 K (100 °C) and 473 K (200 °C) would be recommended. As just mentioned, sample outgassing often occurs under vacuum and this vacuum should

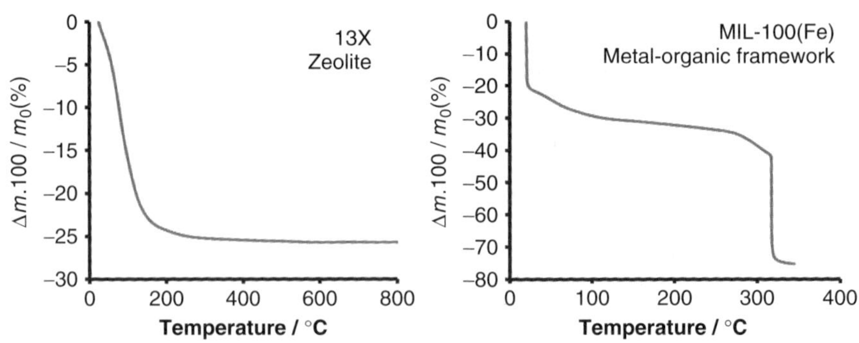

Figure 4.8 Example of a thermogravimetric analysis obtained with a 13X zeolite sample (left) and MIL-100Fe sample (right).

be applied slowly so as not to suck any powder into the vacuum pump. This option is now implemented in all commercial equipment.

A far more elegant solution is to outgas using a sample-controlled thermal analysis (SCTA) approach.[25,26] In its most appropriate form for outgassing, the heating of the sample to its final temperature is adjusted so that the vacuum pressure above the sample is maintained constant. This ensures that the heating rate decelerates when physisorbed species evolve from the sample and accelerates when there is no evolution of such species. This is not often employed, although an intermediate solution can be found in some adsorption equipment in which the sample is only heated if the vacuum pressure is below a fixed limit.

After outgassing, the sample is re-weighed and then placed onto the adsorption manifold, a simplified schematic diagram of which is given in Figure 4.9. The experimental protocol varies more or less around the same steps. Thus, an initial evaluation of the dead space is made using helium – the dead space includes the sample cell and part of the manifold. As the different sample cells are not identical and the sample volume is

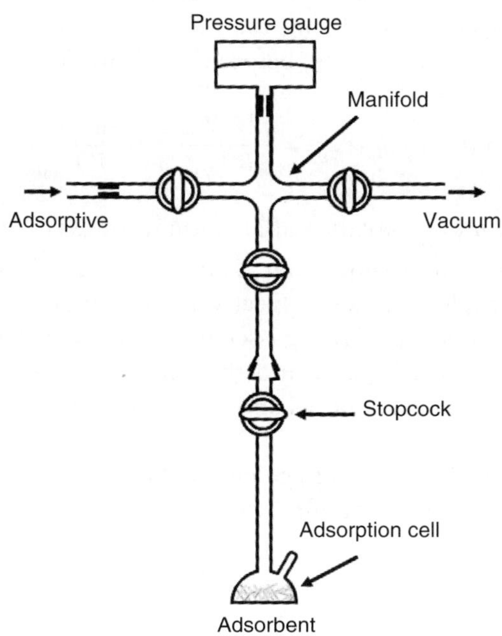

Figure 4.9 Schematic representation of a manometric device to measure adsorption isotherms. Adapted with permission from [10] Copyright (1999) Elsevier Ltd.

different for each experiment, the dead space should be evaluated prior to each experiment. Helium is used for this because it is considered that this gas does not adsorb, although this may not be the case with microporous solids. The actual measurement of the dead space by helium is similar to that of the isotherm with the adsorptive gas.

The adsorption measurement itself consists in dosing a known amount of adsorptive gas to the reference volume, V_{ref}. The reference volume is then opened to the sample volume, V_{sample}, and the system is followed until equilibrium between the sample and the gas is attained. Various criteria can be used to evaluate this equilibrium. The pressure above the sample can be followed, assuming that equilibrium is established when the pressure change is below a certain limit for a given time. Alternatively, a simple time limit can be set. This equilibrium is an important part of the experiment as it is easy to misinterpret results whose data points are taken after a poor equilibration step.

For each adsorption dose, the amount of gas in the reference volume and in the sample volume is calculated using either the simple ideal gas law relationship or a more complicated expression if gas non-ideality must be taken into account, even though departure from non-ideality is small for gases at 77 K. Using the ideal gas relationship, the initial amount of gas in the system, n^{i}, before the adsorption dose is:

$$n^{i} = n^{i}{}_{V_{ref}} + n^{i}{}_{V_{sample}} = \frac{p^{i}{}_{ref} V_{ref}}{RT_{ref}} + \frac{p^{i}{}_{sample} V_{sample}}{RT_{sample}} \tag{4.4}$$

Where $n^{i}{}_{V_{ref}}$ and $n^{i}{}_{V_{sample}}$ are the initial amounts of gas in the reference and sample volumes, respectively, and $p^{i}{}_{ref}$ and $p^{i}{}_{sample}$ are the pressures in the reference and sample volumes, respectively. Note that for the initial dose, $p^{i}{}_{sample}$ is zero if the sample is outgassed under vacuum. The temperatures of the sample volume, T_{sample}, and that of the reference volume, T_{ref}, should be taken into account, as should as the gradient in temperature between them.

A similar expression is used to calculate the final amount of gas phase, n^{f}, after adsorption equilibrium:

$$n^{f} = n^{f}_{V_{ref}} + n^{f}_{V_{sample}} = \frac{p^{f} V_{ref}}{RT_{ref}} + \frac{p^{f} V_{sample}}{RT_{sample}} \tag{4.5}$$

Note that as the reference volume and sample volume are in communication, the measured final pressure p^{f} is the same in each term. When following doses, Equation 4.6 applies:

$$\frac{p^{\mathrm{f}} V_{\mathrm{sample}}}{RT_{\mathrm{sample}}} = \frac{p^{\mathrm{i+1}}_{\mathrm{sample}} V_{\mathrm{sample}}}{RT_{\mathrm{sample}}} \tag{4.6}$$

Thus, to calculate the amount adsorbed per dose:

$$n^{\mathrm{ads}} = n^{\mathrm{i}} - n^{\mathrm{f}} \tag{4.7}$$

The amount adsorbed is then given per gram of sample for the equilibrium pressure. This pressure can be given in relation to the saturation vapour pressure, $p°$, which is measured separately. The adsorption isotherm is then represented as n^{a} (in mmol g^{-1}) as a function of p or $p/p°$. Most commercial equipment still represents the amount adsorbed in terms of gas volume, v^{a} (in $\mathrm{cm^3}_{\mathrm{STP}}\,\mathrm{g}^{-1}$), which is not advisable as it leads to difficulties when comparing with isotherms measured using other methods (*e.g.* gravimetric, chromatographic). Nevertheless, it is possible to relate the amount adsorbed, n^{a} (in mmol g^{-1}), to the gas volume adsorbed (v^{a}, in $\mathrm{cm^3}_{\mathrm{STP}}\,\mathrm{g}^{-1}$) *via* the expression:

$$n^{\mathrm{a}}\,(\mathrm{mmol\,g^{-1}}) = v^{\mathrm{a}}\,(\mathrm{cm^3}_{\mathrm{STP}}\,\mathrm{g^{-1}})/22\,414\,(\mathrm{cm^3\,mol^{-1}}) \tag{4.8}$$

It can be appreciated that adsorption manometry is a quite straightforward method, although systematic errors can occur due to small errors in the reference and sample volumes. Calculation of the amount adsorbed is based on the difference in the terms n^{i} and n^{f}, so that if this difference is small, *i.e.* n^{f} is relatively large, then the error in n^{a} may be significant. This is even more critical when the sample volume is estimated prior to the adsorption experiment *via* helium calibration. One assumes that helium does not adsorb on the sample but this assumption rarely holds in the case of microporous samples as it can adsorb in pores of the smallest dimensions at 77 K. Thus, the sample cell volume is overestimated while the amount adsorbed is underestimated. As with other techniques, a blank experiment without sample gives an excellent means of estimating the errors for a given apparatus.

Experiment gives the amount of fluid retained on the surface of an adsorbent. This is known as the amount adsorbed and depends on the following parameters:

- the extent of the interface;
- the fluid concentration (gas pressure, liquid concentration);
- the temperature.

In the case of the adsorption of nitrogen at its normal temperature of liquefaction (77.4 K), the **adsorption isotherm** (Figure 4.10) therefore represents the totality of equilibrium states for pressures from zero to the saturation vapour pressure of the adsorptive. Nevertheless, it is worth bearing in mind that each isotherm is characteristic of the adsorbate–adsorbent pair under study and, as mentioned above, specificities such as gas polarity and solid surface chemistry (hydroxyl groups, cations, free metal sites *etc.*) should be considered. In this case, however, the focus is on characterisation of pore size and pore volume.

IUPAC[20] recommends that the isotherm be shown graphically as the amount adsorbed per gram of adsorbent as a function of the equilibrium pressure. If the experiment occurs at below the triple point then the amount adsorbed is plotted as a function of the relative equilibrium pressure, denoted $p/p°$, where $p°$ is the saturation vapour pressure of the adsorptive. An example of an isotherm with an adsorption branch and a desorption branch is given in Figure 4.10. This isotherm has several significant points that can provide some initial ideas about the adsorption. Point A is the back extrapolation to $p/p° = 0$ of the quasi-linear region B–C. The amount adsorbed at this point, A, can, in some cases, be related to the volume of micropores (Figure 4.4). Point B is the start of a quasi-linear region of the isotherm to point C. Point B is interpreted as the moment at which a monolayer of gas forms on the surface of the solid (see Figure 4.5a). In a first instance, therefore, one can use the amount adsorbed at this point to calculate an approximate surface area (see below). The quasi-linear region B–C corresponds to the formation of

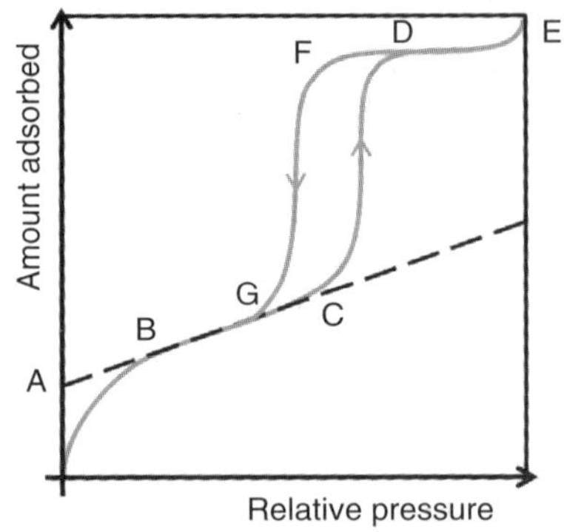

Figure 4.10 Adsorption (open circles)–desorption (closed circles) isotherm obtained with nitrogen on a porous glass at 77 K. The various points A–G are discussed in the text.

the liquid-like multilayer (Figure 4.5b,c). Region C–D corresponds to an upswing in the isotherm due to a pore filling *via* capillary condensation, which is typical of mesoporous solids (Figures 4.5e–g). Region D–E corresponds to the end of pore filling and the value of the amount adsorbed can be related to the total pore volume. Point F on the desorption branch of the isotherm corresponds to the start of pore emptying, and the relative pressure at which this occurs can be related to the pore size. Point G is where the desorption isotherm rejoins the adsorption branch. There is a minimum value for this point for a given gas at a given temperature, which corresponds to the point of meniscus instability. That is to say, for a given gas at a given temperature, there is a minimum pore size at which the liquid meniscus will spontaneously break, allowing complete desorption from any mesopores. For nitrogen at 77 K, this corresponds to a $p/p°$ of 0.42 and for argon at the same temperature, $p/p° = 0.28$.

IUPAC classifies six types of isotherm (Figure 4.11).[20] These correspond to limiting cases for non-porous solids (Type II), non-porous and non-wetting solids (Type III), mesoporous solids (Type IV), mesoporous, non-wetting solids (Type V) and extremely well ordered, non-porous systems such as graphitic systems in which adsorption occurs layer by layer (Type VI). Once again, the hysteresis loops observed for the Type IV and Type V isotherms correspond to a capillary condensation phenomenon associated with the mesporosity.

Microporous solids such as zeolites usually give rise to Type I isotherms, characterised by an increase in the amount adsorbed at low pressure followed by a distinct plateau. The low-pressure adsorption corresponds to filling of the micropore volume.

IUPAC equally considers four limiting cases for hysteresis (Figure 4.12).[20] The H1 hysteresis loop is characterised by parallel

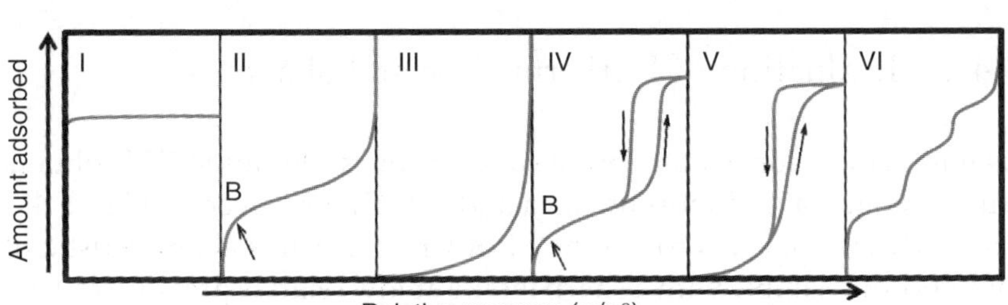

Figure 4.11 Representation of the six isotherm types classified by the IUPAC.[20] The point B is discussed in the above text. Reprinted with permission from [20] Copyright (1985) IUPAC.

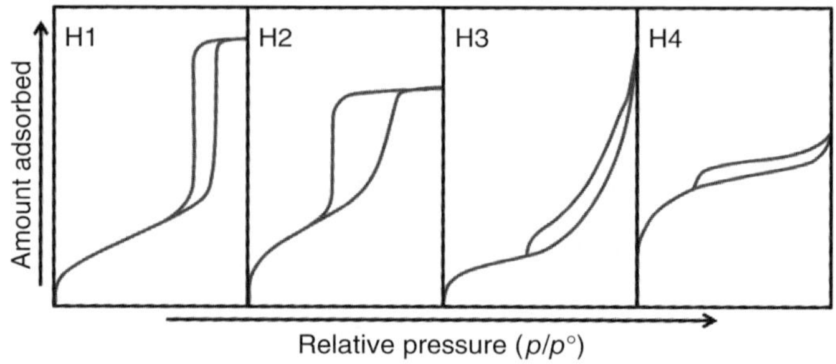

Figure 4.12 Representation of the different hysteresis loops classified by the IUPAC. Reprinted with permission from [20] Copyright (1985) IUPAC.

adsorption and desorption branches, and can be considered to result from the adsorption in unconnected mesopores with a relatively narrow pore size distribution. This type of hysteresis can typically be observed for MCM-41- and SBA-15-type materials, for example. The H2-type hysteresis is almost triangular in shape and is typically found in cases where the pores are interconnected, often with smaller entrance widths with respect to their internal 'body' widths; these are often referred to as 'ink bottle-shaped' pores (see Figure 4.7). Hysteresis types H3 and H4 result from non-rigid pore structures, which can be between particle grains (H3) or flat plates (H4). The non-rigidity in each of these cases is characterised by a lack of a defined flat plateau region near $p/p° = 1$, without any overlapping adsorption and desorption branches.

4.4 INTERPRETATION OF ISOTHERMS TO ESTIMATE POROUS SOLID CHARACTERISTICS

4.4.1 Evaluation of Isotherm Type and Shape

Examples of nitrogen adsorption–desorption isotherms at 77 K obtained with a zeolite, a mesoporous silica and a MOF are given in Figure 4.13. The NaX zeolite leads to a Type I isotherm, which is characteristic of a microporous sample with a sharp intake at low relative pressure followed by a 'plateau' region. This can be interpreted simply by an initial pore filling followed by adsorption on the external surface of the sample. To highlight the pore-filling process, the same data can be plotted in semi-log form (Figure 4.13, lower curves).

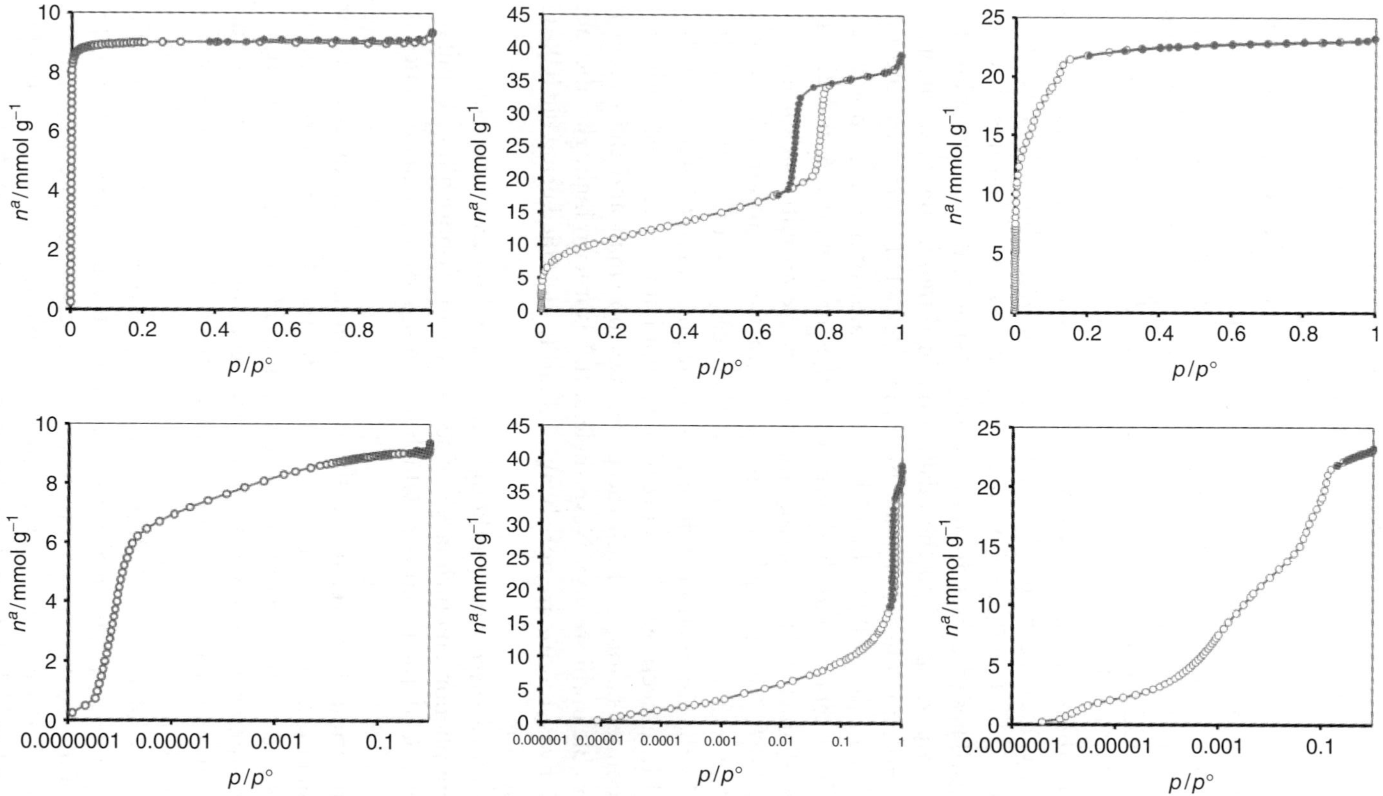

Figure 4.13 Adsorption of nitrogen at 77.4 K on NaX (left), SBA-15 (middle) and MIL-100(Fe) (right). The upper figures show the normal plots and the lower figures the semi-log plots. The adsorption branches are represented by open symbols and the desorption branches by closed symbols.

The mesoporous silica SBA-15[8] leads to an isotherm that can be classified as Type IV and is characteristic of a gradual uptake up to $p/p° = 0.7$. This gradual uptake is more visible in the semi-log plot, which, since no upswing is visible, indicates the absence of any microporosity. A sharp upswing in the isotherm is visible at around $p/p° = 0.75$, followed by a final plateau region that corresponds to the filling of all the available porosity. The sharp upswing in the adsorption branch is accompanied by a hysteresis loop on desorption; this hysteresis is typical of rigid mesoporosity and can be classified as H1.

The isotherm obtained with MIL-100(Fe) is not easily identified using the IUPAC classification. The long plateau at $p/p°$ above 0.2 resembles that of a Type I isotherm but there seem to be several features at lower loadings. These features are more readily visible in the semi-log plot and appear to correspond to the adsorption in different regions. Interpretation of this isotherm is aided by knowledge of the structure and the chemistry of this solid.[9] Indeed, it can contain specific adsorption sites at the iron centres and there are two cages of around 2.6 and 2.8 nm internal diameter. There seems, therefore, to be an initial adsorption at $p/p° = 10^{-6}$, which may be due to the interaction of nitrogen at the iron sites. There then seems to be a gradual coverage of the pore walls, followed by the completion of the filling of the pores at $p/p°$ of around 0.05 and 0.12, respectively. This pore filling should not be confused with capillary condensation, as both the pores are too small and the isotherm is reversible in this region. Nevertheless, this isotherm highlights the fact that many isotherms are not always so readily identified using the IUPAC classification.

In each of these cases, although the pore filling can be attributed to different phenomena (primary and secondary micropore filling, capillary condensation), the pressure at which pore filling occurs can be related to the pore size.

As mentioned above, IUPAC[16] recommends the use of nitrogen at its temperature of liquefaction (*i.e.* 77.4 K) for the characterisation of porous and powder materials by adsorption. However, it can be argued that the quadrupole moment of nitrogen can interact with specific surface groups such as hydroxyls or extra-framework, charge-compensating cations. Thus, other probes such as argon and krypton are also used, either at 77.4 K or at slightly higher temperatures such as 87 K – the liquefaction temperature of argon. Krypton is often involved in the evaluation of low surface areas, although some discrepancies occur regarding the cross-sectional area of the adsorbed molecule due to uncertainties of the adsorbed state and wetting capacity of this adsorbate at 77 K.

Argon can be used as a compromise as it is a non-polar molecule that has no specific interaction with a surface; ideally, it would be of interest to characterise solids with this adsorbate at 87 K. This is not often the case, however, due to the expense of liquid argon. Use of argon at 77 K can be of interest in probing microporosity, surface area and small mesoporosity, but it is not possible to characterise large mesoporosity due to the solidification of argon. Where possible, however, it is of great interest to compare the adsorption of both nitrogen and argon at 77 K, due to the complementary information that can be obtained.

This can be seen in Figure 4.14 for the adsorption of argon and nitrogen in MCM-41, where two points of comparison can be noted. First, more nitrogen is adsorbed at pressures below $p/p° = 0.1$, due to the interaction of the nitrogen with the surface silanol groups in MCM-41. A comparison of the BET surface areas (see Section 4.4.2) of the two samples would suggest a surface area of $614\,m^2\,g^{-1}$ in the case of nitrogen and of $522\,m^2\,g^{-1}$ in the case of argon. Due to the specific interaction induced by the nitrogen quadrupole, the result given by argon adsorption would seem to be more significant, because the diatomic nitrogen does not always sit flat on the surface but can orient itself at various angles depending on the surface chemistry. Thus, it is possible to recalculate the cross-sectional area occupied by the nitrogen molecule on the mesoporous silica, with a value of $0.140\,nm^2$ being found.[28] This value is close to that found on other silica surfaces.[29–31] The second point to note is that no hysteresis is visible for the adsorption of nitrogen, whereas a significant hysteresis loop is observed in the case of argon.

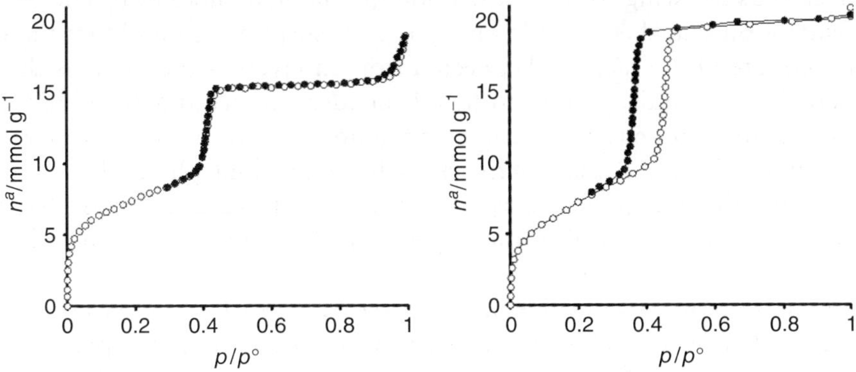

Figure 4.14 Adsorption of nitrogen (left) and argon (right) on MCM-41 at 77.4 K. Reprinted with permission from [27] Copyright (1994) Elsevier Ltd.

This loop closes above the limiting pressure for argon ($p/p° = 0.28$) and can therefore be used for the estimation of the pore size distribution.

4.4.2 Evaluation of Specific Surface Area using the BET Model

The BET equation is used commonly for the evaluation of specific surface area. Proposed by Brunauer, Emmet and Teller in 1938,[32,33] this model is based on the assumption of multilayer adsorption on a flat surface; strictly, the BET model does not apply to porous solids. In reality, however, the BET equation can be used correctly for adsorption on mesoporous solids, *i.e.* before the onset of capillary condensation, although it does not hold for microporous solids. However, as the BET method for the evaluation of surface area is accepted as the standard, it is possible to apply it to microporous solids; the term '*equivalent* surface area' is used to imply that the value given is unity if the adsorption process in micropores occurs in the same manner as on a planar surface.

The BET model is based on that which proposed by Langmuir in 1915 and 1917,[34,35] which considers the limiting case of the edification of a single monolayer of strongly sorbed species on a flat surface. Whilst the Langmuir model can be used to estimate surface areas, it is not strictly applicable to the case of physisorption. Brunauer, Emmet and Teller hypothesised that after the formation of a statistical monolayer, there is the possibility for formation of supplementary layers, the molecules adsorbed in one layer being considered adsorption sites for the next layer, thus allowing for the adsorption of a large number of layers. The assumption that there is only one type of adsorption site and that there are no lateral interactions between adsorbed layers is the same as that made by Langmuir. Further to this, Brunauer, Emmet and Teller added the assumption that the second and following layers are considered as having the same characteristics as the bulk fluid phase. Thus, for example, the energy of adsorption of the first layer, E_1, is considered to be different to that of further layers, which are in turn considered to adsorb with the enthalpy of liquefaction, E_L.

From this, at a given value of the pressure, the surface can be covered with 0, 1, 2 or more layers. The surface can, therefore, be represented by the sum of surfaces, s_0, s_1, s_2, s_3 *etc.* with 0, 1, 2, 3 *etc.* layers of molecules, respectively (Figure 4.15). At each equilibrium pressure, p,

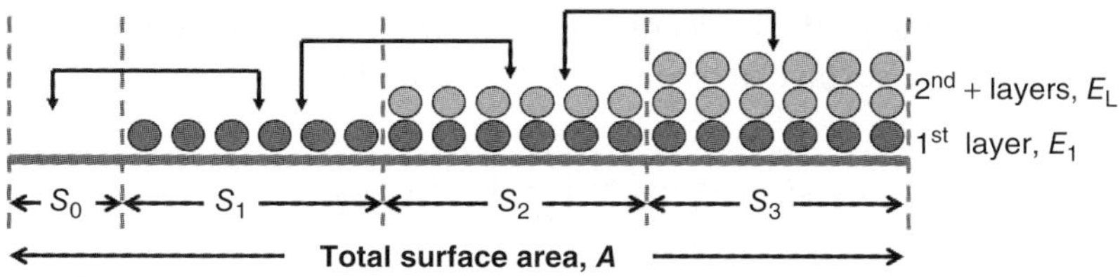

Figure 4.15 Schematic representation, used in the BET model, of an adsorbed film on an adsorbent surface.

Langmuir theory applies between the fraction of uncovered solid s_0/A and the fraction of solid with one layer, s_1/A; between the fraction of solid covered with one layer, s_1/A, and the fraction of solid covered with two layers, s_2/A; and so on. Thus, with this schema and the above hypotheses, the following expression relating the amount adsorbed, n^a, and the relative equilibrium pressure, $x = p/p°$:

$$\frac{n^a}{n^a_m} = \left(\frac{Cx}{1-x}\right)\left(\frac{1-(N+1)x^N + Nx^{N+1}}{1+(C-1)x - Cx^{N+1}}\right) \tag{4.9}$$

where n^a_m is the amount adsorbed at the statistical monolayer, N is the number of layers and C is a constant related to the energy of adsorption, given by:

$$C \approx \exp\left[\frac{E_1 - E_L}{RT}\right] \tag{4.10}$$

Normally, one assumes that N tends towards infinity, giving the simplified expression:

$$\frac{n^a}{n^a_m} = \left(\frac{Cx}{(1-x)\cdot(1-x+Cx)}\right) \tag{4.11}$$

It can be shown that Equation 4.11 is valid for $N > 4$ and for $p/p° < 0.35$. This equation is more often used in the linear form:

$$\frac{p/p^o}{n^a(1-p/p^o)} = \frac{1}{n^a_m C} + \left[\frac{C-1}{n^a_m C}\right]\left(\frac{p}{p^o}\right) \tag{4.12}$$

The BET plot is thus a plot of $[(p/p°)/(n^a\{1-p/p°\})]$ *versus* $p/p°$ and one looks for a linear region below $p/p° = 0.35$. The slope $[(C-1)/n^a_m\,C)]$ and intercept $[1/(n^a_m\,C)]$ can thus be used to obtain the values of C and n^a_m.

The value of the monolayer capacity, n^a_m, can then be used for the calculation of the specific surface area by taking into account the equivalent area, σ_m, that the probe molecule takes up when in the monolayer. Thus the BET specific surface area of a solid is given by:

$$a_{\text{BET}} = \frac{\text{A}}{m^s} = \left[\frac{n_m^a}{m^s}\right] \cdot N_A \cdot \sigma_m \qquad (4.13)$$

As stated above, in the case of non-porous and mesoporous solids the BET equation is valid. One will often look for a linear region in the BET plot between $p/p° = 0.05$ and 0.35. In the case of microporous solids, this will be found at a much lower region of $p/p°$ and often below $p/p° = 0.05$.

The example given in Figure 4.16 shows the adsorption of nitrogen at 77 K on the zeolite NaX sample whose isotherm is shown in Figure 4.13. The left-hand figure shows the BET transformation in the accepted domain of validity, $0.05 < p/p° < 0.35$, where it can be seen that there is no linearity. Furthermore, the intercept at $p/p° = 0$ is negative, which has no physical sense. A number of criteria can be used to aid the identification of a linear region. The first, initially proposed by Rouquerol,[10,36] is that the term $n^a(p° - p)$ should increase with increasing $p/p°$. The BET treatment can then be validated by ensuring that the relative pressure at which the n_m^a term is found is within the region under consideration and that this relative pressure is close to that calculated *via* the term $\frac{1}{\sqrt{C+1}}$. A linear region in the BET transformation can be found at lower relative pressures in the domain $1.4 \times 10^{-5} < p/p° < 0.016$, in which the above criteria hold (Figure 4.16, right). An *equivalent* BET surface area of $685\,\text{m}^2\,\text{g}^{-1}$ is thus obtained for a n_m^a term at a $p/p°$ of 0.0048 and a C value of approximately 43 000. The very high value of C translates the strong attraction of nitrogen in this microporous solid, although such a high number should not be over interpreted.

4.4.3 Evaluation of External Surface Area and Pore Volume using the t- or α_S-Method

The t-[37] and α_S-[38]methods can be used to estimate the external surface area and micropore volume of microporous solids. In some cases, the α_S-method can additionally be used to calculate the total surface area of microporous solids.

The t-method of Lippens and de Boer[37] consists in a comparison of the amount adsorbed for a given isotherm with the thickness of the

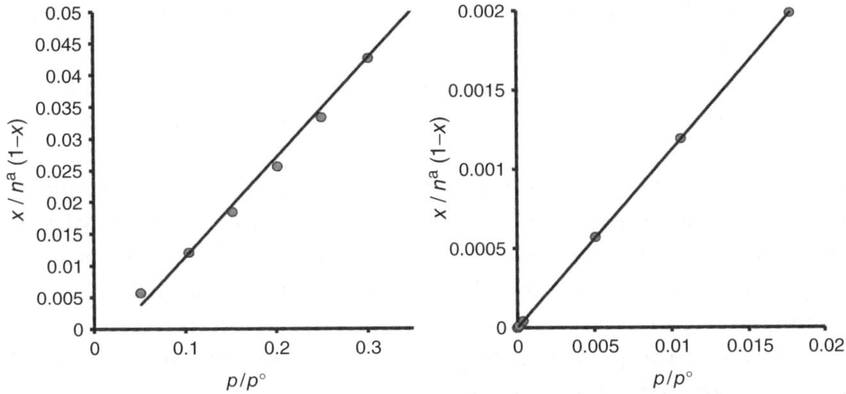

Figure 4.16 BET transformation, obtained for the nitrogen isotherm of NaX shown in Figure 4.13, in the accepted domain of $0.05 < p/p° < 0.35$ (left) and the linear region between $1.4 \times 10^{-5} < p/p° < 0.016$ (right), selected *via* the criteria noted in the text.

adsorbed layer of a known reference isotherm at the same relative pressure. The reference isotherm is obtained with a non-porous solid. In this method, the thickness, t, of the adsorbed multi-molecular layer is calculated as the sum of the number of layers, N, and the thickness of each layer, l:

$$t = Nl \tag{4.14}$$

The following hypotheses are then made:

- The adsorbed layer is assumed to be a liquid of molecular mass, M, density, ρ^λ and of volume, v^λ:

$$v^\lambda = n^a (M/\rho^\lambda) \tag{4.15}$$

- The surface area, $a(t)$, per gram of solid is covered by an adsorbed layer and is therefore given by:

$$a(t) = \mathrm{v}^\lambda / t \tag{4.16}$$

- Each molecular layer has the same thickness, l:

$$l = (M/\rho^\lambda)/N_A\sigma \tag{4.17}$$

where N_A is Avogadro's constant and σ is the cross-sectional area of a single molecule of adsorptive. Thus, for liquid nitrogen at 77 K where $M = 28\,\mathrm{g\,mol^{-1}}$, $\rho^\lambda = 0.809\,\mathrm{g\,cm^{-3}}$ and $\sigma = 0.162\,\mathrm{nm^2}$, the thickness of a monomolecular layer, l, is 0.354 nm.

- The number of monomolecular layers, N, in an adsorbed film can be calculated using $N = n^a/n^a_m$ where n^a is measured for each relative pressure and the monolayer capacity, n^a_m, is calculated using the BET method. For nitrogen at 77 K, $t/\text{nm} = 0.354\ (n^a/n^a_m)$.

Thus, the isotherm obtained with a non-porous adsorbent can be plotted with t as a function of $p/p°$. De Boer[37] showed that the plots of t as a function of $p/p°$ for the adsorption of nitrogen at 77 K on a number of nonporous oxides are practically the same, leading to a 'universal' plot. A number of expressions attempt to fit this universal plot, and it can be shown that of the most widely proposed, that by Harkins and Jura[39] fits the original data most closely:

$$t/nm = \left(\frac{0.1399}{0.034 - \log\left(p/p^0\right)}\right)^{0.5} \tag{4.18}$$

The construction of the t-plot is obtained by plotting n^a for the unknown solid as a function of t, as schematically shown in Figure 4.17.

Microporous and mesoporous solids lead to t-plots similar to those given in Figure 4.18 for the zeolite NaX (left) and SBA-15 (right), respectively. It can be seen that these plots show distinct linear regions whose slopes, S_t, are related to the surface area according to:

$$a(\text{t}) = \text{v}^{\lambda} \cdot S_t \tag{4.19}$$

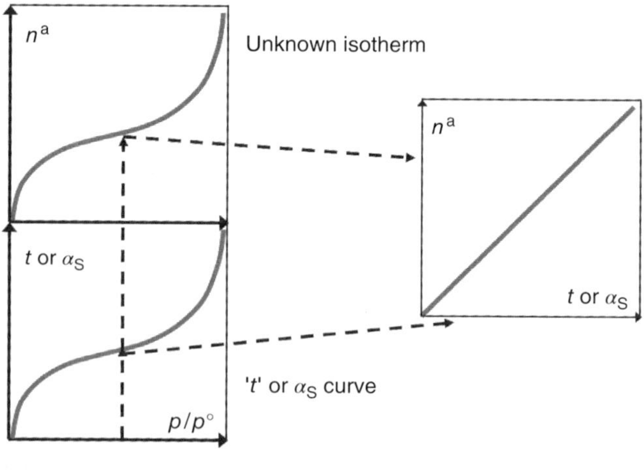

Figure 4.17 Schematic representation of the construction of the t- or α_S-plot.

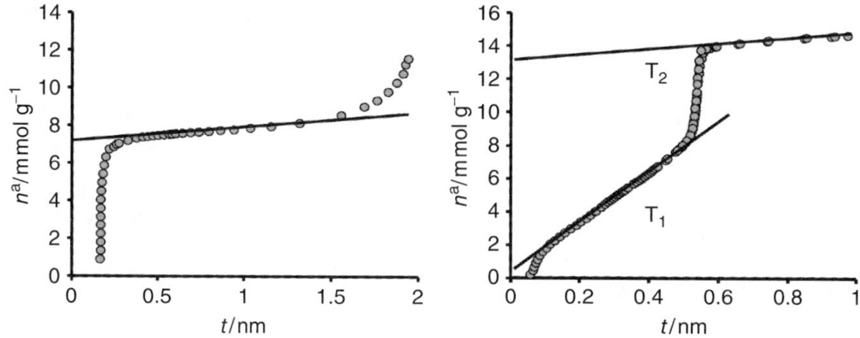

Figure 4.18 *t*-plot obtained from the nitrogen isotherm for NaX zeolite (left) and MCM-41 (right).

or:

$$a(t) = 34.6 \times (S_t/\text{mmol}\,g^{-1}\,nm^{-1})$$

In the case of the zeolite sample, a single linear region with a positive intercept can be observed, which can be related to an external surface area without taking any microporosity into account. The intercept at the origin can be used to calculate the micropore volume.

In the case of the mesoporous silica, two linear regions can be distinguished. The first, denoted t_1, can be related to the total surface area, although a small positive intercept might be related to a small amount of microporosity. The second linear region, t_2, occurs after the mesopore filling and so can be related to the external surface area. The difference in the values of the intercepts of t_1 and t_2 can be related to the mesopore volume.

In both of these examples, an initial linear region at low t can be observed. However, any linear region that crosses the y-axis at negative n^a cannot be exploited as it is unphysical. To remedy this, Sing's plot can be used instead.[38]

Sing noted a number of inconveniences in the t-plot method.[38] The first is that the BET method for the calculation of the surface area is not always applicable. This is especially the case for nitrogen, which does not always take up the cross-section of 0.164 nm^2 normally assumed. The second is that the 'universal' curve is not so universal, so one should ideally compare a given adsorbent to a non-porous material of the *same chemical nature*. In order to overcome these points, Sing[38] proposed modifying the t-method by replacing the monolayer capacity,

n^a_m, with the value n^a_s, which is the amount adsorbed on the non-porous reference at a given value of $p/p°$. To simplify the comparison, the value at $p/p° = 0.4$ is normally used, as this is a region well after any micropore filling and before capillary condensation. The ratio $n^a/n^a_{0.4}$ is known as α_S. The α_S curve of *versus* $p/p°$ for the non-porous reference solid is taken as the reference α_S curve. It is evident that such a curve is obtained without any hypothesis and that it relates directly to the chemical nature of the adsorbate–adsorbent couple under investigation. The α_S plot is, therefore, a plot of n^a *versus* α_S for each value of $p/p°$, by analogy with the *t*-plot.

Any linear region, s_S, in the α_S plot can be related to an unknown surface, a_x, according to the expression:

$$a_x = [a_{ref}/n^a_{0.4(ref)}] \cdot s_S \tag{4.20}$$

where a_{ref} is the area of the reference material and $n^a_{0.4(ref)}$ is the amount adsorbed on the reference material at $p/p° = 0.4 \cdot a_{ref}$ can be estimated using the BET equation or by another method, *e.g.* immersion calorimetry.

4.4.4 Evaluation of Micropore Size: The Horváth–Kawazoe Methodology

An estimation of micropore size can be obtained using the methodology proposed by Horváth and Kawazoe,[40] originally put forward to estimate the pore sizes of microporous carbon, with further extensions being proposed for other microporous solids. One can question the relevance of such a methodology in the case of ordered materials such as zeolites or MOFs, but it may be of some interest to use these methods to monitor any pore size modification resulting from post-synthetic treatments such as de-alumination.

The basic methodology relies on a simple representation of the pore geometry (slit-shaped, cylindrical), such as that given in Figure 4.19. The width of the pore, W, is given as the distance between atom centres. The interaction of a molecule inside the pore can be given, in the case of a simple slit-shaped pore, as the sum of interactions between each pore wall.

A Lennard–Jones-type expression is then used for the interaction potential, ϕ, of a molecule in a pore:

$$\phi(z)_{pore} = k\left[\left(\frac{\sigma}{z}\right)^a - \left(\frac{\sigma}{z}\right)^b + \left(\frac{\sigma}{W-z}\right)^a - \left(\frac{\sigma}{W-z}\right)^b\right] \tag{4.21}$$

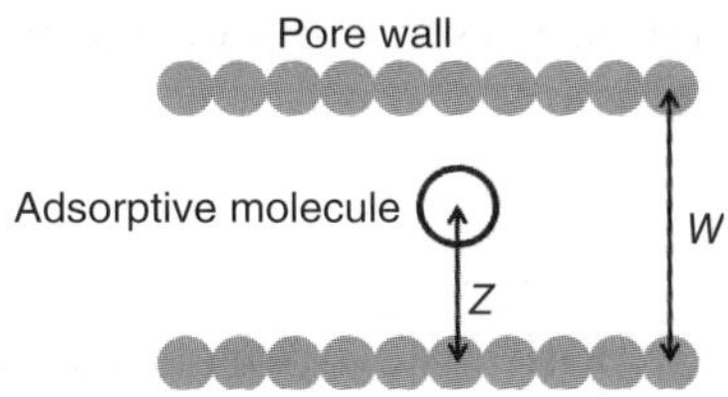

Figure 4.19 Schematic of slit-shaped micropores used as the basic assumption for the original Horváth–Kawazoe method.

where k is a constant for each adsorbate–adsorbent system, σ is the interaction strength and z is the distance between the adsorbed species and the first pore wall. The terms a and b are constants that vary with respect to the pore geometry. A simple relationship between this interaction potential and the adsorption potential $[-RT \ln (p/p^o)_{pore}]$ is then used to relate the relative pressure at which pore filling occurs with the pore size. It has been shown that such an expression can be used to estimate mesopore sizes, but this methodology is generally limited to pores smaller that 2 nm.

In the case of the NaX zeolite, this methodology does not give the expected pore size (Figure 4.20). Nevertheless, as mentioned above,

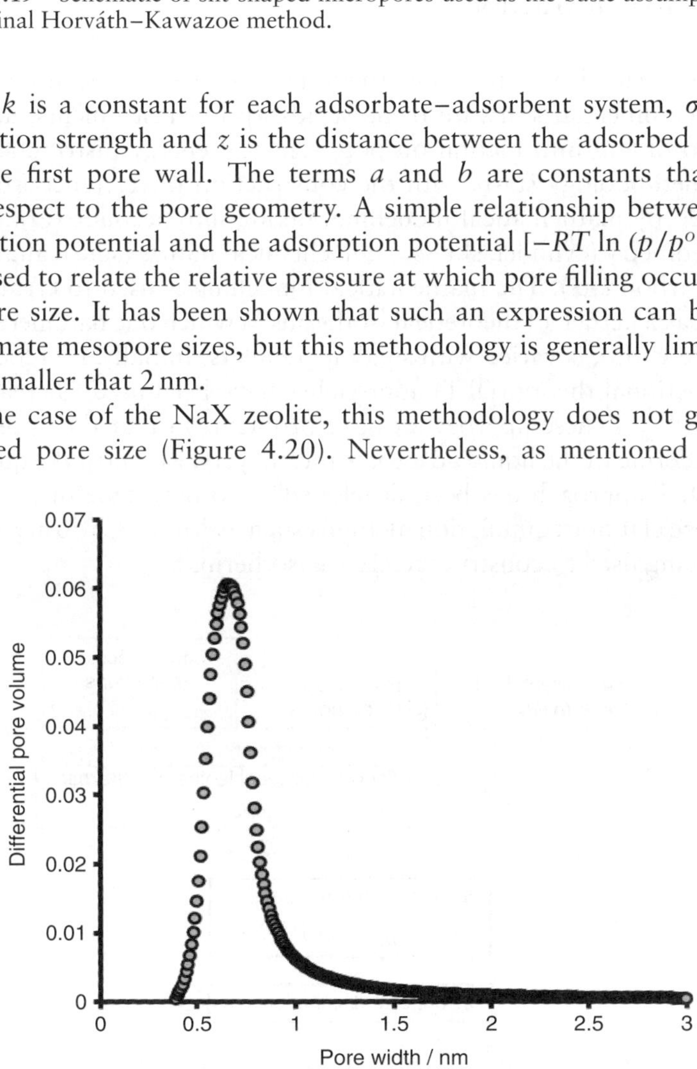

Figure 4.20 The Horváth–Kawazoe method exploited for the nitrogen isotherm on NaX zeolite.

Horváth–Kawazoe analysis presents an interesting way to monitor pore size variations in similar solids.

4.4.5 Evaluation of both Micropore and Mesopore Size using DFT/GCMC Treatment and Isotherm Reconstruction

In recent years, isotherm reconstruction methods have become available in the commercial software of most adsorption instruments, allowing users to obtain information on pore size and energy distribution. The basic methodology starts with the construction of reference isotherms obtained by mathematical modelling, taking into account variations in pore geometry (cylinder, slit *etc.*) and chemical nature (silica, aluminosilicate, carbon *etc.*). The mathematical modelling is used to create a set, known as a kernel, of theoretical isotherms in which one parameter, often the pore width, is varied with small increments. Initially, non-local density functional theory (DFT) approaches have been used to create such kernels.[41,42] These approaches have problems in that they predict an over ordering of the liquid adsorbate multilayers. Recently, the quenched solid DFT approach has been developed[43] to overcome this problem, but more elaborate simulation methods such as Monte Carlo are increasingly being used to construct reference isotherms.

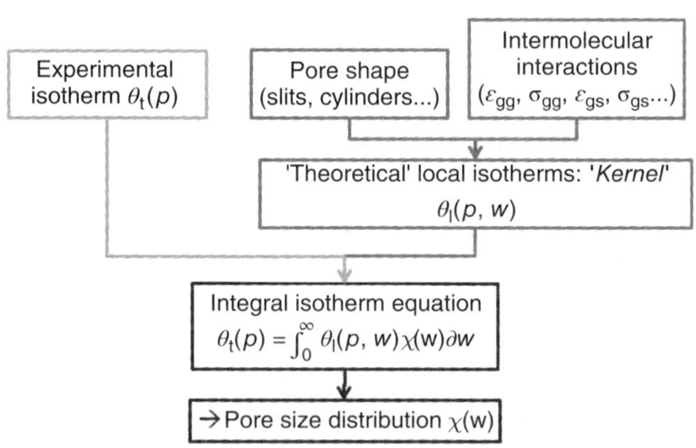

Figure 4.21 Schematic flow of the calculation of pore size distribution using isotherm reconstruction methods. The theoretical kernels (top right) are constructed assuming a given pore shape and given intermolecular interactions. The experimental isotherm (top left) is compared to the kernels *via* the use of the integral isotherm equation (bottom), in which the only variable is the pore size distribution.

It is then possible to reconstruct the 'measured' isotherm from the reference isotherms, assuming a given pore geometry and chemical nature using the integral isotherm equation, as schematised in Figure 4.21. The percentage of each individual isotherm from the kernel that is used to construct the theoretical isotherm is then used to calculate the pore size distribution.

An example is given for the treatment of the SBA-15 isotherm from Figure 4.13. Here the NL-DFT kernel is used for nitrogen at 77 K on a cylindrical silica surface using the desorption branch of the isotherm. From Figure 4.22, it can be seen that the comparison between the

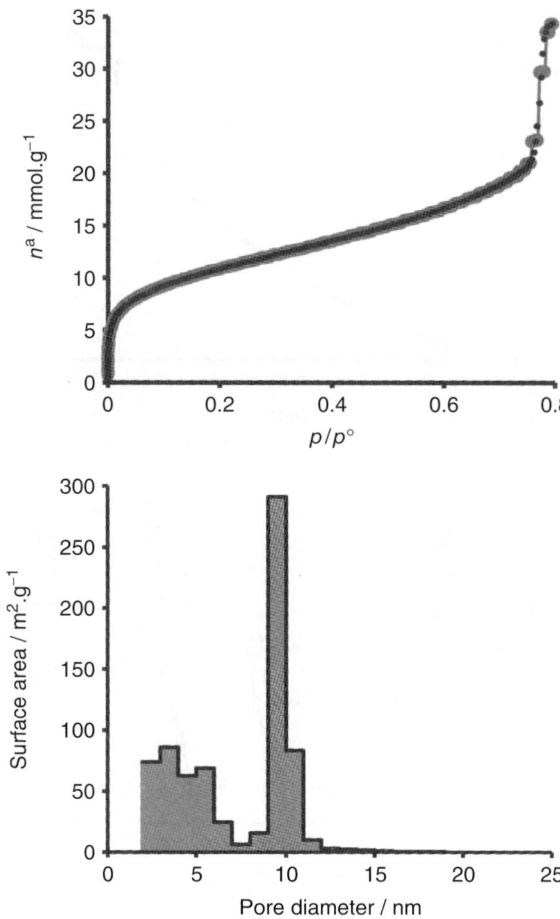

Figure 4.22 N_2 adsorption on SBA-15 at 77 K. Top: comparison of the experimental isotherm (black) and fit from DFT analysis (light grey). Bottom: histogram of pore diameter.

experimental isotherm and the reconstructed isotherm is very good. It is possible to extract results in different manners, and in this case a histogram of pore size distribution is given, suggesting a majority of the pores are centred on a diameter of 10 nm.

A second example is given in Figure 4.23 for the adsorption of nitrogen at 77 K on MIL-100(Fe),[9] whose isotherm is given in Figure 4.13. In this case, as the kernels for MOFs do not exist (at the time of writing) the choice was made to use spherical carbon reference isotherms obtained using the quenched solid DFT method, since the majority of the accessible surface of a MOF is organic in nature and the energies

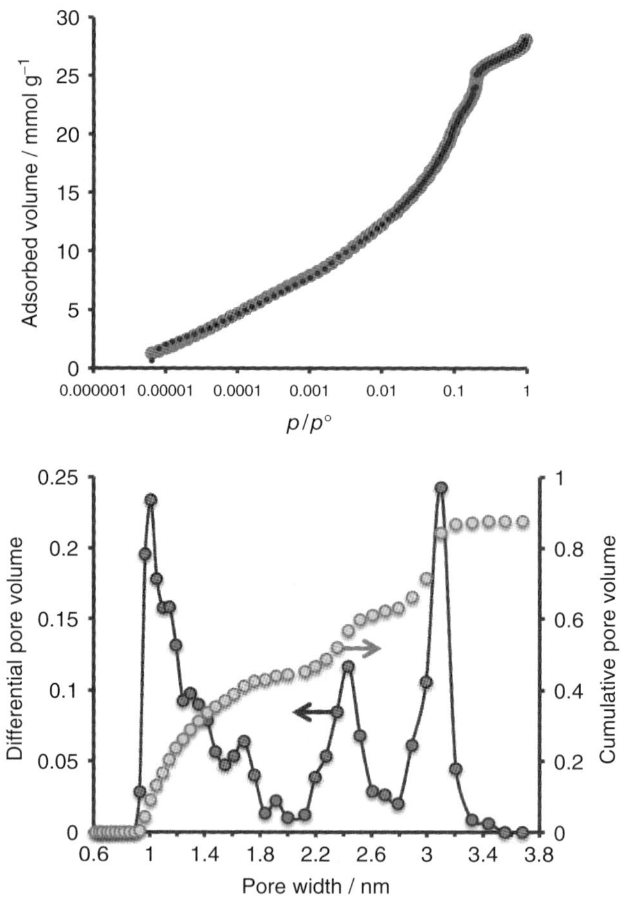

Figure 4.23 N_2 adsorption on MIL-100(Fe) at 77 K. Top: comparison of the experimental semi-logarithmic isotherm (black dots) and fit from DFT analysis (grey). Bottom: results in the form of differential/cumulative pore volume as a function of pore width.

of adsorption of many gases on most MOFs tend towards those of activated carbons rather than inorganic zeolites.

With this assumption made, the comparison between the reconstructed isotherm and the experimental isotherm is surprisingly good. Note that in this case, a semi-log plot is used as much of the pore filling occurs at quite low pressures for this sample. The ensuing results are given as differential and cumulative pore volume as a function of half pore width in the form of curves. Three distributions are visible; interestingly, these correspond quite well to the crystallographic estimations of the pore entrances (0.7–0.9 nm) and the two pore sizes of 2.5 and 2.9 nm.[9]

To a great extent, this isotherm reconstruction approach to the analysis of porous solids is very much like the isotherm comparison methods, such as the t- and the α_S-methods, with the advantage that it is possible to obtain pore size distributions. However, much like the α_S-method, it is crucial to select correct parameters of pore shape and pore surface chemistry, as it is possible to calculate sizes in the micropore range.

In summary, this approach offers the advantage of being the only one that can provide a single analysis of both microporosity and mesoporosity. The main disadvantage lies in the fact that kernels of reference data are required for each type of pore structure and each chemical nature. These kernels can be found in the programs provided by instrument manufacturers. However, for novel materials such as MOFs and some disordered materials, these kernels do not yet exist, so an 'exact' estimation of pore size or pore size distribution is difficult.

4.5 CONCLUSION

This chapter aimed to give the reader a basic idea of how adsorption can be used to characterise porous solids. Simple illustrations have been made to allow the reader to understand the adsorption phenomena in various porous media. The analytical methods used to analyse isotherms that are found in the software of commercial adsorption equipment are numerous and it is not easy to make the correct choice as each method has its own hypothesis and range of validity. In this chapter, an effort has been made to identify the easiest approaches to the analysis of solids in terms of:

- total surface area using the standard BET method;
- external surface areas using the t- (or α_S-) plot;
- micropore size distribution using the Horváth–Kawazoe method;

- mesopore size distribution using the DFT/grand canonical Monte Carlo (GCMC) reconstruction method; this method can equally be used to calculate micropore sizes, although particular care must be made as to the kernel chosen.

While this chapter relates well to most nano- and mesoporous solids, as in many areas of research there are specific cases in which it is essential to combine the above-mentioned methods with complementary characterisation techniques. Historically, this has been the case for the adsorption of simple molecules on well-defined graphite, where 2D epitaxial phases are formed to give rise to Type VI isotherms.[44,45] Similar, well-ordered phases have equally been observed with the pure silica zeolite, silicalite-1. Here, gases such as nitrogen and carbon monoxide show two successive, well-ordered phases[46] and others (argon, oxygen, krypton) equally show one well-ordered phase.[47] In these cases, one often uses the term **commensurate phases** when the adsorbed species seem to be influenced strongly by the pore structure. The phase transitions associated with this adsorption are extremely well highlighted with direct measurements of the energies of adsorption using calorimetry. The structures have been determined using diffraction methods, often in large-scale facilities such as synchrotron X-ray diffraction and neutron scattering.

The family of metal-organic frameworks (MOFs, ZIFs *etc.*) provides a huge challenge to those interested in their characterisation and evaluation with respect to gas separation and storage. There are several reasons for this, including the fact that the surface energy of the organic parts of these materials is not taken into account in the DFT and GCMC models. Furthermore, an interesting feature of some MOFs is their great flexibility during gas adsorption.[48–51] This poses some interesting problems for their characterisation, again requiring the use of complementary techniques such as calorimetry (see Section 2.3) and X-ray or neutron diffraction. Other techniques have also been used successfully to characterise the host–guest interactions in this fascinating family of materials: spectroscopic methods (*e.g.* infrared and electronic spectroscopy) have been used to explore the adsorption on specific sites and Mössbauser spectroscopy has been used to follow the variation in valence of the metal centre. All of these experimental methods are complemented by various modelling approaches (*ab initio*, DFT, Monte Carlo, molecular dynamics).

The majority of materials can thus be characterised in terms of available surface area using the BET method, external surface area and pore volume using the *t*-method and pore size distribution using the DFT/GCMC isotherm reconstruction approaches. Each of these methods comes with a series of assumptions and it is, therefore, impossible to consider the values obtained as absolute. In some cases, it is of interest to combine adsorption measurements with various complementary *in situ* techniques.

REFERENCES

[1] A. D. McNaught and A. Wilkinson (Eds), *Compendium of Chemical Terminology*, 2nd edition (the 'Gold Book'), Blackwell Scientific Publications, Oxford (1997).
[2] R. T. Yang, *Adsorbents: Fundamentals and Applications*, John Wiley, New Jersey, 2003.
[3] http://adsorption.org/org/indexcom.htm (last accessed 28 October 2013).
[4] F. Rodriguez-Reinoso and M. Molina-Sabio, *Adv. Coll. Interf. Sci.*, **76**, 271 (1998).
[5] D. H. Olson, *J. Phys. Chem.*, **74**, 2758 (1970).
[6] http://www.iza-structure.org/databases/ (last accessed 7 October 2013).
[7] J. S. Beck, J. C. Vartuli, W. J. Roth, M. E. Leonowicz, C. T. Kresge, K. D. Schmitt, C. T. W. Chu, D. H. Olson, E. W. Sheppard, S. B. McCullen, J. B. Higgins and J. L. Schlenker, *J. Am. Chem. Soc.*, **114**, 10834 (1992).
[8] D. Zhao, J. Feng, Q. Huo, N. Melosh, G. H. Fredrickson, B. F. Chmelka and G. D. Stucky, *Science*, **279**, 548 (1998).
[9] P. Horcajada, S. Surblé, C. Serre, D.-Y. Hong, Y.-K. Seo, J.-S. Chang, J.-M. Grenèche, I. Margiolakid and G. Férey, *Chem. Commun.*, **2820** (2007).
[10] F. Rouquerol, J. Rouquerol and K. S. W. Sing, *Adsorption by Powders and Porous Solids: Principles, Methodology and Applications*, Academic Press, London, 1999.
[11] S. J. Gregg and K. S. W. Sing, *Adsorption, Surface Area and Porosity*, 2nd edition, Academic Press, London, 1982.
[12] D. W. Breck, *Zeolite Molecular Sieves*, John Wiley, New York, 1974.
[13] D. M. Ruthven, *Principals of Adsorption and Adsorption Processes*, Wiley Interscience, New York, 1984.
[14] J. Cejka, H. van Bekkum, A. Corma and F. Schüth (Eds), *Introduction to Zeolite Science and Practice*, 3rd edition, Elsevier, Amsterdam, 2007.
[15] J. W. Gibbs, *Collected Works*, Longmans Green and Co., New York, 1877.
[16] R. M. Barrer, *J. Coll. Interf. Sci.* **21**, 415 (1966).
[17] R. M. Barrer, *Zeolites & Clay Minerals as Sorbents & Molecular Sieves*, Academic Press, London, 1978.
[18] J. E. Lennard-Jones, *Physica Eindhoven*, **4**, 9411937 (1937).
[19] J. O. Hirschfelder and A. D. Crowell, *Physical Adsorption of Gases*, Butterworths, London, 1962.

[20] K. S. W. Sing, D. H. Everett, R. A. W. Haul, L. Moscou, R. A. Pierotti, J. Rouquerol and T. Siemieniewska, *Pure Appl. Chem.*, **57**, 603 (1985).
[21] T. J. Mays, *Studies in Surface Science and Catalysis*, **160**, 57 (2007).
[22] P. A. Monson, *J. Chem. Phys.*, **128**, 084701 (2008).
[23] http://www.ecs.umass.edu/~monson/ (28 October 2013).
[24] M. R. Baklanov, K. P. Mogilnikov, V. G. Polovinkin and F. N. Dultsev, *J. Vac. Sci. Technol. B*, **18**, 1385 (2000).
[25] J. Rouquerol, *Thermochim. Acta*, **144**, 209 (1989).
[26] O. Toft Sorensen and J. Rouquerol, *Sample Controlled Thermal Analysis*, Kluwer Acad., Dordrecht, 2003.
[27] P. L. Llewellyn, Y. Grillet, F. Schuth, H. Reichert and K. K. Unger, *Micro. Mat.*, **3**, 345 (1994).
[28] P. L. Llewellyn, C. Sauerland, C. Martin, Y. Grillet, J.-P. Coulomb, F. Rouquerol and J. Rouquerol, in *Characterisation of Porous Solids IV*, B. McEnaney, T. Mays, J. Rouquerol, F. Rodriguez-Reinoso, K. S. W. Sing and K. Unger (Eds), Royal Society of Chemistry, London, 1997.
[29] J. Rouquerol, F. Rouquerol, C. Peres, Y. Grillet and M. Boudellal, in *Characterisation of Porous Solids*, S. J. Gregg, K. S. W. Sing and H. F. Stoeckli (Eds), Society of Chemical Industry, London, 1979.
[30] M. K. Ismail, *Langmuir*, **8**, 360 (1992).
[31] J. L. Jelinek and E. Kovats, *Langmuir*, **10**, 4225 (1994).
[32] S. Brunauer, P. H. Emmett and E. Teller, *J. Am. Chem. Soc.*, **60**, 309 (1938).
[33] S. Brunauer, *The Adsorption of Gases and Vapors*, Princeton University Press, NJ, 1945.
[34] I. Langmuir, *Phys. Rev.*, **6**, 79 (1915).
[35] I. Langmuir, *Proc. Nat. Acad. Sci. U.S.A.* **3**, 141 (1917).
[36] J. Rouquerol, P. L. Llewellyn, F. Rouquerol, in *Characterisation of Porous Solids VII*, P. L. Llewellyn, J. Rouquerol, F. Rodriguez-Reinoso and N. A. Seaton (Eds.), Studies in Surface Science and Catalysis, No. 160, Elsevier, Amsterdam, 2006.
[37] B. C. Lippens and J. H. de Boer, *J. Catalysis*, **4**, 319 (1965).
[38] K. S. W. Sing, *Chem. & Ind.*, 1520 (1968).
[39] W. D. Harkins and G. Jura, *J. Am. Chem. Soc.*, **66**, 1362 (1944).
[40] G. Horváth and K. Kawazoe, *J. Chem. Eng. Jpn.*, **16**, 470 (1983).
[41] C. Lastoskie, K. E. Gubbins and N. Quirke, *J. Phys.Chem.*, **97**, 4786 (1993).
[42] J. Z. Jagiello and E. D. Tolles, in *Fundamentals of Adsorption* 6, F. Meunier (Ed.), Elsevier, Paris, 1998.
[43] A. V. Neimark, Y.Lin, P. I. Ravikovitch and M. Thommes, *Carbon*, **47**, 1617 (2009).
[44] A. Thomy and X. Duval, *J. Chim. Phys.*, **67**, 110 (1970).
[45] J. Rouquerol, S. Partyka and F. Roquerol, *J. Chem. Soc., Faraday Trans.*, **73**, 306 (1977).
[46] P. L. Llewellyn, J.-P. Coulomb, Y. Grillet, J. Patarin, H. Lauter, H. Reichert and J. Rouquerol, *Langmuir*, **9**, 1846 (1993).
[47] P. L. Llewellyn, J.-P. Coulomb, Y. Grillet, J. Patarin, G. André and J. Rouquerol, *Langmuir*, **9**, 1852 (1993).
[48] S. Bourrelly, P. L. Llewellyn, C. Serre, F. Millange, T. Loiseau and G. Férey, *J. Am. Chem. Soc.* **127**, 13519 (2005).

[49] P. L. Llewellyn, G. Maurin, T. Devic, S. Loera-Serna, N. Rosenbach, C. Serre, S. Bourrelly, P. Horcajada, Y. Filinchuk and G. Férey, *J. Am. Chem. Soc.*, **130**, 12808 (2008).

[50] S. Bourrelly, B. Moulin, A. Rivera, G. Maurin, S. Devautour-Vino, C. Serre, T. Devic, P. Horcajada, A. Vimont, G. Clet, M. Daturi, J. C. Lavalley, S. Loera-Serna, R. Denoyel, P. L. Llewellyn and G. Férey, *J. Am. Chem. Soc*, **132**, 9488 (2010).

[51] F. Salles, G. Maurin, C. Serre, P. L. Llewellyn, C. Knofel, H. J. Choi, Y. Filinchuk, L. Oliviero, A. Vimont, J. R. Long and G. Férey, *J. Am. Chem. Soc*, **132**, 13782 (2010).

5

Dynamic Light Scattering

Erika Eiser
Cavendish Laboratory, University of Cambridge, Cambridge, CB3 0HE, UK

5.1 INTRODUCTION

As the name suggests, **dynamic light scattering** (DLS) deals with the study of how coherent light, scattered by a solution containing diffusing particles, changes with time. The time-correlation function of these scattered intensities provides information about not only the diffusivity of the suspended particles but also their size.

In the 1960s, Hanbury Brown and Twiss were the first to realise that the time-correlation function of scattered laser light could be used to determine the radii of radio stars.[1−3] The rapid improvement in and availability of continuous lasers and fast photon detectors then promoted the rapid development of DLS for measurement of the size or radius of gyration of colloids, macromolecules, viruses or synthetic and biological molecules in the range of a few nm up to a few μm in diameter.

The essential set-up of a typical DLS experiment is shown in Figure 5.1. A well-defined parallel beam of coherent laser light is directed onto a dilute colloidal suspension. Most of the beam will travel through the sample and hit the screen, placed a large distance away from the sample (far-field realisation). One might assume that the intensity imaged on the screen will decay away smoothly from the centre, but this is not what is seen for a fluid sample containing scattering particles. Instead, there will be many speckles, which are tiny but macroscopically visible light dots

Multi Length-Scale Characterisation, First Edition. Edited by Duncan W. Bruce, Dermot O'Hare and Richard I. Walton.

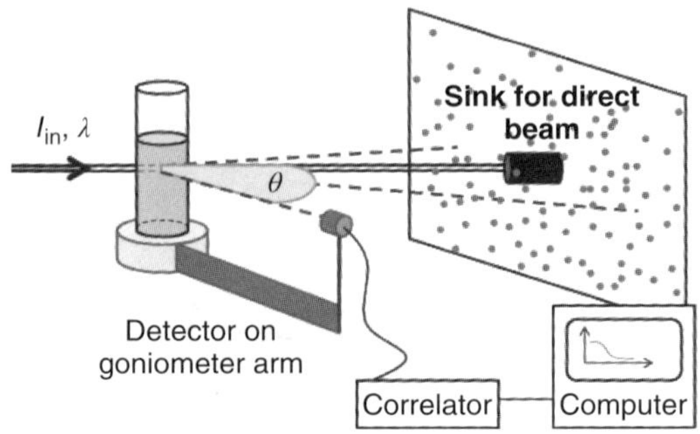

Figure 5.1 Schematic of a general DLS set-up. Monochromatic laser light of intensity I_{in} falls onto a cuvette holding the sample. For dilute, colloidal suspensions, much of the light passes through unaltered and only a small fraction is scattered. The light scattered from the colloids will interfere and form a speckle pattern, indicated as a snapshot by the dots on the screen. The detector is typically mounted on a goniometer, allowing for measurement of the scattered intensities, $I_s(\theta, t)$, with scattering angles, θ, ranging from a few degrees up to about 170° (depending on the instrument used).

that move continuously around the screen. Moreover, these speckles will be visible at all scattering angles, θ, that lie in the plane of the incoming and the unscattered outgoing beam, and also out of plane. The appearance of speckles is caused by constructive interference of the incoming light scattered by individual particles. As the particles diffuse in time, the conditions for constructive interference change, making the speckles move. The heart of DLS lies in measurement of the intensity of a speckle in time with a fast, 'point-like' detector under a fixed scattering angle, θ. Note that depending on how the intensity or frequency of the scattered light is detected, DLS can also be referred to as **photon-correlation spectroscopy** (PCS) or **quasi-elastic light scattering** (QUELS), and sometimes even **intensity fluctuation spectroscopy** (IFS).[4]

In Figure 5.2, a typical trace of the scattered intensity, $I_s(q, t)$, is presented as it fluctuates in time around a time-averaged intensity, $\langle I_s(q) \rangle$. In an DLS experiment, it is necessary to measure the intensity of the autocorrelation function, $g^{(2)}(\tau) = \langle I_s(\vec{q}, 0) I_s(\vec{q}, \tau) \rangle / \langle I_s(\vec{q})^2 \rangle$. Figure 5.2 shows the decay of this autocorrelation function and, as expected, it starts with a fully correlated signal that decays in time to the time-averaged square intensity. It is the decaying part that holds the information about the diffusivity of the diffusing object of interest. It will be shown that for very dilute monodisperse samples, the decay is exponential and written as $g^{(2)}(\tau) = 1 + \exp(-q^2 D \tau)$, where $q = |\vec{q}| \propto \sin(\theta/2)$ is the scattering wave vector and D is the diffusion coefficient of the scattering object.

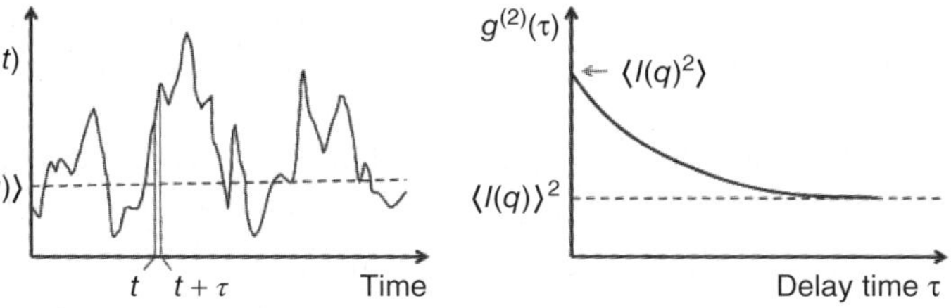

Figure 5.2 Left: Scattered intensity measured in time, showing typical fluctuations due to thermal motion of scattering particles in solution. Right: Intensity-autocorrelation function plotted as a function of delay time, τ.

This chapter provides a more detailed background to DLS, starting with static light scattering (SLS), as the tools derived will be valuable for the subsequent derivation of the relevant electric field and intensity autocorrelation functions. Note that the derivations and optical background will be kept to a minimum. For a very thorough discussion, the classic book by Berne and Pecora is recommended,[5] as are subsequent publications by Chu,[4] Cumming and Pike[6] and the series edited by Lindner and Zemb.[7] Much of what follows is based on the Pusey's contributions to the Lindner and Zemb series.[7] The theoretical background will be followed by a more detailed discussion of particle sizing and the evaluation of size distributions. The main section on DLS will be followed by a description of new instrumental approaches to DLS based mainly on fibre optics, which allow DLS measurements to be made in more difficult environments. The chapter concludes with the introduction of a very new video-microscopy technique (**differential dynamic microscopy**, DDM), which may replace the costly and complex instrumental set-up of DLS and could very soon be adopted by many laboratories.

5.2 THEORETICAL BACKGROUND

Fundamental to all DLS experiments is the use of continuous laser light that fulfils a number of criteria. First, the light striking the sample must be a monochromatic plane wave. The time and spatially dependent amplitude of such a plane wave is given by Equation 5.1:

$$E_{\text{in}}(\vec{r},t) = E_0 \exp\left[\mathrm{i}\left(\vec{k}_{\text{in}} \cdot \vec{r} - \omega t\right)\right] \tag{5.1}$$

with electric field vector $\mathbf{E}_0$ and wave vector $\boldsymbol{k}_{in}$, which points in the propagation direction of the incident beam. The magnitude of the wave vector is $k_{in} \equiv |\mathbf{k}_{in}| = 2\pi/\lambda$, where λ is the wavelength.

Figure 5.3 shows the top view of a basic experimental DLS set-up. The incident beam must be in the same plane as the detected scattered and non-scattered photons and the incident electric field, $\vec{E}_0$, must be linearly polarised perpendicular to the scattering plane. For DLS experiments on isotropic particles, there is no need to place an analyser between the scattered light and the detector. Therefore, in many systems that are equipped with a polariser and an analyser, both are aligned in a parallel fashion – hence the name, **polarised DLS** – and the translational diffusion coefficient is measured. However, when studying suspensions of anisotropic colloids, such as clays (disc-like particles) or fd-viruses (rod-like particles), both polarised and depolarised scattering are measured, as the latter can give information on the rotational diffusion coefficient of the colloids. Depolarised DLS is discussed further in Section 5.3.3.1.

In order to be able to interpret the scattering data recorded based on mathematical analysis, it is necessary to make a few more assumptions concerning both the sampling beam and the sample of interest: (i) the spread of the wavelength of the linearly polarised laser light must be minimal, implying that the light is monochromatic; (ii) the light must be coherent spatially and temporally; (iii) the intensity of the light used must be very stable over time; (iv) the sample must be dielectric, meaning that it is transparent to the laser beam and non-adsorbing or fluorescent – this applies to both the sample medium (mostly liquid continuum with an average electric permittivity, ε_m) and the scattering

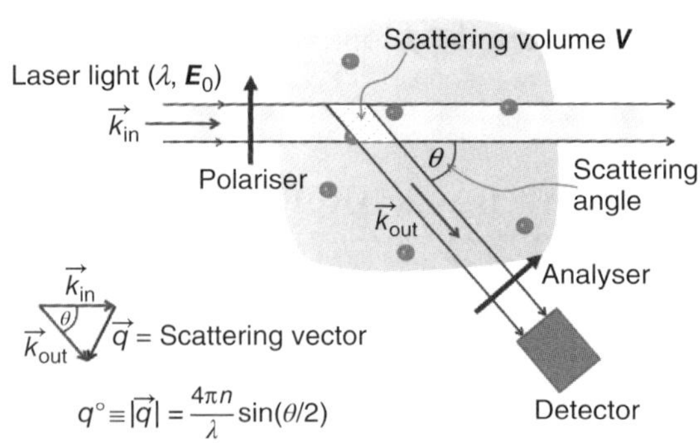

Figure 5.3 Top view of a DLS set-up. A parallel beam of coherent laser light with intensity I_{in} passes through a dilute colloidal suspension from the left. Most of this incoming intensity will pass through unaltered, while a small fraction will be scattered and falls onto a detector (usually a photomultiplier) with a very small area.

Figure 5.4 Photograph of a relatively parallel beam of laser light passing through a quartz cuvette filled with 'dirty' water containing bubbles and dust particles (left), a container filled with hydrogel particles of 1 cm diameter immersed in water (middle) and a similar container filled with the same type of hydrogel particles but with no water added (right). The laser light comes from a laser pointer with a wavelength of $\lambda \approx 532\,\text{nm}$. As expected if there are no scattering objects, the beam remains invisible as it passes through the cuvette. However, there is static scattering at the walls of the cuvette, which lights the air bubbles in the water. Most of the laser beam passes through and enters the middle container. Because the hydrogel is mostly water, just like the continuous phase in which the particles are suspended, only traces of scattering are seen. In this case, the particles are almost completely refractive-index-matched. In the right container, the refractive index of the spherical hydrogel particles is about 1.34, which is much higher than that of air (≈ 1). This leads to very strong, multiply scattered light, as evidenced by the 'lighting up' of the entire sample, although the actual laser beam has a width of only 0.5 mm.

objects (such as colloidal particles or molecules with a local electric permittivity, $\varepsilon(\vec{r}, t)$, that will be time dependent as the colloids diffuse freely); and (v) the concentration of the colloid suspension must be sufficiently dilute, such that the incoming light is scattered only once before reaching the detector (Figure 5.4).

If all these conditions are met, the electric field of the light scattered by an infinitesimal scattering volume, dV, falling onto the detector surface some distance, $\vec{R}$, away from the sample can be written as Equation 5.2:[8]

$$E_s(\vec{R}, t) = -\frac{k^2 E_0}{4\pi} \frac{\exp\left[i\left(\vec{R}\cdot\vec{r} - \omega t\right)\right]}{R} \int \frac{\varepsilon(\vec{r}, t) - \varepsilon_m}{\varepsilon_m} \exp\left[-i(\vec{q}\cdot\vec{r})\right] d^3r \tag{5.2}$$

where $\vec{q}$ is the scattering vector, which is defined as the vectorial subtraction of the incoming from the scattered wave vector:

$$\vec{q} = \vec{k}_s - \vec{k}_{in}, \qquad q \equiv |\vec{q}| = \frac{4\pi}{\lambda}\sin(\theta/2) \tag{5.3}$$

where θ is the scattering angle.

This expression for the scattered electric field, derived by D'Aguanno and Klein,[8] is correct only in the so-called **far-field approximation,** meaning that the detector is far away from the sample ($\vec{R} \gg \vec{r}$), a condition that is easily met. It can be seen that, according to Huygens' principle, the scattered wave is a spherical wave ($\vec{E}_s \propto \vec{E}_0 e^{i(\vec{R}\cdot\vec{r}-\omega t)}/R$) multiplied by the Fourier transform (FT) of the time-dependent local electric permittivity variation that takes place in the illuminated scattering volume, V. As the electric permittivity is directly proportional to the refractive index, n, of a dielectric medium ($\varepsilon \propto n$), the FT of the permittivity variation can be interpreted in terms of local changes in the refractive index. This means that if the refractive index of the colloidal particles is the same as that of the liquid medium in which they are dispersed then the sample does not scatter at all. In other words, the sample will appear completely transparent and no speckled pattern will be seen, except for the transmitted direct beam. Such a procedure is known as **refractive-index matching** and is used to make dense colloidal systems transparent. This also means that the van der Waals interactions are screened completely.

To obtain a more workable expression for $\vec{E}_s(\vec{R}, t)$, it is necessary to realise that it is composed of all contributions from the $\tilde{n}$ particles lying in the scattering volume, V. At time t, the positions of these $\tilde{n}$ particles can be described by a set of vectors, $\{\vec{R}_j(t)\}$, in which each vector points from some arbitrary origin to the centre of mass of the scattering objects (Figure 5.5). Assuming that spherical colloids are being considered, it is necessary to integrate over the volume, dV_j, of the colloids. The integration of the colloidal volume enters the expression describing the variation of the electric susceptibility, $(\varepsilon(\vec{r}, t) - \varepsilon_m)$. This expression

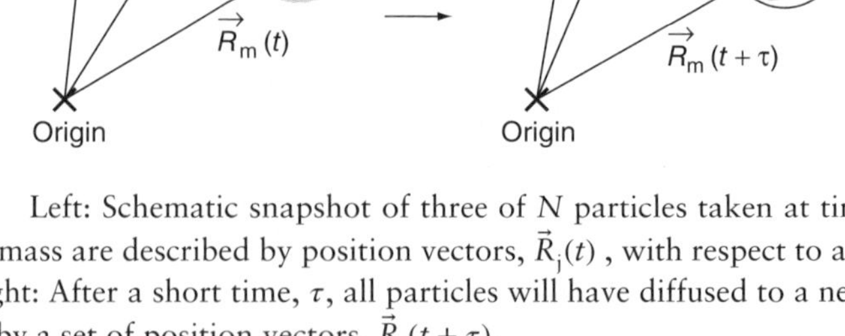

Figure 5.5 Left: Schematic snapshot of three of N particles taken at time t. Their centres of mass are described by position vectors, $\vec{R}_j(t)$, with respect to an arbitrary origin. Right: After a short time, τ, all particles will have diffused to a new position described by a set of position vectors, $\vec{R}_j(t+\tau)$.

can be rewritten in terms of the particles and pure liquid susceptibilities, $\varepsilon_p(\vec{r},t)$ and $\varepsilon_l(\vec{r},t)$, in the following way:

$$\varepsilon(\vec{r},t) - \varepsilon_m = [\varepsilon_p(\vec{r},t) - \varepsilon_l] + [\varepsilon_l(\vec{r},t) - \varepsilon_l] + [\varepsilon_l - \varepsilon_m] \quad (5.4)$$

The first term on the right-hand side describes the susceptibility difference inside the particles, the second describes the variations in the liquid and the last expresses the susceptibility anywhere in the scattering volume, V. In general, we need only consider the first term. Assuming that the density fluctuations in the liquid are extremely short-lived, the second term can be neglected. The susceptibilities in the third term are independent of r, so their contribution to $\vec{E}_s$ will be a constant multiplied by $\int_v \exp(-i\vec{q}\cdot\vec{r})d^3r$, which is simply the FT of the total scattering volume. This constant factor is the background, which can be measured. For DLS measurements, this contribution can also be neglected.

The susceptibility due to the $\tilde{n}$ particles lying inside V can now be written as:

$$\varepsilon_p(\vec{r},t) = \sum_{j=1}^{\tilde{n}} \delta(\vec{r} - \vec{R}_j - \vec{r}_j)\varepsilon_p(\vec{r}_j,t)d^3r_j \quad (5.5)$$

Substituting $\varepsilon_p(\vec{r},t)$ into Equation 5.4, an expression is obtained for the total scattered electric field at position $\vec{R}$ and time t:

$$\vec{E}_s(\vec{R},t) = -\frac{k^2E_0}{4\pi}\frac{e^{i(\vec{R}\cdot\vec{r}-\omega t)}}{R}\sum_{j=1}^{N}\left[\int \frac{\varepsilon_p(\vec{r},t) - \varepsilon_l}{\varepsilon_m} e^{-i\vec{q}\cdot\vec{r}_j}d^3r_j\right] e^{-i\vec{q}\cdot\vec{R}_j} \quad (5.6)$$

Then, defining $\Delta\rho(\vec{r}_j,t)$ as the 'optical density' contrast of the sample:

$$\Delta\rho(\vec{r}_j,t) = \frac{k^2}{4\pi}\frac{\varepsilon_p(\vec{r},t) - \varepsilon_l}{\varepsilon_m} \quad (5.7)$$

we obtain:

$$E_s(\vec{R},t) = -E_0\frac{e^{i(\vec{R}\cdot\vec{r}-\omega t)}}{R}\sum_{j=1}^{N}\left[\int \Delta\rho(\vec{r}_j,t)\, e^{-i\vec{q}\cdot\vec{r}_j}d^3r_j\right] e^{-i\vec{q}\cdot\vec{R}_j}$$
$$= -E_0\frac{e^{i(\vec{R}\cdot\vec{r}-\omega t)}}{R}\sum_{j=1}^{N} b_j(\vec{q},t)e^{-i\vec{q}\cdot\vec{R}_j} \quad (5.8)$$

The coefficient $b_j(\vec{q},t)$ is called the scattering length of particle, j. These coefficients are simply the Fourier transforms of the optical density or refractive index contrast between the particles and the liquid background. Note that the above calculations apply to scattering for all types of radiation, whether it be X-ray, neutron or light scattering. Hence, equivalent scattering lengths will appear in X-ray or neutron scattering, only the density contrast will be due to electronic density and to differences in the masses of the nuclei of different atoms and their isotopes, respectively.

In SLS experiments, it is not the scattered electric field, $E_s(\vec{R},t)$, that is measured with a detector but the ensemble-averaged intensity: $I_s(\vec{q}) = \langle I_s(\vec{q},t)\rangle = \langle |E_s|^2\rangle = \langle E_s(\vec{q},t)E_s^*(\vec{q},t)\rangle$. The intensity is given by:

$$I_s(\vec{q},t) = \left(\frac{E_0}{R}\right)^2 \left(\sum_{j=1}^{N} b_j\left(\vec{q},t\right) e^{-i\vec{q}\cdot\vec{R}_j}\right)\left(\sum_{k=1}^{N} b_k^*\left(\vec{q},t\right) e^{+i\vec{q}\cdot\vec{R}_k}\right) \tag{5.9}$$

The ensemble average is then:

$$\begin{aligned}\langle I_s(\vec{q})\rangle &= \langle I_s(\vec{q},t)\rangle = \lim_{N\to\infty}\frac{1}{N}\sum_{j=1}^{N} I_s^{(j)}(\vec{q},t)\\ &= \left(\frac{E_0}{R}\right)^2 \left\langle \sum_{j=1}^{N}\sum_{k=1}^{N} b_j\left(\vec{q},t\right) b_k^*(\vec{q},t) e^{-i\vec{q}\cdot(\vec{R}_j-\vec{R}_k)}\right\rangle\end{aligned} \tag{5.10}$$

So how is this ensemble-averaged expression of the scattered intensity to be understood? It is convenient to sort the terms in the double sum into those for which $j = i$ and those for which:

$$\begin{aligned}\langle I_s(\vec{q},t)\rangle = \left(\frac{E_0}{R}\right)^2 \Bigg\langle &\sum_{j=1}^{N} b_j\left(\vec{q},t\right) b_j^*(\vec{q},t)\\ &+ \sum_{j=1}^{N}\sum_{k=1}^{N} b_j\left(\vec{q},t\right) b_k^*(\vec{q},t) e^{-i\vec{q}\cdot(\vec{R}_j-\vec{R}_k)}\Bigg\rangle\end{aligned} \tag{5.11}$$

Because summation and the averaging process (which is an integral operation) are linear operations, the order in which they are performed can be swapped so that:

$$\begin{aligned}\langle I_s(\vec{q})\rangle = \left(\frac{E_0}{R}\right)^2 \Bigg[&\sum_{j=1}^{N} \left\langle b_j\left(\vec{q},t\right) b_j^*(\vec{q},t)\right\rangle\\ &+ \sum_{j\neq k}^{N}\sum_{k=1}^{N} \left\langle b_j\left(\vec{q},t\right) b_k^*(\vec{q},t) e^{-i\vec{q}\cdot(\vec{R}_j-\vec{R}_k)}\right\rangle\Bigg]\end{aligned} \tag{5.12}$$

For very dilute colloidal solutions, the averaging in the second sum can be done on the two separate terms such that:

$$\langle I_s(\vec{q})\rangle = \left(\frac{E_0}{R}\right)^2 \left[\sum_{j=1}^{N} \left\langle b_j(\vec{q},t)\, b_j^*(\vec{q},t)\right\rangle + \sum_{j\neq k}^{N}\sum_{k=1}^{N} \left\langle b_j(\vec{q},t)\, e^{-i\vec{q}\cdot\vec{R}_j}\right\rangle \left\langle b_k^*(\vec{q},t)\, e^{+i\vec{q}\cdot\vec{R}_k}\right\rangle\right] \tag{5.13}$$

As a first approximation, it can be assumed that any inter-particle correlations can be neglected and therefore that the second sum averages to zero. Thus, the term for the scattered intensity reduces to:

$$\langle I_s(\vec{q})\rangle = \left(\frac{E_0}{R}\right)^2 \sum_{j=1}^{N} \left\langle \left|b_j(\vec{q},t)\right|^2\right\rangle \tag{5.14}$$

This assumption is, strictly speaking, correct only at the limit of very dilute solutions and is wrong when there is only one particle present. From the definition of the scattering lengths, b_j, we know that these are simply the Fourier transforms of the shape of the scattering object. For a very dilute solution of identical colloidal spheres with homogenous density, the scattered intensity is thus given by:

$$\langle I_s(q)\rangle = \left(\frac{E_0}{R}\right)^2 N\langle |b(q,t)|^2\rangle = \left(\frac{E_0}{R}\right)^2 N\langle |b(0)|^2\rangle\, P(q) \tag{5.15}$$

where $P(q) \equiv \langle |b(q)|\rangle^2/\langle |b(0)|^2\rangle$ is the so-called **form factor** as it reflects purely the actual shape of the scattering particles (see Figure 5.6). It is defined such that $P(q) \to 1$ for $q \to 0$.

When dealing with more concentrated systems, it can no longer be assumed that the cross-terms will average out, as the particle motions will be correlated one to another. However, for identical particles with homogenous density, the averaged scattered intensity can be written as:

$$\begin{aligned}\langle I_s(\vec{q})\rangle &= \left(\frac{E_0}{R}\right)^2 \left\langle \sum_{j=1}^{N}\sum_{k=1}^{N} b(q)\, b^*(q)\, e^{-i\vec{q}\cdot(\vec{R}_j-\vec{R}_k)}\right\rangle \\ &= \left(\frac{E_0}{R}\right)^2 \langle |b(q)|^2\rangle \sum_{j=1}^{N}\sum_{k=1}^{N} \langle e^{-i\vec{q}\cdot(\vec{R}_j-\vec{R}_k)}\rangle \end{aligned} \tag{5.16}$$

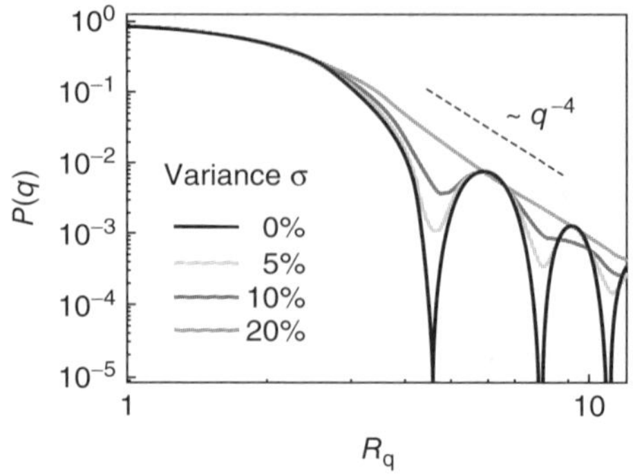

Figure 5.6 Form factor $P(q)$ as a function of the scattering vector, q, multiplied by the radius, R, of a spherical colloid. The 'smearing' of the sharp oscillations expected for a single colloid size is shown for colloids with a symmetric Gaussian size distribution for three different widths, $\sigma = 5, 10$ and 20%.

Inserting the expression for the form factor $P(q)$ and defining the double sum as the 'static structure factor' $S(q)$ gives:

$$S(q) \equiv \frac{1}{N} \sum_{j=1}^{N} \sum_{k=1}^{N} \left\langle e^{-i\vec{q}\cdot\left(\vec{R}_j - \vec{R}_k\right)} \right\rangle \tag{5.17}$$

the scattered intensity can be written as:

$$\langle I_s(\vec{q}) \rangle = \left(\frac{E_0}{R}\right)^2 N\langle |b(0)|^2 \rangle \, P(q)S(q) \tag{5.18}$$

Hence, the ensemble-averaged scattered intensity is a product of $Nb^2(0)P(q)$, which is a result of the uncorrelated particles, and the static structure factor $S(q)$, which describes the averaged spatial distribution of the particles'centres of mass. Evidently, in the limit of very dilute solutions, $S(q) \to 1$, and Equation 5.15 results. An excellent account of the form and structure factors for various colloidal shapes and structures is given by Pedersen in reference [9].

Before going on to introduce the autocorrelation function of the scattered intensity, there is one last technical detail that needs to be considered, namely the averaging process. The ensemble average of the intensity was taken above; this was defined as:

$$\begin{aligned} \left\langle I_s\left(\vec{q}\right)\right\rangle_E = \left\langle I_s\left(\vec{q}, t_0\right)\right\rangle &= \lim\nolimits_{N\to\infty} \frac{1}{N} \sum_{j=1}^{N} I_s^{(j)}(\vec{q}, t_0) \\ &= \int_{-\infty}^{\infty} I_s(\vec{q}) p_I(\vec{q}, t_0) \mathrm{d}\vec{q} \end{aligned} \tag{5.19}$$

Note that the ensemble average, denoted as subscript, E, is performed over the scattered intensities or electric-field amplitudes generated by all possible spatial configurations of the N particles in the system at time $t = t_0$. Even in dilute systems, N can be very large. Hence the ensemble average can also be written in integral form, in which the intensities are multiplied by a probability function, $p_I(\vec{q}, t)$. However, in DLS the time average of $I_s(\vec{q}, t)$ is measured as:

$$\langle I_s(\vec{q}, t)\rangle_T = \lim_{T\to\infty} \frac{1}{T}\int_{-T/2}^{T/2} I_s(\vec{q}, t)\mathrm{d}t \tag{5.20}$$

and the autocorrelation function is computed as:

$$\langle I_s(\vec{q}, t)\, I_s(\vec{q}, t+\tau)\rangle = \lim_{T\to\infty} \frac{1}{T}\int_{-T/2}^{T/2} I_s(\vec{q}, t) I_s(\vec{q}, t+\tau)\mathrm{d}t \tag{5.21}$$

It is important to note that for systems that are ergodic, the time average equals the ensemble average and the two can be interchanged. An example of an ergodic system is a colloidal gas (dilute) or liquid (concentrated). A colloidal crystal, on the other hand, is not ergodic.

Equipped with these general expressions for the scattered electric fields and intensities, the intensity–intensity correlation function can now be derived, which corresponds to what is measured in DLS experiments.

5.2.1 Scattering Intensities and the Autocorrelation Function

In standard light, X-ray or neutron scattering experiments, information about the shapes of the scattering objects and their spatial distributions (structure) is extracted from the static (time-averaged) scattered intensity, $\langle I_s(\vec{q})\rangle$ or $\langle I_s(\theta)\rangle$, which is usually plotted as a function of the scattering vector, $\vec{q}$, or angle, θ, respectively. In DLS, however, it is the time-variant contribution of the scattered intensity that is used to extract useful information about the sample.

It has been shown that the scattered intensity is proportional to the intensity of the incoming intensity, $E_{in}(t)$, multiplied by a time-dependent variation of a function $\delta\alpha(\vec{q}, t)$, such that:

$$E_s(q) = E_{in}(t)\delta\alpha(\vec{q}, t), \text{ with } \delta\alpha(\vec{q}, t) = \frac{1}{R}\sum_{j=1}^{N} b_j(\vec{q}, t)\mathrm{e}^{-i\vec{q}\cdot\vec{R}_j} \tag{5.22}$$

The scattered intensity is then simply:

$$I_s(t) = |E_s|^2 = E_{in}(t)E^*_{in}(t)\delta\alpha(\vec{q},t)\delta\alpha^*(\vec{q},t) \qquad (5.23)$$

Information is only obtained about the sample if the change in intensity with time is considered. This means that the time-averaged intensity–intensity correlation function, also called the autocorrelation function of the scattered intensity, must be considered:

$$\langle I_s(0)I_s(t)\rangle = \langle E_{in}(0)E^*_{in}(0)E_{in}(t)E^*_{in}(t)\rangle \times \langle \delta\alpha(\vec{q},0)\delta\alpha^*(\vec{q},0)\delta\alpha(\vec{q},t)\delta\alpha^*(\vec{q},t)\rangle \qquad (5.24)$$

The first term in the pointed brackets on the right-hand side of Equation 5.24 is the time average over the electric-field amplitudes of the incoming beam, which reflects the intensity fluctuations of the laser. For a continuous-wave laser, these fluctuations decay typically within 10^{-12} to 10^{-9} seconds and, therefore, do not influence the information required from the sample. This means that at time $t = 0$, the normalised scattered intensity observed under an angle corresponding to the scattering vector, $\vec{q}$, is fully correlated with itself ($\langle I_s(0)I_s(t)\rangle = \langle I_s^2(0)\rangle = 4$) and falls off to a factor 2 within fs (or in other words, $\langle I_{in}^2\rangle \rightarrow \langle I_{in}\rangle^2$).

The second term in Equation 5.24 is then the actual quantity of interest, as it describes the fluctuations in the optical density of the sample, reflecting the diffusive motion of the scatterer in the suspending fluid. As colloidal particles and macromolecules of sizes ranging from a few nm to μm in diameter are under study, this term will decay on timescales of the order of 10^{-9} to 10^{-4} seconds. We will see later that for truly diffusive systems, $\langle I_s(0)I_s(t)\rangle/\langle|I_s(0)|^2\rangle$ then falls off further from factor 2 to factor 1.

In order to compute the intensity–intensity correlation function, the time averages of the scattered electric fields are first considered alone and then those of the electric-field products are taken into account. We continue in our assumption that a dilute solution of N identical, spherical colloids is under study, meaning that their electric permittivity, ε_p, is a constant. Therefore, the scattering lengths $b_j(q) = b$ are time-independent, allowing the scattered electric field to be written as:

$$\vec{E}_s(\vec{q},t) = -\vec{E}_0\frac{e^{i(\vec{R}\cdot\vec{r}-\omega t)}}{R}b\sum_{j=1}^{N} e^{-i\vec{q}\cdot\vec{R}_j} \qquad (5.25)$$

The scattered electric field has thus reduced to a spherical wave multiplied by the sum of phase factors, which can be represented by vectors in the complex plane. At time $t = 0$, the resulting electric field will also be a vector, $\vec{E}_s(\vec{q}, t = 0)$, in the complex plane. As we only consider scattering in the plane containing the incoming beam we can drop the vectorial sign for the electric fields and simply write them as $E_s(\vec{q}, t)$ or E_s. Because colloidal particles perform a random walk in the fluid as time evolves, the length and direction of $\boldsymbol{E}_s$ will change constantly. But because this thermal diffusion of the particles is a completely random process (ergodicity), the time average over these scattered vectors will be zero, since the phase angles are equally distributed over the range $[0, 2\pi]$ – Equation 5.26:

$$\langle E_s(\vec{q}, t)\rangle = -\lim_{T\to\infty}\frac{1}{T}\int \mathrm{d}t \sum_{j=1}^{N} \mathrm{e}^{-\mathrm{i}\vec{q}\cdot\vec{R}_j(t)} = \sum_{j=1}^{N}\langle \mathrm{e}^{-\mathrm{i}\vec{q}\cdot\vec{R}_j(t)}\rangle = 0 \qquad (5.26)$$

For now the pre-factor is omitted, as it drops out when intensities are normalised later. The time-averaged scattered intensity is then given by:

$$\begin{aligned}\langle I_s(\vec{q}, t)\rangle = \langle |E_s(\vec{q}, t)|^2\rangle &= \left\langle \sum_{j=1}^{N} \mathrm{e}^{-\mathrm{i}\vec{q}\cdot\vec{R}_j(t)} \sum_{k=1}^{N} \mathrm{e}^{+\mathrm{i}\vec{q}\cdot\vec{R}_k(t)} \right\rangle \\ &= \sum_{j=1}^{N}\sum_{k=1}^{N}\left\langle \mathrm{e}^{-\mathrm{i}\vec{q}\cdot\left(\vec{R}_j(t)-\vec{R}_k(t)\right)}\right\rangle \end{aligned} \qquad (5.27)$$

Sorting the sums for factors $j = k$ and $j \neq k$, the following is obtained:

$$\langle I_s(\vec{q}, t)\rangle = \sum_{j=1}^{N} 1 + \sum_{j\neq k}^{N}\sum_{k=1}^{N}\left\langle \mathrm{e}^{-\mathrm{i}\vec{q}\cdot\left(\vec{R}_j(t)-\vec{R}_k(t)\right)}\right\rangle \qquad (5.28)$$

Finally, this expression can be simplified further by making use of the fact that the positions of the colloids in dilute solutions are uncorrelated. Hence, the time average in the cross-terms can be taken over the individual phase factors, leading to:

$$\langle I_s(\vec{q}, t)\rangle = N + \sum_{j\neq k}^{N}\sum_{k=1}^{N}\langle \mathrm{e}^{-\mathrm{i}\vec{q}\cdot\vec{R}_j(t)}\rangle\langle \mathrm{e}^{-\mathrm{i}\vec{q}\cdot\vec{R}_k(t)}\rangle = N \qquad (5.29)$$

Note that a quantity whose time or spatial average is zero, while its averaged square is proportional to the number of particles or steps taken, is known to be described by a Gaussian distribution. In fact, all random processes are Gaussian distributed in the limit of long enough averaging (tying into the discussion in Section 5.1, where the equivalence between time and ensemble averages for ergodic systems was discussed).

As normalised intensities are being considered, the autocorrelation function of the scattered electric field is defined as:

$$g^{(1)}(q,\tau) \equiv \frac{\langle E(\vec{q},0)E^*(\vec{q},\tau)\rangle}{\langle I_s(\vec{q},0)\rangle} \tag{5.30}$$

This notation is used widely. The superscript in brackets does not indicate a power law but rather refers to the type of correlation function. In fact, it is a first-order correlation function, whereas the intensity autocorrelation function is a second-order correlation function, as will be shown in a moment. From the fact that $\langle I_s(\vec{q},0)\rangle = N$, and given that the phase factors $\exp(-i\vec{q}(\vec{R}_j(0) - \vec{R}_j(\tau)))$ do not simply reduce to 1 while the mixed terms vanish again because the positions are uncorrelated, we obtain:

$$g^{(1)}(q,\tau) \equiv \frac{\sum_{j=1}^{N}\left\langle e^{-i\vec{q}\cdot\left(\vec{R}_j(0)-\vec{R}_j(\tau)\right)}\right\rangle}{N} \tag{5.31}$$

Because uncorrelated identical particles are under study, the time-averaged motion of all particles is identical, so that:

$$\sum_{j=1}^{N}\left\langle e^{-i\vec{q}\cdot\left(\vec{R}_j(0)-\vec{R}_j(\tau)\right)}\right\rangle = N\left\langle e^{-i\vec{q}\cdot\left(\vec{R}(0)-\vec{R}(\tau)\right)}\right\rangle \tag{5.32}$$

or:

$$g^{(1)}(q,\tau) = \left\langle e^{-i\vec{q}\cdot\left(\vec{R}(0)-\vec{R}(\tau)\right)}\right\rangle \tag{5.33}$$

The electric-field autocorrelation function, $g^{(1)}(\vec{q},\tau)$, is also called the **intermediate scattering function,** $f(\vec{q},t)$.

Now it remains to compute the normalised intensity autocorrelation function, which is defined as:

$$g^{(2)}(\vec{q},\tau) \equiv \frac{\langle I_s(\vec{q},0)I_s(\vec{q},\tau)\rangle}{\langle I_s(\vec{q},0)^2\rangle} \tag{5.34}$$

Fully expanded, the intensity–intensity correlation can be written as:

$$\begin{aligned}\langle I_s(\vec{q},0)I_s(\vec{q},\tau)\rangle &= \left\langle \sum_{j=1}^{N} e^{-i\vec{q}\cdot\vec{R}_j(0)} \sum_{k=1}^{N} e^{+i\vec{q}\cdot\vec{R}_k(0)} \sum_{l=1}^{N} e^{-i\vec{q}\cdot\vec{R}_l(\tau)} \sum_{m=1}^{N} e^{+i\vec{q}\cdot\vec{R}_m(\tau)} \right\rangle \\ &= \left\langle \sum_{j=1}^{N}\sum_{k=1}^{N} e^{-i\vec{q}\cdot\left(\vec{R}_j(0)-\vec{R}_k(0)\right)} \sum_{l=1}^{N}\sum_{m=1}^{N} e^{-i\vec{q}\cdot\left(\vec{R}_l(\tau)-\vec{R}_m(\tau)\right)} \right\rangle \\ &= \sum_{j}^{N}\sum_{k}^{N}\sum_{l}^{N}\sum_{m}^{N}\left\langle e^{-i\vec{q}\cdot\left(\vec{R}_j(0)-\vec{R}_k(0)+\vec{R}_l(\tau)-\vec{R}_m(\tau)\right)}\right\rangle\end{aligned} \tag{5.35}$$

This quadruple sum is rather heavy to compute, but there is a simpler way of evaluating the sum from a number of equations that have been derived already. Thus:

$$\langle E_s(0)E_s^*(0)E_s(t)E_s^*(t)\rangle = \langle E_s(0)E_s^*(0)\rangle\langle E_s(t)E_s^*(t)\rangle + \langle E_s(0)E_s^*(t)\rangle\langle E_s^*(0)E_s(t)\rangle \tag{5.36}$$

From Equation 5.29, we know that $\langle I_s(\vec{q},t)\rangle = \langle E_s(0)E_s^*(0)\rangle = \langle E_s(t) E_s^*(t)\rangle = N$, while the mixed term is simply the electric-field autocorrelation function, $g^{(1)}(q,\tau) = \langle E_s(0)E_s^*(t)\rangle/N$. Therefore, the intensity autocorrelation function becomes:

$$g^{(2)}(q,\tau)\langle I_s(\vec{q},0)\rangle^2 = \langle I_s(\vec{q},0)\rangle^2 + N^2(g^{(1)}(q,\tau))^2 \tag{5.37}$$

or:

$$g^{(2)}(q,\tau) = 1 + (g^{(1)}(q,\tau))^2 \tag{5.38}$$

This is the main result used in all DLS measurements – it relates the intensity autocorrelation function to the electric-field autocorrelation function and is known as the **Siegert relation**. It is important to note that this result is applicable only to variables that describe complex Gaussian random processes such as thermally driven fluctuations in the colloid density in solution. This is also why only optically transparent (dispersive) systems that do not adsorb light can be studied: materials that adsorb light in the range of wavelengths used in DLS will convert part of the adsorbed energy into heat (excitation of molecular vibration modes inside the colloid) or into additional kinetic energy, thereby leading to localised heating, which in turn leads to convection. Transport due to convection is not a Gaussian process, however, so experiments should be conducted at constant temperature and should avoid any local temperature gradients.

In Section 5.3 it will become apparent that the expression for the intensity autocorrelation function is much more general. It therefore applies not only to dilute solutions of identical spherical particles but also to concentrated colloidal solutions with a given size distribution and to solutions of non-spherical particles.

5.2.2 Homodyne *versus* Heterodyne Detection

So far, only scattered intensities have been considered, which were generated exclusively by the random diffusive motion of the colloidal particles in solution, giving an expression for the intensity autocorrelation function (Equation 5.38). This type of measurement is called a **homodyne** arrangement. In such homodyne experiments, the measured intensity correlation function $g^{(2)}(\vec{q}, t)$ is proportional to the square of the electric-field correlation function. This means that $g^{(2)}(\vec{q}, t)$ must always be positive (Figures 5.2 and 5.7). If this is not the case, there will be something wrong in the experimental set-up.

In general, the total scattered intensity arriving at the detector surface may also contain a static component, $E_{\text{local}}(\vec{q}, t) = E_{\text{l}}(\vec{q}, t)$, such that:

$$E_{\text{s,tot}}(\text{q}, t) = E_{\text{l}}(\vec{q}, t) + E_{\text{s}}(\vec{q}, t) \tag{5.39}$$

This additional intensity can arise from the scattering of the container walls or of dispersed larger objects such as dust particles that move so slowly that their scattered light appears as quasi-static over the periods of single measurements. Such static intensity can be added deliberately by mixing a small fraction of the incoming intensity with the scattered one. This type of arrangement is called a **heterodyne** measurement. In this case, the total scattered intensity is written as:

$$I(\vec{q}, t) = |E_{\text{l}}(\vec{q}, t) + E_{\text{s}}(\vec{q}, t)|^2 \tag{5.40}$$

It is convenient to rewrite the time- and $\vec{q}$-dependent scattered intensity arising from the particle motion (Equation 5.39) as $E_{\text{s}}(\vec{q}, t) = E_0$ $exp(-i\omega t)\alpha(\vec{q}, t)$, such that the total intensity at the detector becomes:

$$I(\vec{q}, t) = |E_{\text{l}}(\vec{q})\text{e}^{-i\omega t} + E_0\text{e}^{-i\omega t}\alpha(\vec{q}, t)|^2 \tag{5.41}$$

Recognising that $|E_{\text{s}}(\vec{q}, t)|^2 = I_{HOM}$ results from homodyne scattering, we can write:

$$I(\vec{q}, t) = |E_{\text{l}}|^2 + 2E_{\text{l}}E_0\alpha(\vec{q}, t) + I_{\text{HOM}} \tag{5.42}$$

The first part is simply a constant. The second term will be the dominant one, because in general the amplitude from a static scatterer will be much larger than the scattered intensity from the diffusing particles in solution: $E_l \gg E_0\alpha(\vec{q}, t)$. Therefore, the third term can be neglected. The heterodyne intensity autocorrelation function thus becomes:

$$\langle I(\vec{q}, t)I^*(\vec{q}, t)\rangle_{\text{HET}} \propto |E_{\text{l}}|^2 E_0\langle\alpha(\vec{q}, 0)\alpha^*(\vec{q}, t)\rangle = |E_{\text{l}}|^2 E_0\, g^{(1)}(q, \tau) \tag{5.43}$$

This result is very interesting as it shows that in heterodyne DLS the electric-field correlation function is measured directly; this can be of advantage. For instance, the correlation function may become negative, which is relevant when studying the rotational diffusion of an anisotropic particle. Further, the diffusion of a single particle can be measured, meaning that the work is carried out under conditions of infinite dilution. From an inspection of the field correlation function:

$$g^{(1)}(q,\tau) = \left\langle e^{-i\vec{q}\cdot\left(\vec{R}(0)-\vec{R}(\tau)\right)} \right\rangle = \left\langle e^{-i\vec{q}\cdot(\vec{v}(0)-v(\tau))/\tau} \right\rangle \tag{5.44}$$

it can be seen that it depends on the velocity of that particle measured. Thus, the diffusion velocity of particles in solution can be measured directly, which is not really possible in homodyne detection.

5.2.3 Relations between the Correlation Functions and Static Light Scattering

The electric-field correlation function is often referred to as the intermediate scattering function, $f(q,\tau)$, because in homodyne detection we measure the intensity correlation function. However, in order to obtain any actual information about the system, we need to extract $f(q,\tau)$ from $g^{(2)}(q,\tau)$. This seems straightforward, but in realistic experiments the expression for $g^{(2)}(q,\tau)$ must be modified to Equation 5.45:

$$g^{(2)}(q,\tau) = 1 + \beta\left(g^{(1)}(q,\tau)\right)^2 \tag{5.45}$$

The constant factor, β, is the ratio between the area of the detector and the area of a typical speckle hitting it. When deriving the electric field and the intensity-correlation function, the scattered field arriving at the detector was considered to be point-like. In reality, the detector surface has a finite area: if the detector area chosen is too large, then β will be very different from unity and the detector may 'see' more than one speckle at the time. Such a situation is not desirable in a DLS set-up. Therefore, the aperture of the detector is chosen such that $\beta \approx 0.8$. When the detector surface becomes larger than a typical speckle size, several independent fluctuating events are recorded simultaneously and $\beta \to 0$.

In homodyne measurements, where the intensity correlation function is measured, information is obtained about the intermediate function:

$$\beta^{1/2} f(q,\tau) = \sqrt{g^{(2)}(q,\tau) - 1} \tag{5.46}$$

by treating β as a fitting parameter. A detailed discussion of the relation between speckle size and detector area and the importance of the coherence of the laser light used can be found in reference [10].

Further, it is useful to realise that there is a direct link between the electric-field correlation, or measured intermediate scattering function, and the static structure factor, *S*(*q*), derived in Equation 5.17. To show this, Equation 5.30 is re-written as:

$$f(q,\tau) \equiv \frac{\langle E(\vec{q},0)E^*(\vec{q},\tau)\rangle}{\langle I_s(\vec{q},0)\rangle} = \frac{\sum_{j=1}^{N}\sum_{k=1}^{N}\left\langle b_j(\vec{q},0)\, b_k^*(\vec{q},\tau)\mathrm{e}^{-\mathrm{i}\vec{q}\cdot\left(\vec{R}_j(0)-\vec{R}_k(\tau)\right)}\right\rangle}{\sum_{j=1}^{N}\sum_{k=1}^{N}\left\langle b_j(\vec{q},0)\, b_k^*(\vec{q},0)\mathrm{e}^{-\mathrm{i}\vec{q}\cdot\left(\vec{R}_j(0)-\vec{R}_k(0)\right)}\right\rangle} \tag{5.47}$$

For suspensions of identical spheres, the scattering lengths, $b_{j,k}$, are time-independent and can be taken in front of the sum in both the denominator and numerator. They thus cancel, so we can write:

$$f(q,\tau) = \frac{\sum_{j=1}^{N}\sum_{k=1}^{N}\left\langle \mathrm{e}^{-\mathrm{i}\vec{q}\cdot\left(\vec{R}_j(0)-\vec{R}_k(\tau)\right)}\right\rangle}{\sum_{j=1}^{N}\sum_{k=1}^{N}\left\langle \mathrm{e}^{-\mathrm{i}\vec{q}\cdot\left(\vec{R}_j(0)-\vec{R}_k(0)\right)}\right\rangle} = \frac{\sum_{j=1}^{N}\sum_{k=1}^{N}\left\langle \mathrm{e}^{-\mathrm{i}\vec{q}\cdot\left(\vec{R}_j(0)-\vec{R}_k(\tau)\right)}\right\rangle}{NS(q)}$$

$$= \frac{F(q,\tau)}{S(q)} \tag{5.48}$$

F(q, τ) is known as the dynamic structure factor but is also referred to as the intermediate scattering function.

5.3 APPLICATIONS

5.3.1 Particle Sizing

One great advantage of DLS over SLS is perhaps the fact that the size of a particle can be determined over a very large scale, ranging from a few μm down to a few nm, in a very quick and easy way. It is, of course, also possible to determine the size of a particle from SLS measurements in dilute solutions, for which the intensity is simply the form factor, $P(\vec{q})$, multiplied by a constant (Equation 5.15). However, this requires great care and time, as the scattered intensity has to be measured as a function of the scattering vector, $\vec{q}$, and subsequent normalisation, which makes it difficult to follow growth processes in which the particle size or structure is changing over the course of a measurement.

The advantage of SLS measurements is their greater accuracy in determining the real size and the size distribution (also called size polydispersity) of the particles dispersed, simply because the form factor has sharp minima and maxima in a logarithmic plot of the static scattered intensity *versus* q (see Figure 5.6). The sizes determined in DLS measurements are referred to as **hydrodynamic diameters**, because they are generally somewhat larger than the true sizes obtained from static measurements, as will be explained later.

5.3.2 Identical Spherical Colloids in Dilute Suspensions

What is measured in DLS is not the sizes of the particles directly but rather their diffusion constants. To see this, consider the electric-field autocorrelation function (Equation 5.44) obtained for dilute solutions, where it is assumed that the particles do not interact and that therefore their motion is uncorrelated:

$$g^{(1)}(q,\tau) = \left\langle \mathrm{e}^{-\mathrm{i}\vec{q}\cdot\left(\vec{R}(0)-\vec{R}(\tau)\right)} \right\rangle = \left\langle \mathrm{e}^{-\mathrm{i}\vec{q}\cdot\Delta\vec{R}(\tau)} \right\rangle \qquad (5.49)$$

where $\Delta\vec{R}(\tau)$ is the displacement a particle undergoes in time interval, τ.

In one of Einstein's seminal publications from 1906,[11] he showed that the translational diffusion coefficient, D_0, of non-interacting spherical particles dispersed in a solvent of viscosity η is given by what is known

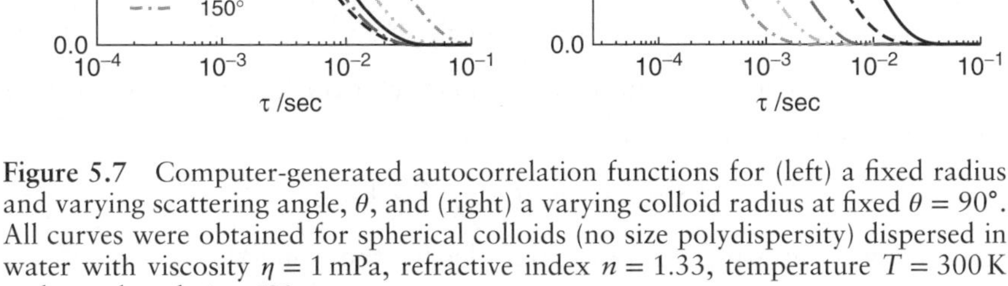

Figure 5.7 Computer-generated autocorrelation functions for (left) a fixed radius and varying scattering angle, θ, and (right) a varying colloid radius at fixed $\theta = 90°$. All curves were obtained for spherical colloids (no size polydispersity) dispersed in water with viscosity $\eta = 1\,\text{mPa}$, refractive index $n = 1.33$, temperature $T = 300\,\text{K}$ and wavelength $\lambda = 622\,\text{nm}$.

today as the Stokes–Einstein relation:

$$D_0 = \frac{k_B T}{\zeta} = \frac{k_B T}{6\pi\eta R_H} \quad (5.50)$$

where, as usual, k_B is the Boltzmann constant and T the temperature (in Kelvin). Stokes derived the friction coefficient, ξ, that a sphere of hydrodynamic radius, R_H, experiences in a viscous fluid. The Stokes–Einstein relation is also known as the **generalised fluctuation-dissipation theorem.** In the same publication, Einstein also showed that the distribution of the position in time of such particles is described by a Gaussian function. The normalised distribution in one dimension is given by:

$$p(x,\tau) = \frac{1}{\sqrt{4\pi D_0}} \frac{e^{-\Delta x^2/4D_0\tau}}{\sqrt{\tau}} \quad (5.51)$$

Furthermore, the average displacement of a particle in time is given by $\langle \Delta x^2(\tau)\rangle = 2D_0\tau$ in one dimension and by $\langle \Delta R^2(\tau)\rangle = 6D_0\tau$ in three. Using the 1D distribution function and inserting it into the 1D electric-field correlation function, Equation 5.52 is obtained:

$$g^{(1)}(q,\tau) = \langle e^{-iq\Delta x(\tau)}\rangle = \frac{1}{\sqrt{4\pi D_0 t}} \int_{-\infty}^{\infty} e^{-\Delta x^2/4D_0\tau} e^{iqx} dx \quad (5.52)$$

The product of the two exponentials can then be rewritten as:

$$e^{\frac{1}{4D_0 t}[\Delta x^2 - 4D_0\tau iqx + (2D_0\tau iq)^2] - \frac{(2D_0\tau iq)^2}{4D_0 t}} = e^{\frac{(\Delta x - 2D_0\tau iq)^2}{4D_0 t}} e^{-q^2 D_0\tau} \quad (5.53)$$

Inserting this back into Equation 5.52, we obtain:

$$g^{(1)}(q,\tau) = \langle e^{-i\vec{q}\cdot\Delta\vec{R}(\tau)}\rangle = \frac{1}{\sqrt{4\pi D_0\tau}} e^{-q^2 D_0 t} \int_{-\infty}^{\infty} e^{-(\Delta x - 2D_0\tau iq)^2/4D_0\tau} dx \quad (5.54)$$

Now all that is required is to evaluate the integral over a Gaussian distribution, which is simply $(4\pi D_0\tau)^{1/2}$. Thus the field-correlation function simplifies to:

$$g^{(1)}(q,\tau) = e^{-q^2 D_0\tau} \quad (5.55)$$

So, when measuring the intensity autocorrelation function for a dilute solution of identical hard spheres:

$$g^{(2)}(q,\tau) = 1 + \left(g^{(1)}(q,\tau)\right)^2 = 1 + e^{-2q^2 D_0\tau} \quad (5.56)$$

Thus, from the single exponential decay, the spheres' self-diffusion coefficients and, *via* the Stokes–Einstein relation, their hydrodynamic

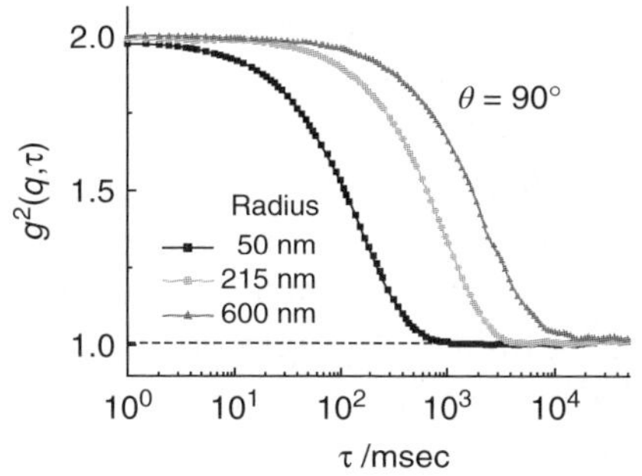

Figure 5.8 Intensity-autocorrelation functions measured for three dilute aqueous solutions of spherical polystyrene particles with different diameters. Measurements were performed at room temperature, using a Malvern particle sizer. The legend lists the manufacturer's values of the commercially obtained colloids. The CONTIN[12,13] fitting algorithm of the particle sizer provides hydrodynamic radii $R_h = 58.10 \pm 15.22$ nm for the smallest colloids, $R_h = 324.1 \pm 56.6$ nm for the intermediate size colloids and $R_h = 1036 \pm 171.4$ nm for the largest colloids.

radii, R_H, can be obtained, as demonstrated in Figure 5.7. Note that this expression is measured in homodyne detection. Should the choice be made to measure with a heterodyne detection arrangement, the intensity-correlation function would change to:

$$\langle I(\vec{q},t)I^*(\vec{q},t)\rangle_{\mathrm{HET}} \propto \text{constant} \times e^{-q^2 D_0 \tau} \tag{5.57}$$

5.3.3 Particle Sizing in Realistic Systems: Size Distributions

In general, synthetically produced colloids are never perfectly equal, as it is very difficult to make colloids of exactly one size. The colloids in any suspension will thus have some spread in diameter, and sometimes in shape or density.

A way to take the size distribution into account is to multiply the electric-field correlation function with a normalised distribution function, $p(D)$, for the diffusivities of differently sized particles, such that:

$$|g^{(1)}(q,\tau)| = \int_0^\infty p(\Gamma)e^{-\Gamma\tau}d\Gamma \equiv \sum_{j=1}^{N} p(\Gamma_j)e^{-\Gamma_j\tau} \tag{5.58}$$

where $\Gamma = q^2 D$.

Hence, $g^{(1)}(q,\tau)$ is simply the Laplace transform of $p(\Gamma)$ with respect to the diffusion coefficient, D. In order to obtain the precise form of the distribution $p(\Gamma)$, it is essential to measure $g^{(1)}(q,\tau)$ with very high accuracy and then perform an inverse Laplace transform on the field correlation. This method is prone to large errors and is not really fit for purpose. Nevertheless, a widely used software package called CONTIN[12,13] is used to assess not only the polydispersity (size distribution) of a given set of colloids but also cluster growth in diffusion- and reaction-limited colloidal aggregation.[14] Others have addressed this problem of polydispersity by assuming a known distribution, such as a Gaussian. However, the size distribution of synthesised colloids or macromolecules is not always symmetric. In this case a different distribution needs to be used, which may not be known in advance.

Koppel realised that the electric-field autocorrelation function can be related to the moments of the Taylor expansion of $\langle \exp(-\Gamma\tau)\rangle$ and the cumulants in the Taylor expansion of $\ln\langle \exp(-\Gamma\tau)\rangle$.[15] The method of cumulants is a general mathematical method for expanding the logarithm of a function, first introduced by Thiele in 1889[16] and then reintroduced by Fisher and Wishart in 1932.[17] By comparing the coefficients in both expansions, a relationship between the moments and cumulants can be found, which allows us to express $g^{(1)}(q,\tau)$ in terms of the width of the distribution $(\langle D^2\rangle - \langle D^2\rangle)$ around the mean, $\langle D\rangle$. The procedure for obtaining these relations is outlined below.

The starting point is Equation 5.49. For size distributions that are not too large, the electric-field autocorrelation function can be expanded in a Taylor series:

$$\langle e^{-\Gamma\tau}\rangle = \left\langle \sum_{j=1}^{N} p(\Gamma_j)\, e^{-\Gamma_j\tau} \right\rangle = \left\langle \sum_{j=1}^{N} p(\Gamma_j)\left(1 - \Gamma_j\tau + \frac{(\Gamma_j\tau)^2}{2!} - \ldots\right)\right\rangle$$
$$= \left\langle \sum_{j=1}^{N} p(\Gamma_j)\right\rangle - \left\langle \sum_{j=1}^{N} p(\Gamma_j)\,\Gamma_j^1\right\rangle\tau + \left\langle \sum_{j=1}^{N} p(\Gamma_j)\,\Gamma_j^2\right\rangle\tau^2 - \ldots \qquad (5.59)$$

The averages are simply the moments of this expansion, where the first term must be 1, as the distribution $p(\Gamma)$ function is normalised. Thus, the general form of the field-autocorrelation function in terms of moments is given by:

$$\langle e^{-\Gamma\tau}\rangle = 1 - \langle\Gamma\rangle\tau + \frac{\langle\Gamma^2\rangle\tau^2}{2!} - \ldots = \sum_{m=0}^{\infty}\langle\Gamma^{(m)}\rangle\frac{(-\tau)^m}{m!} \qquad (5.60)$$

Similarly, we can expand the logarithm of the field-correlation function:

$$\ln\langle e^{-\Gamma\tau}\rangle = \sum_{m=1}^{\infty} K_{\mathrm{m}} \frac{(-\tau)^m}{m!} \tag{5.61}$$

The coefficients, K_{m}, are the cumulants. By comparing the two expansions, one can show that the cumulants are a combination of the moments. To find the cumulants, it is important to realise that the argument in the field-autocorrelation function can also be expanded around the mean diffusion coefficient, $\langle D\rangle$ or $\langle\Gamma\rangle$:

$$\langle e^{-\Gamma\tau}\rangle = e^{k_1\tau + k_2\tau^2 + k_3\tau^3 + \dots} \tag{5.62}$$

Then: $\ln\langle e^{-\Gamma\tau}\rangle = k_1\tau + k_2\tau^2 + k_3\tau^3 + \dots$. The trick now is to see that the nth derivative of the expanded logarithm with respect to the delay time, τ, around $\tau = 0$ will give only one term, as all others will be zero at the initial time:

$$\left(\frac{\partial^n}{\partial\tau^n}\ln\left\langle e^{-\Gamma\tau}\right\rangle\right)_{\tau=0} = n!k_{\mathrm{n}} \Rightarrow k_{\mathrm{n}} = \frac{1}{n!}\left(\frac{\partial^n}{\partial\tau^n}\ln\left\langle e^{-\Gamma\tau}\right\rangle\right)_{\tau=0} \tag{5.63}$$

Hence, the first two derivatives give:

$$\left(\frac{\partial}{\partial\tau}\ln\left\langle e^{-\Gamma\tau}\right\rangle\right)_{\tau=0} = \left.\frac{\left\langle \Gamma e^{-\Gamma\tau}\right\rangle}{\langle e^{-\Gamma\tau}\rangle}\right|_{\tau=0} = \langle\Gamma\rangle \tag{5.64}$$

and:

$$\begin{aligned}\frac{1}{2}\left(\frac{\partial^2}{\partial\tau^2}\ln\left\langle e^{-\Gamma\tau}\right\rangle\right)_{\tau=0} &= \frac{1}{2}\frac{\partial}{\partial\tau}\left[\frac{\left\langle \Gamma e^{-\Gamma\tau}\right\rangle}{\langle e^{-\Gamma\tau}\rangle}\right]_{\tau=0} \\ &= \frac{1}{2}\left[\frac{\left\langle \Gamma^2 e^{-\Gamma\tau}\right\rangle}{\langle e^{-\Gamma\tau}\rangle} - \frac{\langle \Gamma e^{-\Gamma\tau}\rangle^2}{\langle e^{-\Gamma\tau}\rangle^2}\right]_{\tau=0} = \frac{1}{2}(\langle\Gamma^2\rangle - \langle\Gamma\rangle^2)\end{aligned} \tag{5.65}$$

Thus, the cumulants are combinations of the moments about the mean $\langle\Gamma\rangle$. The first three cumulants are:

$$K_1 = \langle\Gamma\rangle = q^2\langle \mathrm{D}\rangle \tag{5.66}$$

$$K_2 = \langle\Gamma^2\rangle - \langle\Gamma\rangle^2 = q^4(\langle \mathrm{D}^2\rangle - \langle \mathrm{D}\rangle^2) \tag{5.67}$$

$$K_3 = \langle\Gamma^3\rangle - 3\langle\Gamma^2\rangle\langle\Gamma\rangle + 2\langle\Gamma\rangle^3 = q^6(\langle \mathrm{D}^3\rangle - 3\langle\Gamma^2\rangle\langle \mathrm{D}\rangle + 2\langle \mathrm{D}\rangle^3) \tag{5.68}$$

K_1 describes the average decay rate of the exponential in the field-autocorrelation function. The second cumulant describes the width or variance of the distribution of diffusivities and thus the spread in particle size. Finally, the third cumulant corresponds to the degree of skew or asymmetry of the distribution about the mean diffusivity. Note that the first two cumulants must be positive by construction, as $\langle \Gamma^2 \rangle > \langle \Gamma \rangle^2$, while the third can be negative.

We can then express the field-correlation function in terms of a single exponential function around the mean:

$$\ln\left(g^{(1)}(q,\tau)\right) = \ln\langle e^{-\Gamma\tau}\rangle = -\langle \Gamma \rangle \tau + \frac{K_2}{2!}\tau^2 - \frac{K_3}{3!}\tau^3 + \ldots \tag{5.69}$$

In a standard DLS equipment, we measure the intensity-correlation function:

$$g^{(2)}(q,\tau) = 1 + \beta\left(g^{(1)}(q,\tau)\right)^2 \tag{5.70}$$

which is given in terms of a cumulant expansion as:

$$\begin{aligned} \ln\left(g^{(2)}(q,\tau) - 1\right) &= \ln(\beta\langle e^{-\Gamma\tau}\rangle^2) \\ &= \ln(\beta) + 2\left(-\langle \Gamma \rangle \tau + \frac{K_2}{2!}\tau^2 - \frac{K_3}{3!}\tau^3 + \ldots\right) \end{aligned} \tag{5.71}$$

This equation is the basis for many commercial DLS systems that use non-linear fitting routines to fit the measured data directly. However, as Frisken[18] has pointed out, in order to obtain correct results it is essential to measure the baseline (factor unity in Equation 5.71, which is sometimes written as factor B) to a very high precision. This means that the intensity-autocorrelation function must decay to exactly 1. Any value below unity makes theoretically no sense, as the exponential decay of the correlations cannot be negative. Measurements to such precision are almost impossible as instrument fluctuations (thermal, mechanical, electronic *etc.*) will always lead to some noise. These seemingly small fluctuations can lead to rather large errors in the determination of the size polydispersity of the colloidal suspension of interest.[18]

There is an additional problem, which becomes apparent when writing the intensity-correlation function as:

$$g^{(2)}(q,\tau) = B + \beta e^{2\left(-\langle \Gamma \rangle \tau + \frac{K_2}{2!}\tau^2 - \frac{K_3}{3!}\tau^3 + \ldots\right)} \tag{5.72}$$

The second term in the exponential is positive and grows in proportion to the square of the delay time. For sufficiently long times, the entire exponential can then become positive, causing the exponential to diverge, which is unphysical.

Pusey and Frisken[18–20] proposed a way of circumventing this problem by writing the electric-field autocorrelation function as:

$$g^{(1)}(q,\tau) = \mathrm{e}^{-\langle\Gamma\rangle\tau} \int_0^\infty p(\Gamma)\mathrm{e}^{-(\Gamma-\langle\Gamma\rangle)\tau}\mathrm{d}\Gamma \tag{5.73}$$

where account is taken of the fact that for small particle-size distributions:

$$\exp(-\Gamma\tau) = \exp(-\langle\Gamma\rangle\tau)\exp(-(\Gamma-\langle\Gamma\rangle)\tau) \tag{5.74}$$

Expanding the exponential inside the integral, we obtain:

$$g^{(1)}(q,\tau) = \mathrm{e}^{-\langle\Gamma\rangle\tau} \int_0^\infty p(\Gamma)\left[1 - (\Gamma - \langle\Gamma\rangle)\,\tau + \frac{(\Gamma-\langle\Gamma\rangle)^2}{2!}\tau^2 + \frac{(\Gamma-\langle\Gamma\rangle)^3}{3!}\tau^3 + \ldots\right]\mathrm{d}\Gamma \tag{5.75}$$

Integrating the electric-field autocorrelation function then gives:

$$g^{(1)}(q,\tau) = \mathrm{e}^{-\langle\Gamma\rangle\tau}\left(1 + \frac{\langle\Gamma^2\rangle - \langle\Gamma\rangle^2}{2!}\tau^2 - \frac{\langle\Gamma^3\rangle - 3\langle\Gamma^2\rangle\langle\Gamma\rangle + 2\langle\Gamma\rangle^3}{3!}\tau^3 + \ldots\right) \tag{5.76}$$

Writing this in terms of cumulants, we get:

$$g^{(1)}(q,\tau) = \mathrm{e}^{-\langle\Gamma\rangle\tau}\left(1 + \frac{K_2}{2!}\tau^2 - \frac{K_3}{3!}\tau^3 + \ldots\right) \tag{5.77}$$

and using the Siegert relation we finally obtain:

$$g^{(2)}(q,\tau) = \mathrm{B} + \mathrm{e}^{-2\langle\Gamma\rangle\tau}\left(1 + \frac{K_2}{2!}\tau^2 - \frac{K_3}{3!}\tau^3 + \ldots\right)^2 \tag{5.78}$$

This is a converging function even for long delay times, τ, that allows direct fitting of the baseline, B.

5.3.3.1 Anisotropic Particles: Depolarised DLS

So far only spherical particles have been considered, which, because of their isotropic shape, will scatter in all directions in the same way. It has thus been possible to reduce the intermediate scattering function, leading us to the exponential dependence of the averaged translational diffusion coefficient. This type of measurement is usually done with polarised coherent light. Many biological and other colloidal samples have shape anisotropy. Examples include *E.coli*, a rod-shaped bacterium with a small aspect ratio (tobacco mosaic and fd virus are also rod-shaped and have a very large aspect ratio[21–23]), and clays such as laponite, bentonite and montmorillonite, which are all disc-shaped.[24–26] Of course, there are numerous other examples and the references given here are by no means exhaustive.

The important point to make is that all particles also have a rotational diffusion coefficient, D_R. For symmetry reasons, the rotational diffusion coefficient of a sphere is not measureable in DLS, but it is measureable for anisotropic particles (see Figure 5.9) when using depolarised DLS (DDLS). Since anisotropic particles give rise to an anisotropic polarisability tensor, they will display different electric dipole moments when interacting with an external field such as that of the incoming laser light. In particular, the non-diagonal elements can be non-zero, *e.g.* $p_a = \alpha_{ab} E_b$, where p_a is the dipole moment along the long axis, a, of an ellipsoid, E_b is the component of the electric field hitting the short axis, b, and α_{ab} is the polarisability of the particle, which depends on both the particle's electric permittivity and the solvent. Hence, the scattered field will have a non-zero component in the direction perpendicular to the polarisation of the incident field. Here we focus on ellipsoids with arbitrary aspect ratios, for which rods and discs can be considered limiting cases. We consider the case of a particle whose radius is small compared to the wavelength of the laser light ($R/\lambda < 1$) is considered, allowing the use of the Rayleigh approximation. For such particles, the eigenvalues of the polarisability tensor correspond to the directions perpendicular

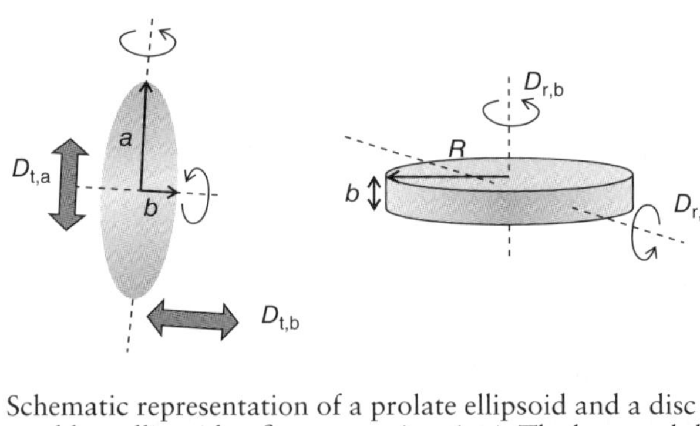

Figure 5.9 Schematic representation of a prolate ellipsoid and a disc (which can be modelled as an oblate ellipsoid to first approximation). The long and short symmetry axes, *a* and *b*, indicate the major axes of the two particles along which rotational and translational diffusion can take place.

(α_b) and parallel (α_a) to the long symmetry axis. The total electric field scattered by the ellipsoidal particle has two components: the vertically polarised component, E_{VV}, has an amplitude proportional to the average polarisability, $\alpha = (\alpha_a + 2\alpha_b)/3$, while the horizontally depolarised component, E_{VH}, has an amplitude proportional to the intrinsic particle anisotropy, $\beta = \alpha_a - \alpha_b$. The electric-field or intermediate correlation functions in polarised (VV) and depolarised (VH) DLS are then:

$$f_{\mathrm{VV}}(q,\tau) \equiv \frac{\langle E_{\mathrm{VV}}(\vec{q},0)E^*_{\mathrm{VV}}(\vec{q},\tau)\rangle}{\langle I_{\mathrm{VV}}(\vec{q},0)\rangle} = \frac{\sum_{j=1}^{N}\sum_{k=1}^{N}\left\langle b_{\mathrm{VV,j}}(\vec{q},0)\, b^*_{\mathrm{VV,k}}(\vec{q},\tau) e^{-i\vec{q}\cdot\left(\vec{R}_j(0)-\vec{R}_k(\tau)\right)}\right\rangle}{\sum_{j=1}^{N}\sum_{k=1}^{N}\left\langle b_{\mathrm{VV,j}}(\vec{q},0)\, b^*_{\mathrm{VV,k}}(\vec{q},0) e^{-i\vec{q}\cdot\left(\vec{R}_j(0)-\vec{R}_k(0)\right)}\right\rangle} \tag{5.79}$$

and:

$$f_{\mathrm{VH}}(q,\tau) \equiv \frac{\langle E_{\mathrm{VH}}(\vec{q},0)E^*_{\mathrm{VH}}(\vec{q},\tau)\rangle}{\langle I_{\mathrm{VH}}(\vec{q},0)\rangle} = \frac{\sum_{j=1}^{N}\sum_{k=1}^{N}\left\langle b_{\mathrm{VH,j}}(\vec{q},0)\, b^*_{\mathrm{VH,k}}(\vec{q},\tau) e^{-i\vec{q}\cdot\left(\vec{R}_j(0)-\vec{R}_k(\tau)\right)}\right\rangle}{\sum_{j=1}^{N}\sum_{k=1}^{N}\left\langle b_{\mathrm{VH,j}}(\vec{q},0)\, b^*_{\mathrm{VH,k}}(\vec{q},0) e^{-i\vec{q}\cdot\left(\vec{R}_j(0)-\vec{R}_k(0)\right)}\right\rangle} \tag{5.80}$$

Note that the polarisabilities have been replaced by the scattering lengths, b, which are simply the Fourier transforms of the polarisabilities themselves, in order to be consistent with the previous derivations.

Obviously these averages are very difficult to compute. However, if a sufficiently dilute solution is considered, such that the rotational motion and the orientation of the particles are completely uncorrelated from each other, it is possible to arrive at simplified expressions for the intermediate scattering functions. After a lengthy calculation (more details can be found in references [5] and [27]), the following expressions emerge:

$$f_{\mathrm{VV}}(q,\tau) = \frac{\left[\overline{b}^2 + \frac{4}{45}\beta^2 e^{-\overline{D}_r\tau}\right]\left\langle e^{-i\vec{q}\cdot\left(\vec{R}(0)-\vec{R}(\tau)\right)}\right\rangle}{\overline{b}^2 + \frac{4}{45}\beta^2 e^{-\overline{D}_r\tau}} \tag{5.81}$$

$$f_{\mathrm{VH}}(q,\tau) = e^{-D_r\tau}\left\langle e^{-i\vec{q}\cdot\left(\vec{R}(0)-\vec{R}(\tau)\right)}\right\rangle \tag{5.82}$$

where $\overline{b}$ is the average polarisability of the particle and $\langle e^{-i\vec{q}\cdot(\vec{R}(0)-\vec{R}(\tau))}\rangle$ is the contribution due to translational diffusion, identified as $g^{(1)}(q,\tau)$ for spherical particles. Hence, the autocorrelation function for polarised scattered light will be that of the translational case multiplied by a contribution stemming from the rotational diffusion. However, in many cases the anisotropy factor is very small, such that $f_{VV}(q,\tau) \approx g^{(1)}(q,\tau)$. All of the information about the rotational particle diffusion is thus in the autocorrelation function for the depolarised scattered light.

All that remains is to identify the rotational diffusion coefficient, D_r. To do so, it is recognised that there are usually two rotational axes, to which independent diffusion coefficients are associated. Defining D_a as the diffusion coefficient along the long axis and D_b as the rotational diffusion along the short axis, both the difference between them, $\Delta D = D_a - D_b$, and the averaged translation diffusion coefficient, $\overline{D_t} = (D_a + 2D_b)/3$, can be defined.[5] The latter has to be used in polarised DLS. Perrin derived an expression for ΔD, which is also known as the coupling parameter between rotational and translational diffusion coefficients.[21] Defining the aspect ratio between the long and short rotation axes of an ellipsoid as $\rho = a/b$, Perrin obtained:

$$D_r = \frac{3k_B T}{16\pi\eta a}\frac{(2\rho^2-1)G(\rho)-\rho}{\rho^4-1} \tag{5.83}$$

where η is again the viscosity of the solvent in which the colloid diffuses and $G(\rho)$ is a function of the particle aspect ratio. For oblate ellipsoids,[28] which are essentially disc-like particles, $\rho < 1$, so Equation 5.84 is obtained:

$$G(\rho) = \sqrt{1-\rho^2}\tan^{-1}\left(\frac{\sqrt{1-\rho^2}}{\rho}\right) \tag{5.84}$$

Similarly, Shimizu[29] derived general expressions for the two translational diffusion coefficients for ellipsoidal particles:

$$D_{t,a} = \frac{k_B T}{8\pi\eta a}G(\rho) \tag{5.85}$$

$$D_{t,b} = \frac{k_B T}{16\pi\eta a}G(\rho) \tag{5.86}$$

Following a representation by Jabbari-Farouji[28] for discs with $b = 0$ and $a = R$, these translational diffusion coefficients become:

$$D_{t,a} = \frac{k_B T}{16\pi\eta R} \tag{5.87}$$

$$D_{t,b} = \frac{3k_B T}{32\pi\eta R} \tag{5.88}$$

The average translational diffusion coefficient for discs is then $\overline{D}_t = k_B T(12\pi\eta R)^{-1}$, and the average rotational diffusion coefficient is:

$$\overline{D_r} = \frac{3k_B T}{32\pi\eta R^3} \tag{5.89}$$

Again, the intermediate correlation function in depolarised DLS measures only the rotational diffusion coefficient. It is interesting to note that in depolarised DLS, the polariser placed before the sample and the analyser placed between the sample and the detector will 'filter out' most of the scattered intensity. Consequently, long measurements are needed in order to achieve good statistics, particularly as the colloidal suspension must be very dilute in order to allow free rotational motion along the long axis. An interesting study of the ageing of clay suspensions (using laponite, a synthetic clay of discs of 30 nm diameter and 1 nm thickness) with polarised and depolarised DLS was carried out by Jabbari-Farouji *et al.*[24,25,28]

5.3.4 Dense Systems

So far only very dilute systems have been considered, and it has been assumed that only single scattering events take place and that average particle separations are such that any inter-particle interactions can be neglected. However, when the particle volume fraction starts to increase, both multiple scattering and collective motion due to direct and indirect particle interactions will contribute to the overall scattering. Multiple light scattering is easily detected in the form of a diverging laser beam (Figure 5.3). Another indication of multiple scattering is when the value of $\exp(-2q^2 D_0 \tau)$ commences well below unity. Unfortunately, many instruments normalise the intermediate scattering function to unity automatically, making this check impossible. A possible way of reducing multiple scattering is to match the refractive index of the solvent with that of the material of which the particles are made. However, this is not always possible, as it can compromise the chemical reaction or other relaxation mechanisms of interest.

The contributions arising from direct interactions can be caused by nonspecific Coulomb, van der Waals and steric interactions, for example;

indirect interactions are usually a result of hydrodynamics. Any particles or molecules will experience some drag forces due to friction with the molecules of the solvent in which they are diffusing. All particles will thus set up a flow field that dies off slowly with the inverse of the distance from the particles. These interfering interactions lead to an enhanced effective viscosity, which can be described by the Einstein relation: $\eta_{eff} = \eta_s(1 + 2.5\phi + \ldots)$, where η_s is the viscosity of the pure solvent and ϕ the volume fraction of the particles.[11] Higher-order terms can be taken into account, but this approximation will break down for systems with volume fractions larger than about 10%. Moreover, the typical relaxation times in such dense systems will be different to those measured in very dilute dispersions. A number of theories have been developed to compute these effects and new experimental set-ups have been developed in order to overcome the difficulties. In particular, fibre-optics,[30–35] but also cross-correlation techniques using two detectors and two different lasers, have been developed; these are summarised beautifully in references [36] and [37]. Another technique, developed by the groups of Maret and Pine in the 1980s, is diffuse wave spectroscopy (DWS).[38–40] A detailed discussion of these techniques is beyond the scope of this book, but fibre-optic DLS (FODLS) and dense, self-assembling systems will be discussed in more detail in subsequent sections.

5.4 INSTRUMENTAL DEVELOPMENTS AND NEW METHODS

5.4.1 Fibre-Optic DLS

Traditional DLS set-ups are built on a goniometer that allows high-precision adjustment of the scattering angle, which is required when the size and the spread in size of the object of interest are to be measured with high accuracy. A complication arises from the random static scattering that results from impurities in the sample or from scratches on the cuvette holding it; this can be minimised by placing the cuvette in a large glass vessel filled with a density-matching fluid such as toluene or a specialised oil with a refractive index equal to that of the cuvette. These two conditions make the instrumentation very costly. Furthermore, the set-up must be placed in a dark room so that light cannot influence the measurement.

A nontechnical issue faced by traditional DLS is multiple light scattering in concentrated solutions. Scientists therefore began to develop more compact, low-cost systems based on fibre optics once these became commercially available. Initially, multimode fibre optics were used to deliver the laser light to the sample.[30,41] In 1987, Brown *et al.* and Dhadwal and Chu were the first to probe single-mode optical fibres terminating in a lens for both delivery of the laser light and collection of the scattered light at a fixed angle (initially in back-scattering geometry).[32–35,42–47] A detailed summary and account of the technical details involved in FODLS can be found in Chapter 6 of reference [4].

A few very recent developments in FODLS will be discussed here. One notable design was put forward by Pristinski and Chastek,[48] whose set-up is constructed from simple, rather inexpensive components (Figure 5.10). Its main advantage is that it can be used to follow reactions *in situ* inside a reaction vessel, thereby avoiding any unwanted static scattering from the sample holder. Moreover, the closeness of the light-delivering and -detecting fibres makes alignment rather easy. The authors were able to follow the growth of silica particles using Stöber synthesis[49] and they give an excellent description of the instrumentation, including the components.[48] However, it should be noted that detection analysis with PMTs and the correlator is demanding, and requires a good understanding of photon counting and electronics.

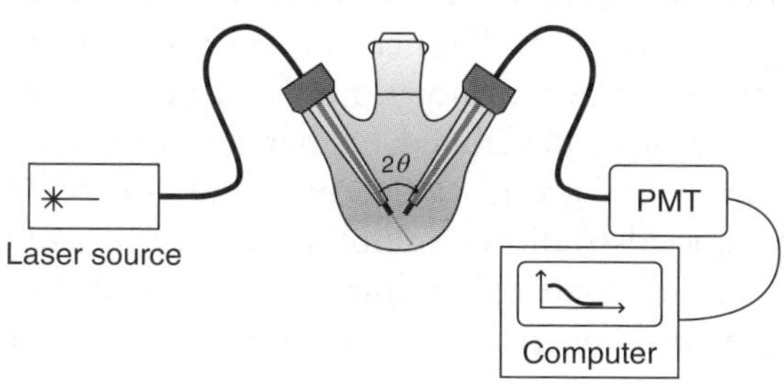

Figure 5.10 Simplified schematic representation of a FODLS set-up developed by Pristinski and Chastek.[48] The laser source is coupled to an *xyz*-stage holding a collimating lens and a fitting for a single-mode optical fibre (required to direct the laser light into the fibre appropriately). The optical fibre is directed into a reaction vessel, where the sample is probed directly. The tip of the fibre holds a collimating lens to ensure the parallel beam hits the sample. A second fibre, arranged under the scattering angle, 2θ, and only a few mm away from the delivering fibre, directs the scattered light into a single-mode fibre splitter in order to perform cross-correlation spectroscopy. The two ends are connected to photo multiplier tubes (PMTs). The amplified signals are fed into a correlator and the resulting data into a computer.

Another exciting development is the miniaturisation of DLS to fit within a microfluidic device. Chastec *et al.* designed such a device both to synthesise block copolymers and to study their size and micellisation behaviour in solution down the production line.[50] As polymerisation synthesis often requires organic solvents and harsh conditions, Chastec *et al.* used a stainless-steel microfluidic device to synthesise poly(methyl methacrylate-β-lauryl methacrylate) and poly(methyl methacrylate-β-octadecyl methacrylate) *via* atom-transfer radical polymerisation. Through controlled mixing of monomers, initiators and other factors, the micellisation behaviour of these block copolymers could be studied as a function of the ratio of the block lengths.[50]

5.4.2 Differential Dynamic Microscopy

Differential dynamic microscopy (DDM) is a newly developed microscopy technique that successfully provides information on the dynamics of 'passive' colloidal solutions and may thus make DLS, with its complex and often rather expensive instrumentation, redundant. Most importantly, DDM enables correct particle-diameter measurements over a much wider range of sizes; in particular, very small scattering angles can be reached, whereas DLS often fails at angles $\lesssim 20°$ due to the interference of the unscattered intensity, which becomes dominant at these small angles. The optical ideas used in DDM were first explored and put forward by Cannell *et al.* as a shadowgraph technique used to study the dynamics of non-equilibrium fluctuations in diffusive systems[51−53] and were developed further and tested successfully in a variety of systems by the group of Cherbino and Trappe.[54,55] Moreover, this new technique also allows sampling of 'active' colloidal systems, such as solutions of bacterial swimmers,[56] and distinction between diffusion-limited and reaction-limited aggregation processes.[57] Even the dynamics of anisotropic particles in external fields can be studied.[58]

The experimental set-up and procedure (see Figure 5.11) are, in principle, straightforward, and they will be explained briefly here. The actual physics is then presented, though a more rigorous discussion can be found in the works by Cannell *et al.*[51] and Cherbino *et al.*[55] The section concludes with a brief discussion of the limitations of DDM and developments in its application.

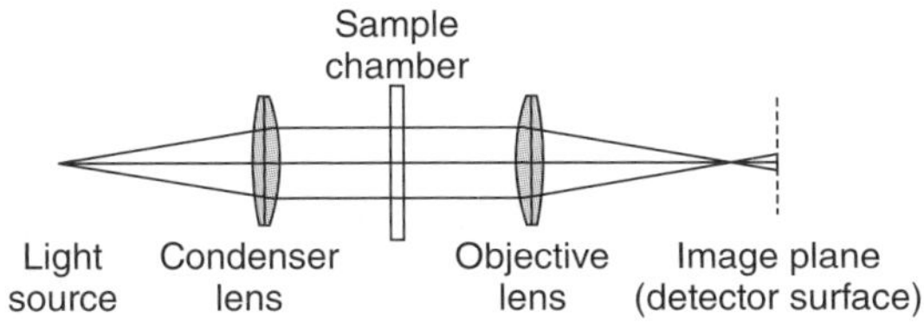

Figure 5.11 Schematic set-up for a DDM measurement, not shown to scale. Using Koehler illumination, white light from a light source is collimated with a condenser onto the sample chamber. The light scattered by any object in the illuminated part of the sample will be directed onto the object with a typical magnification of 40× or 63×. The numerical aperture, *NA*, of the condenser should be high, while the *NA* of the objective should be rather low, in contrast to what is often proposed.[54,56] Typical values might be $NA_{\text{condensor}} = 0.9$ and $NA_{\text{objective}} = 0.1$. The scattered light passing through the objective is then transferred onto the charge-coupled device (CCD) or complementary metal-oxide semiconductor (CMOS) camera with $N \times N$ pixels. The image thus formed in the camera plane reflects a direct image of the density or refractive index variations in the sample.

5.4.2.1 Experimental Set-Up and Functioning

The technique is based on taking a time series of bright-field microscopy images with a fast camera. Both an upright and an inverted microscope can be used. Choosing *e.g.* a frame rate of 100 images per second, the shortest time interval, Δt_{min}, between subsequent images is 10 ms, and the exposure time of each image can be selected to be 1 ms, depending on the camera used. Figure 5.12a shows a typical bright-field image taken from a solution of 1.2 μm large polystyrene colloids in aqueous solution.

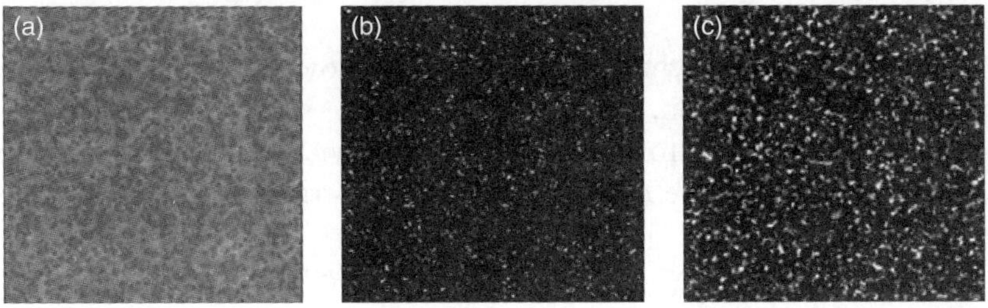

Figure 5.12 (a) Bright-field image of an aqueous solution containing 1.05 μm large polystyrene colloids measured by S.H. Nathan in the group of E. Eiser. Exposure (shutter) time: 1 ms; field of view: 68 × 68 μm; camera pixel size: 132 × 132 nm. (b, c) Difference image taken from two images (b) 75 ms, and (c) 5 s apart.

This image is a superposition of the incoming intensity, $I_0(\vec{r})$, and the scattered light, $I_s(\vec{r}, t)$, which are identified as the heterodyne intensity derived earlier:

$$I(\vec{q}, t) = |E_0(\vec{r}) + E_s(\vec{r}, t)|^2 \cong I_0(\vec{r}) + 2\Re[E_0^*(\vec{r})E_s(\vec{r}, t)] \tag{5.90}$$

The essential step is then to create difference images by subtracting two images from each other for fixed time differences, Δt:

$$D(\vec{r}, \Delta t) = I(\vec{r}, \Delta t) - I(\vec{r}, 0) \tag{5.91}$$

where $I(\vec{r}, 0)$ is *e.g.* the first image taken in the time series. Following a derivation by Wilson *et al.*[56] the difference image can also be written as $D(\vec{r}, \Delta t) = \Delta I(\vec{r}, \Delta t) - \Delta I(\vec{r})$, where $\Delta I(\vec{r}, \Delta t) = I(\vec{r}, t) - \langle I(\vec{r}, t)\rangle$ represents the intensity and thus the density fluctuations in the sample. In Figure 5.12, such difference images are shown for $\Delta t = 75$ ms and 5 s. Two observations can be made. First, subtracting two images taken under constant conditions (illumination should be homogenous throughout and no thermal drift should be present to introduce *e.g.* convection) removes all static contributions that might arise from dust particles on the optical elements or dirt in the sample, but the static incoming light will also be removed, leaving only the dynamically changing intensities (or optical densities) that result from the diffusion of the colloids. Second, subtracting two images taken within the shortest possible time of each other will show very little variation as the particles will not have moved away from their original positions: they are hence still self-correlated to a large extent. As Δt increases, the colloid positions become increasingly decorrelated from their original positions in frame one, resulting in an increasingly coarse-grained difference image (Figure 5.12b,c).

Good statistics are achieved when averages over several hundred difference images are taken.

In order to obtain information about the dynamics and thus the diffusion coefficient of the colloids, the Fourier transform, $D(x, y, \Delta t)$, is used:

$$F_D(\vec{q}, \Delta t) = \int D(\vec{r}, \Delta t) e^{-i(\vec{q}\cdot\vec{r})} d\vec{r} \tag{5.92}$$

The main output we obtain from DDM is then a **differential intensity-correlation function,** $\langle|F_D(\vec{q}, \Delta t)|^2\rangle$ (see top-left image in Figure 5.13). Some authors refer to $\langle|F_D(\vec{q}, \Delta t)|^2\rangle$ as the power spectrum or spectral density, but it should be remembered that we are dealing with a 2D function. This correlation function is related directly to the intermediate

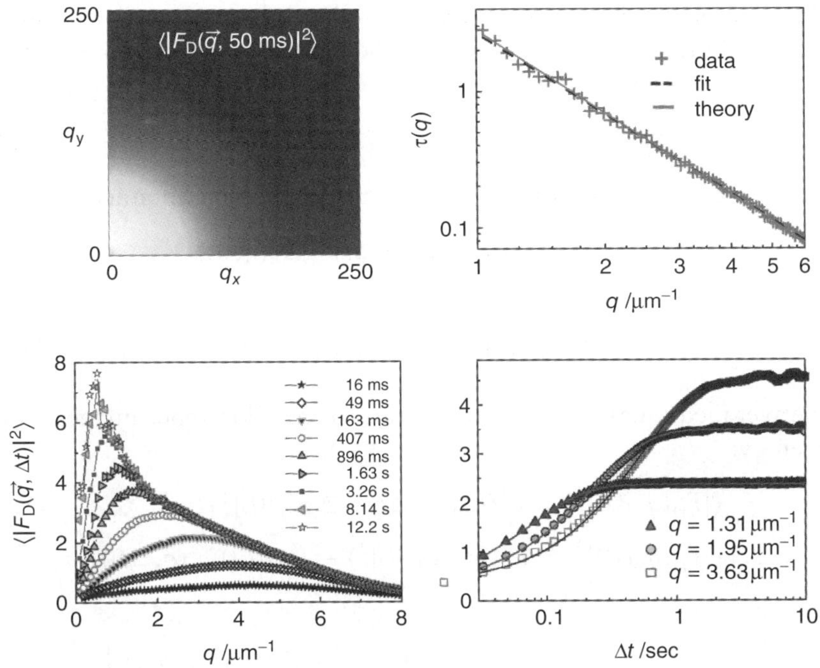

Figure 5.13 Analysis procedure. Top left: The first difference images obtained for given time lags are fast-Fourier-transformed; the image shows such a fast-Fourier-transform (FFT) averaged over all $\Delta t = 55$ ms. The images were taken for an aqueous solution of polystyrene particles with a radius of 527 nm (see Figure 5.8). Bottom left: Subsequently, the $\langle |F_D(\vec{q}, \Delta t)|^2 \rangle$ are integrated radially and then plotted as a function of q. Here only a subset for different Δt's is shown. Bottom right: $\langle |F_D(q, \Delta t)|^2 \rangle$ is then plotted for a given scattering vector, q, and fitted using Equation 5.100. Each fit yields a q-dependent decay time, $\tau(q) = (Dq^2)^{-1}$. Top right: These decay times are plotted as a function of the scattering vector. The slope of the fitted (dashed) line corresponds to the diffusion coefficient of the polystyrene particles. The diffusion coefficient obtained for the given experimental settings matches the theoretical curve almost perfectly.

scattering function, $f(q, \tau)$. This can be understood when we realise that in near-field microscopy, fluctuations in the scattering intensities are measured directly by the camera's pixels,[55] so intensity fluctuations in the pixels reflect the density (equivalent to refractive index) changes in the sample:

$$\Delta\rho(\vec{r}, \Delta t) = \rho(\vec{r}, \Delta t) - \langle \rho(\vec{r}, \Delta t) \rangle \tag{5.93}$$

that occur when the colloids diffuse.[56] Thus, we can write:

$$\Delta I(\vec{r}, \Delta t) = k\Delta\rho(\vec{r}, \Delta t) \tag{5.94}$$

The proportionality constant, k, will depend on a complex optical transfer function given by the specific optical set-up. Expressing the difference images as:

$$D(\vec{r}, \Delta t) = \Delta I(\vec{r}, \Delta t) - \Delta I(\vec{r}) = k[\Delta\rho(\vec{r}, \Delta t) - \Delta\rho(\vec{r}, 0)] \tag{5.95}$$

the Fourier transform for this new expression can be obtained:

$$F_{\rm D}(\vec{q}, \Delta t) = k[\Delta\rho(\vec{q}, \Delta t) - \Delta\rho(\vec{q}, 0)] \tag{5.96}$$

with:

$$\Delta\rho(\vec{q}, \Delta t) = \int \Delta\rho(\vec{r}, \Delta t)\mathrm{e}^{-\mathrm{i}(\vec{q}\cdot\vec{r})}\mathrm{d}\vec{r} \tag{5.97}$$

The averaged square of the Fourier-transformed difference images is then given by:

$$\langle |F_{\rm D}(q, \Delta t)|^2\rangle = k^2\langle |\Delta\rho(\vec{q}, \Delta t) - \Delta\rho(\vec{q}, 0)|^2\rangle \tag{5.98}$$

$$\langle |F_{\rm D}(q, \Delta t)|^2\rangle = k^2[2\langle |\rho(\vec{q}, 0)|^2\rangle - 2\langle \rho(\vec{q}, 0)\rho^*(\vec{q}, \Delta t)\rangle]$$

or:

$$\langle |F_{\rm D}(q, \Delta t)|^2\rangle = 2k^2\langle |\rho(\vec{q}, 0)|^2\rangle \left[1 - \frac{\langle \rho\left(\vec{q}, 0\right)\rho^*(\vec{q}, \Delta t)\rangle}{\langle |\rho(\vec{q}, 0)|^2\rangle}\right] \tag{5.99}$$

Recognising that $\langle |\rho(\vec{q}, 0)|^2\rangle$ is simply the static structure factor, $S(q)$, that describes the averaged distribution of all scattering centres in the illuminated sample (compare Equation 5.17), and that the second term in the square brackets is the immediate scattering function, $f(q, \tau)$ (Equation 5.47), we obtain $\langle |F_{\rm D}(q, \Delta t)|^2\rangle = 2k^2S(q)[1 - f(q, \Delta t)]$. This central result can be written in the more general form:

$$\langle |F_{\rm D}(q, \Delta t)|^2\rangle = A(q)[1 - f(q, \Delta t)] + B(q) \tag{5.100}$$

The added term, $B(q)$, is introduced as a fitting parameter and takes into account any camera noise. It is important to note that in contrast to DLS, where the intensity-autocorrelation function is measured at a single value of q at the time, in DDM the image-correlation function is measured for a whole range of scattering vectors, $\vec{q}$, simultaneously. To illustrate this point, $\langle |F_{\rm D}(q, \Delta t)|^2\rangle$ is shown in Figure 5.13.

When deriving the intensity-autocorrelation function, the expression $f(q, \tau) = \langle \mathrm{e}^{-\mathrm{i}\vec{q}\cdot\Delta\vec{r}(\tau)}\rangle = \mathrm{e}^{-q^2 D\tau}$ arises for independently diffusing spherical particles in solution, where D is the diffusion coefficient. The time-averaged $\langle |F_{\rm D}(q, \Delta t)|^2\rangle$ can be plotted as a function of the scattering vector for different delay times, Δt, which allows the experimental data to be fitted for different q-values (Figure 5.13). This step is computationally rather expensive and time-consuming on a standard PC, although it can be sped up by use of a graphics processing unit (GPU).[59] The resulting characteristic relaxation times, $\tau(q) = (Dq^2)^{-1}$, are then plotted as a function of q, where the slope of the curve obtained in a log–log plot yields the diffusion coefficient. A comparison with the theoretical value is astounding: the nominal value of the particle diameter provided by the producer of the polystyrene particles used here was 1.05 μm, which is very close to the value obtained from the DDM analysis presented in Figure 5.13. When comparing this result to that obtained from a commercial DLS set-up (Figure 5.8), there is a deviation of more than 50% from the manufacturer's size estimate for μm-sized particles. This huge discrepancy observed in DLS measurements is based mainly on the two shortcomings discussed earlier, namely that it is problematic to fit the intensity-autocorrelation function properly, and that larger particles scatter more in the forward direction, but small scattering angles are difficult to access because of static (heterodyne) contributions from the non-scattered beam. It should be noted that these problems can be minimised to a great extent, but they require very careful setting up and alignment of the equipment, as well as fitting.

Another advantage of DDM is that it does not suffer from the multiple scattering that is typical in solutions with high colloidal volume fractions. He *et al.* conducted DDM and DLS studies on aqueous suspensions of both non-fluorescence- and fluorescence-labelled spherical particles with diameters ranging from 100 to 400 nm and different volume fractions.[60] Their conclusions were threefold: (i) fluorescence labelling of the colloids improved the DDM results when studying particles with a size below the diffraction limit or at very low concentrations, at which the detector noise, $(B(q))$, may dominate; (ii) DDM provided much more reliable results for the dynamics in systems of higher volume fractions than DLS, because the latter suffers from multiple scattering effects under these conditions; and (iii) the special accessibility of the sample and the experimental set-up in DDM are greatly enhanced, allowing for dynamic measurements in confined systems such as cells or microfluidic channels.

5.4.2.2 *Further Applications of DDM*

As mentioned in the introduction to this section, DDM has already been extended to applications other than particle sizing. One beautiful extension is the study of the dynamics of *E. coli* bacteria reported by

Wilson *et al.*[56,61] These rod-like bacteria possess flagella that allow them to undergo active swimming. They are known for their 'run and tumble' motion, where the rotating flagella make them swim for several μm (not reliably measureable with DLS, as angles less than 3° need to be accessed), until the active motion stops, at which moment the particle motion becomes purely diffusive (Figure 5.14). Wilson and co-workers showed that in order to assess this 'run and tumble' motion, the intermediate scattering function must be extended appropriately. Their results show that in a solution in which a fraction of the bacteria is not motile, these 'purely diffusive' bacteria experience an enhancement of their diffusivity when motile particles are present, a finding that is extremely important in understanding active transport at low Reynolds numbers. In this context, Lu *et al.* extended the study of motile bacteria using confocal DDM, examining their ballistic motion as a function of concentration and position within the sample.[62] This is particularly interesting because the dynamics of both diffusive and active particles will experience hydrodynamic effects close to the confining walls of the sample chamber.

A second interesting development is the extension of DDM to the study of the dynamics of anisotropic particles. Reufer *et al.* developed a

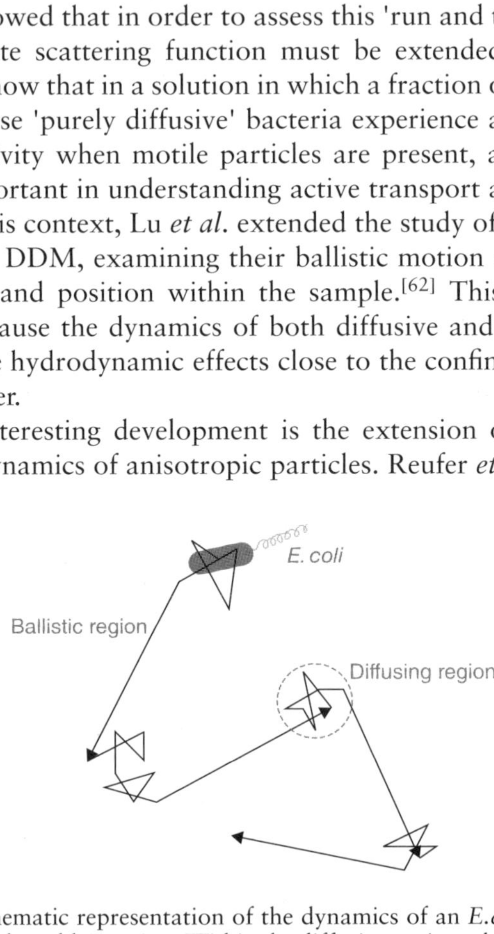

Figure 5.14 Schematic representation of the dynamics of an *E.coli* bacterium performing a 'run and tumble' motion. Within the diffusing regime, the bacterium undergoes regular diffusive motion, with an effective translational diffusion coefficient, D_{eff}, and a relaxation time, $\tau(q) = (D_{\text{eff}}q^2)^{-1}$. In the ballistic region, the flagellum becomes active and causes the bacterium to move in a directed motion for several μm (note that a single bacterium usually has several flagella). In this region, the relaxation time is no longer proportional to q^{-2} and the particle will also experience diffusive motion, decoupled from the ballistic motion. The ballistic path drawn will thus not be completely straight.

theory that allows the interpretation of DDM data taken from a solution of rod-like haematite particles in the presence of an external magnetic field.[58] These authors point out that because the diffusivity of such particles is different along their different axes, the simple integration of difference images is no longer valid, and it is thus necessary to re-compute a modified dynamic and static structure factor that takes into account the distribution of the particle orientations.

Finally, Poon, Weeks and Royall pointed out that DDM is also an effective tool in the measurement of the colloidal volume fraction of a sample of μm-sized colloids, when combined with particle counting.[63] In the early 1980s, Pusey and van Megen presented the first phase diagram of nearly hard spherical colloids using density- and refractive index-matched poly(methyl methacrylate) (PMMA) particles.[64] Their benchmark experiments made it possible to study the crystallisation and glassy behaviour of spherical particles in real time, making colloidal suspensions a model system for the study of liquid-to-crystal phase transitions by analogy with atomic systems, which remain difficult to study *in situ*. Moreover, they also made it possible to test theories and simulations of hard spheres directly. However, since the liquid-to-crystal phase transition in a hard sphere system depends solely on the colloid volume fraction, $\phi = \frac{4\pi R^3 N}{3V}$ (where R is the radius and N the total number of colloids in the total volume, V), it is of paramount importance that this be determined with very high precision. An exact measurement of ϕ, is very difficult, however, and caused a great deal of controversy in the interpretation of experimental data and in comparisons of theories and simulations. Poon and co-workers summarised the problem and suggested the use of DDM in conjunction with confocal microscopy in a recent review.[64] While it is possible to count the number of N μm-sized particles in a given volume, V, with confocal microscopy the diffraction limit does not allow the precise estimation of the colloid radius, R. This problem can now be circumvented using DDM, which provides excellent results for the diffusion coefficient, provided the size polydispersity of the particles is lower than a few per cent. To this end, it should be mentioned that when working with μm-sized colloids it is also possible to obtain their dynamics using video microscopy,[65] which has been applied successfully to obtain the self-diffusion coefficient of particles in very dilute systems and to measure the collective diffusion coefficient in dense systems. Habdas and Weeks provide a very good review of the application of video microscopy to various systems in reference [66].

5.4.2.3 Final Technical Remarks on DDM

To conclude, a few technical details should be mentioned. One big assumption made by Cerbino and Trappe[54] in their derivation of the intensity-correlation function of the difference images, $\langle |F_D(q, \Delta t)|^2 \rangle$, was that density in the sample – and thus refractive-index fluctuations–is imaged directly onto the pixels of the CCD or CMOS camera. Over the past two decades, rapid advances in camera technology and image-analysis procedures have allowed new developments to be made in near-field microscopy.[37] In particular, Giglio and collaborators have developed new small-angle, light-scattering techniques in the near-field region that make it possible to measure the density-correlation function directly.[67] Although it appeared at first that their interpretation of the scattered light in the near field was the same as that obtained in DDM, it is important to note that there is a difference between their approach and the subsequent DDM papers. In reference [59], the authors consider near-field scattering from a sample of colloidal particles and project the resulting near-field speckle – but not the image of the sample – onto a CCD. Importantly, the size of the near-field speckle is, to a good approximation, independent of the vertical distance of scattering particles from the objective lens. In contrast, DDM, as described in references [54–56], images a plane in the sample onto the CCD detector. As particles move in and out of the focal plane, their image sharpens/blurs. This leads to an additional time dependence of the DDM signal that complicates the analysis. For this reason, it is not beneficial to perform DDM experiments when the objective has a high numerical aperture (as this leads to small focal depths).[55] Unfortunately, this point is not always described clearly in the 'Methods' section of publications that use DDM.

5.5 PHYSICAL CHEMISTRY APPLICATIONS

The two main applications of DLS in chemistry are particle sizing and the study of molecular aggregation and growth. The theoretical background to particle sizing for solutions containing only one size of colloid has been discussed already, as has the case in which there is some spread about the mean of the particle diameter. In this section, several practical examples are discussed, with particular focus on nanocolloids for their relevance in catalysis and self-assembled systems such as surfactants and block copolymers, which form micelles, vesicles, lamellae or other aggregates. Other areas, such as polymerisation, dendritic molecules, and some biological systems, will also be discussed briefly. Let us start with a few more notes on particle sizing.

5.5.1 Particle Sizing Revisited

The averaged static part of the scattered intensity is given by:

$$\langle I_s(\vec{q})\rangle = \left(\frac{E_0}{R}\right)^2 N\langle|b(0)|^2\rangle\, P(q)S(q) \tag{5.101}$$

where $P(q)$ and $S(q)$ are the form factor and the static structure factor, respectively, which describe the shape and the spatial distribution of the scattering objects in the sample. It is, however, important to note that the scattered intensity also depends on the factor $\langle|b(0)|^2\rangle$: the scattering length at zero scattering angle. Despite its name, $b(0)$ is actually proportional to the mass of the scattering object. Hence, the scattered intensity is proportional to the number of scatterers, N, and the square of their mass, M: $\langle I_s(\vec{q})\rangle \propto NM^2$, realising that the mass is simply the density divided by the unit volume of the scatterers, $\langle I_s(\vec{q})\rangle \propto R^{-6}$, where R is the average radius of the scattering object. This very strong size dependence means that the larger the particles or molecules, the stronger they will scatter in the forward direction. Consequently, when monitoring *e.g.* a polymerisation process or aggregation of colloids, proteins or other small molecules with time, the scattering intensity will shift continuously to smaller angles, θ. However, since much of the incoming light remains unscattered, the sizes that can be resolved are naturally limited to a few μm. Any larger objects will scatter towards angles smaller than roughly 10°. It is, therefore, important to measure the intensity-autocorrelation function for several angles corresponding to different q-values. Great care must also be taken to filter all solvents and samples used more than once, as any dust particle can dominate the signal completely. At the same time, when working with macromolecules or colloids smaller than 500 nm, one should also take care to de-gas the sample either by centrifuging it or by placing it briefly in a weak vacuum chamber, thus removing possible signals from unwanted nanobubbles. These are found in surfactant systems and viscous samples in particular.

We have seen that the intensity-autocorrelation function for monodisperse spherical particles is expressed as $g^{(2)}(q,\tau) = 1 + \beta e^{-\langle\Gamma\rangle\tau}$. In the left-hand panel of Figure 5.15, relaxation times, $\Gamma(q) = q^2 D$, are plotted

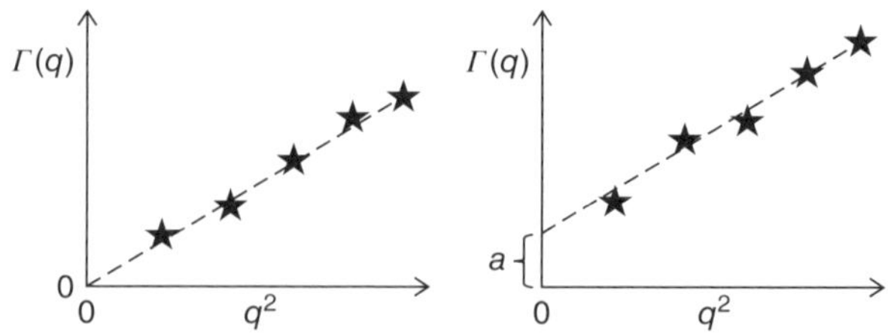

Figure 5.15 Schematic illustration of the dependence of the characteristic relaxation time, $\Gamma(q) = q^2D$, on the scattering vector, q, which can as usual be expressed in terms of the scattering angle, $\sin(\theta/2) \propto |\vec{q}|$. Left: Plotting $\Gamma(q)$ as a function of q^2 shows that the values lie on a straight line that extrapolates to zero at zero scattering angle for dilute solutions of monodisperse isotropic particles or molecules. Right: When the line does not extrapolate to zero but intercepts the zero angle at a positive value a, additional processes will contribute to the intermediate scattering function. These can result from rotational diffusion of anisotropic molecules or chemical transformations.

schematically as a function of the square of the scattering vector. If the object of interest is truly diffusive and isotropic in shape, the measured points must lie on a straight line that extrapolates to the origin. This is an excellent test for probing both the shape isotropy (meaning that spherically symmetric particles or molecules are under investigation) and the question of whether the particles are diffusing freely. In denser solutions, where inter-particle interactions may become important, or in an external field such as gravity that causes sedimentation or thermal drifts, the measured relaxation times may no longer lie on a straight line.

When the line in the $\Gamma(q)$ *versus* q^2 plane intercepts the ordinate at some finite, positive, non-zero value, a, as shown in the right-hand image of Figure 5.15, either shape anisotropy or some internal modes can be present. Shape anisotropy results in a measureable scattering contribution due to rotational diffusion, which can be analysed by cross-polarised DLS as detailed in Section 5.3.3.1. Internal modes can be due to the complexity of the macromolecules, such as DNA or dendritic molecules, but might also stem from chemical reactions that cause internal conformational changes. A number of examples are discussed in detail below.

5.5.2 Quantum Dots, Gold and Other Nanocrystals

In catalysis[68] and a number of biological applications,[69] nanoclusters made of metals and transition metals, and quantum dots made of semiconducting materials have gained increasing interest since the early

1980s. It became obvious very quickly that size and shape[70] were important factors in the catalytic activity of metal nanoclusters, placing them between homogenous and heterogeneous catalysts. In particular, nano-sized gold turned out to be a very active material.[71–76] As catalytic activity depends crucially on the size of the nanoclusters, it is advantageous to measure their size in an effective and inexpensive way either during or immediately after synthesis. The obvious way to obtain a rough estimate of the dimensions of nano-sized Group 11 metals such as copper, silver and gold is to use electronic spectroscopy, as their characteristic plasmon frequency is in the visible range.[77,78] For instance, in aqueous solutions gold clusters of about 5–20 nm diameter will display a strong but relatively broad absorption peak at around 500 nm, which shifts to longer wavelengths as the size of the gold clusters increases (Figure 5.16). Other important metal catalysts such as ruthenium, nickel, palladium and platinum have a plasmon frequency buried in the far ultraviolet (UV) region that is not accessible by UV–visible (UV-vis) spectroscopy. In fact, it remains difficult to determine the actual size and shape of a nanoparticle from UV-vis measurements by fitting the adsorption curves, particularly when the nanocrystals are made of mixed metals that form core–shell particles. Thus, it is appealing to measure their size with DLS, although it needs to be remembered that DLS generally works well only for transparent (also referred to as 'dielectric') samples that are neither adsorbing nor fluorescent. Therefore, DLS will not be suitable for particle sizing of solutions of quantum-dots, where UV-vis combined with transmission electron microscopy (TEM) or even the new technique, DDM, is advisable.[79]

Zimbone *et al.*[80,81] showed a beautiful example of both polarised and depolarised DLS, which they employed to follow the aggregation

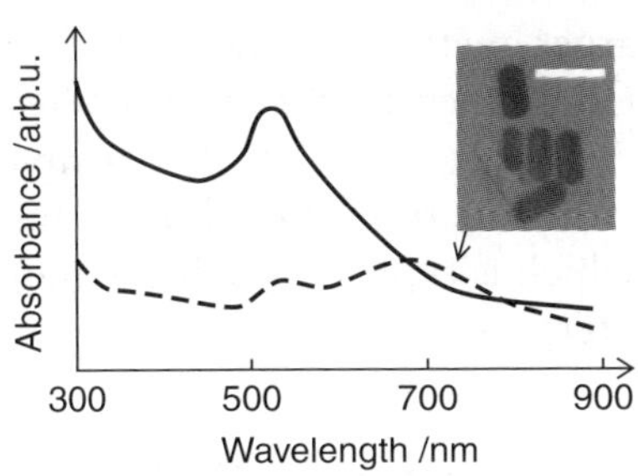

Figure 5.16 Typical UV-vis adsorption curves obtained from two different aqueous solutions of roughly 20 nm large 'spherical' gold colloids (solid line) and of 'rod-like' gold colloids (dashed line). The inset is a TEM of a later solution, where the bar indicates 50 nm. All measurements and sample preparations were conducted by Gaikwad. [79]

of 'spherical' gold colloids in time. They prepared gold nanocrystals by pulsed laser ablation in water[82] and studied their aggregation behaviour in the presence of pyridine, which is known to induce aggregation of gold clusters. Using a home-made DLS set-up that allows for measurements at various scattering angles, Zimbone *et al.* were able to record the intensity-autocorrelation function in polarised, ($g^{(2)}_{\mathrm{VV}}(q,t)$), and depolarised, ($g^{(2)}_{\mathrm{VH}}(q,t)$), geometries. As expected, when plotting the decay times, $\Gamma_{\mathrm{VV}}(q) = q^2 D_t$, obtained from a freshly prepared sample as a function of q^2 for different angles, a straight line could be fitted through the data, demonstrating that mostly spherical Au particles with a small size distribution were present. Simultaneously, the scattering signal in depolarised geometry (DDLS) was very low, again supporting the fact that mostly isotropic particles were present. Once pyridine was added, the scattering contribution in DDLS started to increase significantly. The relaxation time in DDLS is the sum of two terms stemming from the translational and rotational diffusion coefficients:[83]

$$\Gamma_{\mathrm{VH}}(q) = q^2 D_{\mathrm{t}} + 6D_{\mathrm{r}} \tag{5.102}$$

Assuming that the rotational and translational motions are uncorrelated, it is possible to extract D_{r}. The difficulty now lies in the interpretation of the rotational diffusion coefficient. As detailed in reference [81], the concentration of the added pyridine seems to determine whether more 'necklace-like' aggregates or fractal structures will form. If the latter, it is also necessary to distinguish between diffusion-(DLC) and reaction-(RLC) limited clusters, as they will display different dynamic modes.[57,84]

A final remark concerning 'coloured' samples should be borne in mind. When measuring a sample that appears red and hence adsorbs in the region between red and green, it is advantageous to measure using a red laser. Larger metal particles will, however, adsorb in the entire visible range. Consequently, they may be heated by the incoming laser beam leading to convection in the sample. In this case, the intensity-autocorrelation function may show a fast decay, rather than an exponential one, which is typical of ballistic motion. Such results are often impossible to interpret.

5.5.3 Self-Assembling Systems: Micelles, Vesicles and Other Equilibrium Structures

One important application of DLS is in the measurement of the onset of micellisation, also referred to as the **critical micelle concentration** or

CMC. The amphiphilic nature of molecular surfactants is responsible for their spontaneous microphase separation in pure solvents and in binary mixtures such as oil-in-water (o/w) or water-in-oil (w/o). All surfactants will have some degree of solubility in water, despite their hydrophobic aliphatic tail. However, as their concentration rises, the mixing free energy of these systems will be dominated by the unfavourable interaction between the hydrocarbon tail and the water molecules, making the mixing free entropy contribution negligible. As a consequence, surfactants that are added to a solution already at this CMC will self-assemble into micelles, worm-like micelles, lamellae or other, more complex assemblies. The thermodynamic theory describing these self-assembling systems is discussed in great detail in the books by Israelachvili[85] and Evans and Wennerström . [86]

Depending on the application, various types of surfactant are used. Anionic surfactants, such as sodium dodecylsulfate (SDS), have a negatively charged headgroup. These are used primarily in detergents, but have also been the subject of intense academic study, as they show a great variety of distinct phases. However, SDS alone in water forms spherical micelles.[85] Surfactants with a positively charged headgroup are called **cationic surfactants**[87–90] and are used heavily in the cosmetics industry, whereas non-ionic,[85] polymeric and biological surfactants such as phospholipids and sterols are being studied for their importance in nanotechnology[91,92] and other industrial applications. Most of the above-mentioned surfactant systems are employed in aqueous environments. Self-assembly has also been observed in non-aqueous dispersions. An example is the study of aggregation and flocculation of asphaltenes in hydrocarbons, which has been carried out by DLS.[93]

Less explored are metallosurfactants, which self-assemble in aqueous[94–101] and non-polar solvents[101–110] and which show enhanced colour and catalytic activity when self-assembled into micelles or vesicles. The charged headgroup in metallosurfactants is typically replaced by a transition-metal complex. In order to render the complex a surfactant, sufficiently long hydrocarbon chains must be attached to this head group. Dominguez-Gutierrez *et al.* introduced a new $[Ru(bipy)(4,4,\text{-dihexadecyl-}2,2'\text{-bipyridyl})_2]Cl_2$ complex with four $C_{16}H_{33}$ chains attached,[110,111] investigating its self-assembly in both toluene and hexane, using DLS and atomic force microscopy (AFM). DLS measurements in toluene revealed two sizes: one corresponding to the approximate size of a single surfactant molecule and the other suggesting the presence of spherical aggregates of about 500 nm diameter. Using simple geometric arguments introduced by Israelachvili[85,112] and the fact that the head group alone is not soluble in toluene, it became evident that the large aggregates in toluene were not micelles but vesicles. The headgroup size, *a*, was estimated to be 100 Å^2, while the chain length, *l*, was roughly 10 Å, rendering the volume taken up by the chains slightly larger than that occupied by the head group. Using Israelachvili's shape parameter, $p = v/al \approx 1$, Dominguez-Gutierrez hypothesised that the self-assembled structures should be inverted vesicles (Figure 5.17). In hexane, however, inverted micelles of about 11 nm diameter with, on average, 78 ± 15 molecules per aggregate were found. This illustrates how the aggregation process depends on the interactions of the head group with the solvent molecules, or on its solubility.

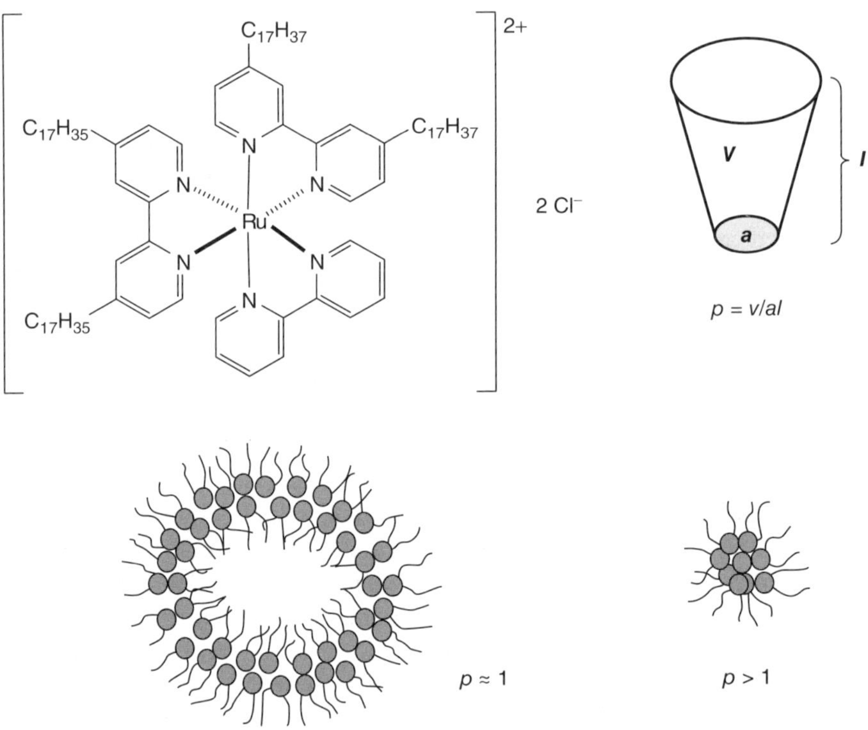

Figure 5.17 General representation of a metallosurfactant introduced by Dominguez-Gutierrez, who used ruthenium as the transition metal and chlorine as the negative counterion.[110] In the top-right corner, the geometric appearance of the metallosurfactants represented is sketched. The geometry factor, *p*, is the ratio between the volume of the truncated cone and the volume computed from the effective headgroup size and the chain length. Below, the cross-sections through an inverted vesicle (left) and an inverted micelle (right) are drawn schematically.[111]

5.6 CONCLUSION

DLS is a very versatile technique that probes particle sizes and proves very valuable in monitoring gelation and aggregation processes. Although the technique is limited in principle to transparent and, in particular, very dilute solutions, fibre-optic set-ups and newly developed DLS techniques such as DWS allow these difficulties to be circumvented. However, it should always be kept in mind that measurements with commercial particle sizers operating at one or two scattering angles only should be cross-checked with a complementary technique such as AFM or scanning electron microscopy (SEM), as inaccurate fitting procedures can provide spurious results.

Finally, DDM, a completely new light-scattering technique, is very useful and may replace DLS in certain applications, particularly due to its easy and rather inexpensive set-up and its efficiency when looking at µm-sized objects.

REFERENCES

[1] R. Hanbury Brown and R. Q. Twiss, *Nature*, **178**, 1046 (1956).
[2] R. Hanbury Brown and R. Q. Twiss, *Nature*, **178**, 1447 (1956).
[3] R. Hanbury Brown and R. Q. Twiss, *Nature*, **177**, 27 (1956).
[4] B. Chu, *Laser Light Scattering*, Dover Publications Inc., 2007.
[5] B. J. Berne and R. Pecora, *Dynamic Light Scattering with Applications to Chemistry, Biology, and Physics*, John Wiley & Sons, New York, 1976.
[6] H. Z. Cummins and E. R. Pike (Eds), *Photon Correlation and Light Beating Spectroscopy*, Plenum Press, New York, London, 1974.
[7] *Neutrons, X-Rays and Light Scattering Methods Applied to Soft Condensed Matter*, Elsevier, Amsterdam, Boston, London, New York, Oxford, Paris, San Diego, San Francisco, Singapore, Sydney, Tokyo, 2002.
[8] R. Klein and B. Daguanno, in *Light Scattering, Principles and Development*, W. Brown (Ed), Clarendon Press, Oxford, 1996.
[9] J. S. Pedersen, *Adv. Colloid Interface Sci.*, **70**, 171 (1997).
[10] P. N. Pusey, *Macromolecular Symposia*, **79**, 17 (1994).
[11] A. Einstein, *Ann. Phys.*, **19**, 371 (1906).
[12] S. W. Provencher, *Comput. Phys. Commun.*, **27**, 213 (1982).
[13] S. W. Provencher, *Comput. Phys. Commun.*, **27**, 229 (1982).
[14] R. T. C. Ju, C. W. Frank and A. P. Gast, *Langmuir*, **8**, 2165 (1992).
[15] D. E. Koppel, *J. Chem. Phys.*, **57**, 4814 (1972).
[16] H. Cramer, in *Mathematical Methods in Statistics*, Princeton University Press, 1946.
[17] R. A. Fisher and J. Wishart, *Proc. London Math. Soc.*, **33**, 195 (1932).
[18] B. J. Frisken, *Appl. Opt.*, **40**, 4087 (2001).

[19] P. N. Pusey, D. E. Koppel, D. W. Schaefer, R. Camerini and S. H. Koenig, *Biochem.*, **13**, 952 (1974).
[20] J. C. Brown and P. N. Pusey, *J. Phys. D: Appl. Phys.*, **7**, L31 (1974).
[21] F. Perrin, *J. Phys. Radium*, **5**, 497 (1934).
[22] S. Broersma, *J. Chem. Phys.*, **32**, 1626 (1960).
[23] T. Maeda and S. Fujime, *Macromolecules*, **18**, 2430 (1985).
[24] S. Jabbari-Farouji, D. Mizuno, M. Atakhorrami, F. C. MacKintosh, C. F. Schmidt, E. Eiser, G. H. Wegdam and D. Bonn, *Phys. Rev. Lett.*, **98** (2007).
[25] S. Jabbari-Farouji, R. Zargar, G. H. Wegdam and D. Bonn, *Soft Matter*, **8**, 5507 (2012).
[26] T. Nicolai and S. Cocard, *J. Colloid Interface Sci.*, **244**, 51 (2001).
[27] J. D. Jackson, *Classical Electrodynamics*, John Wiley, New York, 1998.
[28] S. Jabbari-Farouji, PhD thesis, University of Amsterdam, 2007.
[29] H. Shimizu, *J. Chem. Phys.*, **37**, 756 (1962).
[30] H. S. Dhadwal and D. A. Ross, *J. Colloid Interface Sci.*, **76**, 478 (1980).
[31] H. Auweter and D. Horn, *J. Colloid Interface Sci.*, **105**, 399 (1985).
[32] R. G. W. Brown, *Appl. Opt.*, **26**, 4846 (1987).
[33] R. G. W. Brown, *J. Phys. E-Scient. Instrum.*, **20**, 1312 (1987).
[34] R. G. W. Brown and A. P. Jackson, *J. Phys. E-Scient. Instrum.*, **20**, 1503 (1987).
[35] H. S. Dhadwal and B. Chu, *J. Colloid Interface Sci.*, **115**, 561 (1987).
[36] P. N. Pusey, *Curr. Opin. Colloid Interface Sci.*, **4**, 177 (1999).
[37] F. Scheffold and R. Cerbino, *Curr. Opin. Colloid Interface Sci.*, **12**, 50 (2007).
[38] G. Maret and P. E. Wolf, Z. Phys. *B: Condens. Matter*, **65**, 409 (1987).
[39] D. J. Pine, D. A. Weitz, P. M. Chaikin and E. Herbolzheimer, *Phys. Rev. Lett.*, **60**, 1134 (1988).
[40] P. E. Wolf and G. Maret, *Phy. Rev. Lett.*, **55**, 2696 (1985).
[41] T. Tanaka and G. B. Benedek, *Appl. Opt.*, **14**, 189 (1975).
[42] R. G. W. Brown and A. E. Smart, *Appl. Opt.*, **36**, 7480 (1997).
[43] H. S. Dhadwal, R. R. Ansari and W. V. Meyer, *Rev. Sci. Instrum.*, **62**, 2963 (1991).
[44] H. S. Dhadwal, W. Chi and B. Chu, *Appl. Opt.*, **28**, 4199 (1989).
[45] H. S. Dhadwal and B. Chu, *Rev. Sci. Instrum.*, **60**, 845 (1989).
[46] J. C. Thomas, *Langmuir*, **5**, 1350 (1989).
[47] J. C. Thomas and V. Dimonie, *Appl. Opt.*, **29**, 5332 (1990).
[48] D. Pristinski and T. Q. Chastek, *Meas. Sci. Technol.*, **20** (2009).
[49] W. Stober, A. Fink and E. Bohn, *J. Colloid Interface Sci.*, **26**, 62 (1968).
[50] T. Q. Chastek, K. Iida, E. J. Amis, M. J. Fasolka and K. L. Beers, *Lab on a Chip*, **8**, 950 (2008).
[51] F. Croccolo, D. Brogioli, A. Vailati, M. Giglio and D. S. Cannell, *Ann. NY Acad. Sci.*, **1077**, 365 (2006).
[52] F. Croccolo, D. Brogioli, A. Vailati, M. Giglio and D. S. Cannell, *Appl. Opt.*, **45**, 2166 (2006).
[53] S. P. Trainoff and D. S. Cannell, *Phys. Fluids*, **14**, 1340 (2002).
[54] R. Cerbino and V. Trappe, *Phys. Rev. Lett.*, **100** (2008).
[55] F. Giavazzi, D. Brogioli, V. Trappe, T. Bellini and R. Cerbino, *Phys. Rev. E*, **80** (2009).
[56] L. G. Wilson, V. A. Martinez, J. Schwarz-Linek, J. Tailleur, G. Bryant, P. N. Pusey and W. C. K. Poon, *Phys. Rev. Lett.*, **106** (2011).

[57] F. Ferri, A. D'Angelo, M. Lee, A. Lotti, M. C. Pigazzini, K. Singh and R. Cerbino, *Eur. Phys. J., Spec. Top.*, **199**, 139 (2011).

[58] M. Reufer, V. A. Martinez, P. Schurtenberger and W. C. K. Poon, *Langmuir*, **28**, 4618 (2012).

[59] G. Cerchiari, F. Croccolo, F. Cardinaux and F. Scheffold, *Rev. Sci. Instrum.*, **83** (2012).

[60] K. He, M. Spannuth, J. C. Conrad and R. Krishnamoorti, *Soft Matter*, **8**, 11933 (2012).

[61] V. A. Martinez, R. Besseling, O. A. Croze, J. Tailleur, M. Reufer, J. Schwarz-Linek, L. G. Wilson, M. A. Bees and W. C. K. Poon, *Biophys. J.*, **103**, 1637 (2012).

[62] P. J. Lu, F. Giavazzi, T. E. Angelini, E. Zaccarelli, F. Jargstorff, A. B. Schofield, J. N. Wilking, M. B. Romanowsky, D. A. Weitz and R. Cerbino, *Phys. Rev. Lett.*, **108** (2012).

[63] W. C. K. Poon, E. R. Weeks and C. P. Royall, *Soft Matter*, **8**, 21 (2012).

[64] P. N. Pusey and W. Van Megen, *Nature*, **320**, 340 (1986).

[65] J. C. Crocker and D. G. Grier, *J. Colloid Interface Sci.*, **179**, 298 (1996).

[66] P. Habdas and E. R. Weeks, *Curr. Opin. Colloid Interface Sci.*, **7**, 196 (2002).

[67] M. Giglio, M. Carpineti and A. Vailati, *Phys. Rev. Lett.*, **85**, 1416 (2000).

[68] M. B. Thathagar, J. Beckers and G. Rothenberg, *J. Am. Chem. Soc.*, **124**, 11858 (2002).

[69] T. H. Tran and T. D. Nguyen, *Colloids Surf. B*, **88**, 1 (2001).

[70] R. Fenger, E. Fertitta, H. Kirmse, A. F. Thunemann and K. Rademann, *Phys. Chem. Chem. Phys.*, **14**, 9343 (2012).

[71] G. C. Bond and D. T. Thompson, *Cat. Rev., Sci. Eng.*, **41**, 319 (1999).

[72] M. Haruta and M. Date, *Appl. Catal. A*, **222**, 427 (2001).

[73] M. Haruta, N. Yamada, T. Kobayashi and S. Iijima, *J. Catal.*, **115**, 301 (1989).

[74] R. Meyer, C. Lemire, S. K. Shaikhutdinov and H. Freund, *Gold Bulletin*, **37**, 72 (2004).

[75] A. S. K. Hashmi and G. J. Hutchings, *Angew. Chem. Int. Ed.*, **45**, 7896 (2006).

[76] L. M. Liz-Marzan, M. Giersig and P. Mulvaney, *Langmuir*, **12**, 4329 (1996).

[77] P. Mulvaney, *Langmuir*, **12**, 788 (1996).

[78] S. Sengupta, N. Ganguli, I. Dasgupta, D. D. Sarma and S. Acharya, *Adv. Mater.*, **23**, 1998 (2011).

[79] A. V. Gaikwad, PhD thesis, University of Amsterdam, 2009.

[80] M. Zimbone, P. Musumeci, P. Baeri, E. Messina, S. Boninelli, G. Compagnini and L. Calcagno, *J. Nanopart. Res.*, **14** (2012).

[81] M. Zimbone, L. Calcagno, G. Messina, P. Baeri and G. Compagnini, *Mater. Lett.*, **65**, 2906 (2011).

[82] A. V. Kabashin and M. Meunier, *J. Appl. Phys.*, **94**, 7941 (2003).

[83] A. M. Shetty, G. M. H. Wilkins, J. Nanda and M. J. Solomon, *J. Phys. Chem. C*, **113**, 7129 (2009).

[84] D. A. Weitz, J. S. Huang, M. Y. Lin and J. Sung, *Phys. Rev. Lett.*, **54**, 1416 (1985).

[85] J. N. Israelachvili, *Intermolecular and Surface Forces*, Elsevier, 2011.

[86] D. F. Evans and H. Wennerström, *The Colloidal Domain: Where Physics, Chemistry, Biology, and Technology Meet*, Wiley-VCH, 1999.

[87] C. A. Gracia, S. Gómez-Barreiro, A. González-Pérez, J. Nimo and R. J.R., *J. Colloid Interface Sci.*, **276**, 408 (2004).

[88] D. N. Rubingh and D. P. M. Holland, *Cationic Surfactants: Physical Chemistry*, Dekker, New York, 1991.

[89] J. M. Richmond, *Cationic Surfactants: Organic Chemistry*, Dekker, New York, 1990.

[90] R. Zana, *J. Colloid Interface Sci.*, **78**, 330 (1980).

[91] S. Mashaghi, T. Jadidi, G. Koenderink and A. Mashaghi, *Int. J. Mol. Sci.*, **14**, 4242 (2013).

[92] H. Y. Lee, K. Hashizaki, K. Diehn and S. R. Raghavan, *Soft Matter*, **9**, 200 (2013).

[93] M. A. Anisimov, I. K. Yudin, V. Nikitin, G. Nikolaenko, A. Chernoutsan, H. Toulhoat, D. Frot and Y. Briolant, *J. Phys. Chem. C*, **99**, 9576 (1995).

[94] J. Bowers, K. E. Amos, D. W. Bruce and R. K. Heenan, *Langmuir*, **21**, 5696 (2005).

[95] J. Bowers, K. E. Amos, D. W. Bruce and J. R. P. Webster, *Langmuir*, **21**, 1346 (2005).

[96] J. Bowers, M. J. Danks and D. W. Bruce, *Langmuir*, **19**, 292 (2003).

[97] P. C. Griffiths, I. A. Fallis, D. J. Willock, A. Paul, C. L. Barrie, P. M. Griffiths, G. M. Williams, S. M. King, R. K. Heenan and R. Gorgl, *Chem. Eur. J.*, **10**, 2022 (2004).

[98] J. D. Holbrey, G. J. T. Tiddy and D. W. Bruce, *J. Chem. Soc., Dalton Trans.*, 1769 (1995).

[99] J. T. Kunjappu, V. K. Kelkar and C. Manohar, *Langmuir*, **9**, 352 (1993).

[100] G. Mayer, V. Vogel, B. G. G. Lohmeijer, J. F. Gohy, J. A. Van Den Broek, W. Haase, U. S. Schubert and D. Schubert, *J. Polym. Sci., Part A: Polym. Chem.*, **42**, 4458 (2004).

[101] P. C. Griffiths, I. A. Fallis, T. Tatchell, L. Blishby and A. Beeby, *Adv. Colloid Interface Sci.*, **144**, 13 (2008).

[102] S. C. N. Chandar, D. Sangeetha and M. N. Arumugham, *J. Solution Chem.*, **40**, 608 (2011).

[103] R. Dong and J. Hao, *Chemphyschem.*, **13**, 3794 (2012).

[104] P. Mahato, S. Saha, S. Choudhury and A. Das, *Chem. Commun.*, **47**, 11074 (2011).

[105] M. J. M. Munoz and G. Fernandez, *Chem. Sci.*, **3**, 1395 (2012).

[106] M. J. Danks, M. E. Light and D. W. Bruce, *Eur. J. Inorg. Chem.*, **4232** (2009).

[107] U. Maran, D. Britt, C. B. Fox, J. M. Harris, A. M. Orendt, H. Conley, R. Davis, V. Hlady and P. J. Stang, *Chem. Eur. J.*, **15**, 8566 (2009).

[108] X.-N. Xu, L. Wang and Z.-T. Li, *Chem. Commun.*, **6634** (2009).

[109] J. Zhang, Y.-F. Song, L. Cronin and T. Liu, *Chem. Eur. J.*, **16**, 11320 (2010).

[110] D. Dominguez-Gutierrez, G. De Paoli, A. Guerrero-Martinez, G. Ginocchietti, D. Ebeling, E. Eiser, L. De Cola and C. J. Elsevier, *J. Mater. Chem.*, **18**, 2762 (2008).

[111] D. Dominguez-Gutierrez, M. Surtchev, E. Eiser and C. J. Elsevier, *Nano Lett.*, **6**, 145 (2006).

[112] J. N. Israelachvili, D. J. Mitchell and B. W. Ninham, *J. Chem. Soc., Faraday Trans. II*, **72**, 1525 (1976).

Index

Figures are indicated by *italic numbers*, Tables by **emboldened numbers**.
Abbreviations: AFM = atomic force microscopy; DDM = differential dynamic microscopy; DLS = dynamic light scattering; DSC = differential scanning calorimetry; DTA = differential thermal analysis; NMR = nuclear magnetic resonance; TGA = thermogravimetric analysis

Multi Length-Scale Characterisation, First Edition. Edited by Duncan W. Bruce, Dermot O'Hare and Richard I. Walton.